Algebraic Theory for
Multivariable Linear Systems

This is Volume 166 in
MATHEMATICS IN SCIENCE AND ENGINEERING
A Series of Monographs and Textbooks
Edited by RICHARD BELLMAN, *University of Southern California*

The complete listing of books in this series is available from the Publisher upon request.

Algebraic Theory for Multivariable Linear Systems

Hans Blomberg and Raimo Ylinen

Helsinki University of Technology

Technical Research Centre of Finland

1983

Academic Press

A Subsidiary of Harcourt Brace Jovanovich, Publishers
London New York
Paris San Diego San Francisco
São Paulo Sydney Tokyo Toronto

ACADEMIC PRESS INC. (LONDON) LTD.
24/28 Oval Road,
London NW1 7DX

United States Edition published by
ACADEMIC PRESS INC.
111 Fifth Avenue
New York, New York 10003

British Library Cataloguing in Publication Data
Blomberg, H.
Algebraic theory for multivariable linear
systems.—(Mathematics in science and engineering).
1. Control theory 2. Polynomials
I. Title II. Ylinen, R. III. Series
629.8'32 QA402.3

ISBN 0-12-107150-2

LCCCN 82-72337

Typeset and printed in Great Britain by Page Bros, (Norwich) Ltd.

Preface

The present book has evolved out of research work done by a great many people in the course of the last two decades. The work of the older of the present authors, HB, started back in the late fifties with a series of lecture notes on control theory. A safe method was sought for the determination of the correct characteristic polynomial for various feedback and other configurations with the aid of the then so popular transfer function technique. The considerations led to the introduction of an "uncancelled" form of the ordinary transfer function, but the result was not encouraging. In fact, it became apparent that there was something seriously wrong with the traditional form of the whole transfer function representation.

It was an easy matter to isolate and explore the weak point in the transfer function technique—obviously the source of all the trouble was the fact that the method was based on the assumption of zero initial conditions. On the other hand, if nonzero initial conditions were to be cared for too, then the nice transfer function algebra—essentially a field structure—would no longer work. The classical transfer function technique evidently needed to be improved, but it was certainly not so easy to see how this should and could be done. The improved technique should, in the first place, also make it possible to deal with nonzero initial conditions, but at the same time it should be based on some suitable algebraic structure in order to make the calculations easy to perform.

Fortunately, HB had at the time two bright young assistants, Sampo Ruuth (formerly Salovaara) and Seppo Rickman. Above all they were mathematically oriented, and they pointed out that quite a lot of highly developed algebraic structures existed other than just fields and vector spaces. They also suggested that it might be possible to replace, in a sense, the field structure utilized in the transfer function technique by some other suitable algebraic structure. As a result, an extensive study of set theory, topology, and abstract algebra was taken up within the research group working with HB. At the same time the qualitative properties of linear time-invariant differential and difference equations were being explored.

But it was still not quite clear how these qualitative properties should be interpreted in terms of abstract algebra in order to arrive at the desired result. The key, a suitable "system" concept, was still missing.

Then things started to happen. With respect to the present subject the most important events were the appearance of the early papers by Rudolf Kalman (e.g. Kalman, 1960; notations of this kind refer to the list of references at the end of the book) and the book by Lotfi Zadeh and Charles Desoer: Linear System Theory, The State Space Approach, 1963.

Through Kalman's eminent works the power of the "state-space approach" as a means for structural studies of the properties of linear time-variant and time-invariant differential equations was convincingly shown. The algebraic structure was here induced by the vector space structure of the state space—in the finite dimensional real case an Euclidean vector space. Moreover, the state-space representation also suggested very natural "state" and "system" concepts. The state-space approach made it possible to solve many problems in a rigorous and elegant fashion using known vector space methods. However, the methodology thus obtained does not show much resemblance to the transfer function technique—for instance, the treatment of systems of interacting differential equations by means of state-space methods turns out to be strikingly awkward, and it lacks the simple elegance of the transfer function technique.

The book by Zadeh and Desoer proved to be a real gold mine with respect to future work on an improved transfer function technique. The book contained new and fresh ideas concerning many concepts appearing in the present monograph. Sampo Ruuth (Salovaara, 1967) formalized and developed in set theoretical terms many of the suggestions presented by Zadeh and Desoer. Our system and related concepts rely very much on Ruuth's work. HB and his co-workers Sampo Ruuth, Jyrki Sinervo, Aarne Halme, Raimo Ylinen and Juhani Hirvonen combined Zadeh's and Desoer's ideas concerning representations of linear differential systems and interconnections of such systems with results obtained through their own research work. As a result a well-founded basic mathematical machinery suitable for the present purpose emerged. The main principles of this machinery were published in a number of papers and reports during the period 1968–1972 (e.g. Blomberg and Salovaara, 1968, Blomberg *et al.*, 1969; Sinervo and Blomberg, 1971; Blomberg, 1972a). These early works dealt almost entirely with structural problems—during the years that followed applications were also considered. It also became apparent that further studies were required of the application of the machinery to interconnections of systems. The final results of these studies are published for the first time in the present monograph.

The new transfer function technique—here called the "polynomial systems theory"—thus created relies essentially on a module structure (a module is "almost" a vector space, the difference being that the scalar set of a module is only a ring and not a field as in the vector space case). The set of scalars of this module is a ring formed from polynomials in the differentiation operator $p \triangleq d/dt$ (in the discrete time case p is replaced by a shift operator) interpreted as a linear mapping between suitably chosen signal spaces. It turns out that the polynomial systems theory has many features in common with the classical transfer function technique—it is therefore justified to call it a "generalized transfer function technique". It works rather nicely and it is well suited to the treatment of various problems concerning interconnections of systems. Further, the state-space representation has, interestingly enough, its own natural place within this framework. Note in passing that the module structure introduced and utilized by Kalman (Kalman, Falb, and Arbib, 1969) differs from our module and serves different purposes.

While HB and his co-workers were busy developing a rigorously founded mathematical machinery for the purpose at hand, Howard H. Rosenbrock (*cf.* Rosenbrock, 1970, and further references therein) was creating in an ingenious and rather direct way a methodology that closely resembled the polynomial systems theory developed by HB and his co-workers. Rosenbrock's works contain far-reaching results which are also directly applicable within the framework of the present theory. W. A. Wolovich (*cf.* Wolovich, 1974, and further references therein) also derived a number of new and important results based essentially on the same polynomial interpretations as Rosenbrock's work. Many details in the present text are founded on and inspired by results obtained by Rosenbrock and Wolovich.

In later years new people interested in the subject have joined the above "pioneers" and produced new and valuable results. References to some of them can be found later on in the main part of the text.

The present monograph presents the "state of the art" of the polynomial systems theory and related matters based on results obtained as described above. The organization of the material is as follows.

We start with an introductory chapter 1, which provides a quick survey of some of the main points of the present polynomial systems theory along with examples. Of particular interest from an application point of view is section 1.5, where a new algorithm for the design of a feedback compensator is outlined.

There are a few basic ideas and concepts which are essential to a true understanding of "systems thinking". These are introduced and discussed in a rather concentrated manner in part I (chapters 2 and 3) of the book. The subsequent treatment of the subject relies firmly on the material presented here.

After a very brief and superficial motivation we start, in part II (chapters 4 to 7), a detailed study of the polynomial systems theory as applied to systems governed by ordinary linear time-invariant differential equations. The mathematical machinery needed is discussed in detail in chapters 4 and 5. It is based on a module structure. A survey of the relevant part of the abstract algebra used in this context can be found in appendices A1 and A2. In chapter 6 a number of basic system concepts are interpreted in terms of the module structure chosen. Chapter 7 is probably the most important chapter in the whole book from an application point of view. It contains a great deal of material of significance to the application of the polynomial systems theory to various analysis and synthesis problems. Problems concerning the design of feedback controllers and compensators, as well as of estimators and observers, are discussed in some detail. Considerable space has also been devoted to representation theory, in particular some effort has been made to interpret Rosenbrock's system matrices (Rosenbrock, 1970) and related concepts in terms of the present theory. All this has unfortunately led to a rather oversized chapter.

Part III (chapters 8 and 9) serves a number of purposes, the main one being to give a convincing justification for the polynomial systems theory as developed and applied earlier in the text. For this purpose a rather delicate vector space structure is constructed, also involving generalized functions. Appendices A3 and A4 contain some basic material of relevance in this context. Chapter 8 devises a method for constructing a suitable system (i.e. a family of input-output mappings) from a given input-output relation. The method is called the "projection method" because it is based on a certain projection mapping. In passing, a justification for the engineering use of the Laplace transform method is obtained. In chapter 9 this projection method is then utilized to build up the convincing justification wanted for the polynomial systems theory. The considerations also lead to a number of significant by-results, for instance concerning the so-called realization problem.

Part IV (chapters 10 to 14) is devoted to a study of the polynomial systems theory as applied to systems governed by ordinary linear time-invariant difference equations. Because the theory is very similar to the theory of differential systems we can pass many considerations only by referring to the corresponding considerations in parts II and III. There are, however, some dissimilarities which make the algebraic theory a little more complicated. These are caused by the fact that in the most useful case, where our time set consists of all integers, the unit shift operators are invertible. On the other hand, the difference systems theory is simpler than the differential systems theory in the sense that in using the projection method for constructing finite dimensional input-output systems we do not need any generalized functions.

The presentation of the material within the different paragraphs is organized in the following way.

Each paragraph contains a main part, which includes definitions, theorems, proofs, etc. of fundamental importance to the subject. This part is written in a rather precise way. Numerous subtitles are inserted in order to make it easier for the reader to get a general view of the material presented. Examples are also included—often with suitable details left as exercises for the reader. In order to make the text as readable as possible, the number of special concepts and terms introduced has been kept at a near-minimum. For the same reason we have as a rule avoided presenting things in their most general setting.

The material covered by the main part of the text most often leads to a great deal of useful consequences and implications. Results of this kind are presented in a more informal way as "notes". Most of the items contained in the notes are statements given without, or with only incomplete proofs—the proofs and other considerations relating to the statements are left as exercises for the reader. In view of the exercises implicitly contained in the examples and notes, it has been possible to dispense entirely with separately formulated exercises.

There are also a number of "remarks". These are just remarks of some general interest but without much bearing on the actual presentation of the subject at hand.

For referencing and numbering we have essentially adopted the very flexible and convenient system used in Zadeh and Desoer, 1963:

Theorems, definitions, notes, remarks, as well as other items of significance are numbered consecutively within each section. These numbers appear in the left-hand margin. References and cross references to various items like literature, figures, equations, etc. are given in a rather self-explanatory way. A few examples will further illustrate the simple system. For instance, a cross reference of the form "note (8.2.1), (ii)" refers to item (ii) of note number 1 within section 8.2 of chapter 8. A reference of the form "theorem (A2.69)" correspondingly refers to theorem number 69 within appendix A2. If reference is made in a section to an item within the same section, then only the item number is indicated, for instance "consider (2)" means that we shall consider item (equality, expression, etc.) number 2 of the same section.

The symbol □ indicates the end of theorems, proofs, definitions, notes, remarks, examples etc.

Hans Blomberg and Raimo Ylinen
Helsinki, Summer 1982

Acknowledgements

The work on this monograph started on a suggestion made by Karl Johan Åström in connection with a series of lectures on the subject delivered by HB at Lund Institute of Technology back in 1973. KJÅ even sketched a draft for the contents of the book. Since then KJÅ has closely followed the progress of the authors' work, showing great and continuous interest in the achievements and giving encouraging advice. The authors are very grateful for all this valuable help.

The book could not have been written without the research work carried out by a great many people acting in co-operation with the authors over the years. All these people are remembered with great gratitude and affection.

Parts of the manuscript were included in a graduate seminar at Helsinki University of Technology. The authors are indebted to the students participating in this seminar. They made a careful reading of the material and suggested many significant improvements. The authors further thank Pirkko Mähönen for typing the manuscript, Susan Sinisalo for checking the English language, Eila Meriläinen for preparing the figures, and finally the publisher for a helpful and understanding attitude towards the work.

A task of the present kind, in which a large number of people have been involved, could hardly have been carried out without generous support from a great many foundations and institutions. The authors are grateful for financial and other kinds of support granted in particular by Helsinki University of Technology, the Academy of Finland, the Societas Scientiarium Fennica, the Emil Aaltonen Foundation, the Finnish Cultural Fund, the Foundation for Technology in Finland, the Jenny and Antti Wihuri Fund, and the Oskar Öflund Foundation.

Contents

Preface v
Acknowledgements x
List of Symbols xv

1. Introduction 1
1.1 Differential input-output relations and systems 4
1.2 Compositions of differential input-output relations 7
1.3 Algebraic foundations 13
1.4 Properties of differential input-output relations 22
1.5 Feedback compensation and control 27
1.6 Concluding remarks 44

Part I Basic Concepts of Systems Theory

2. Systems and system descriptions 49
2.1 Parametric input-output mappings. Abstract input-output systems. System descriptions 50
2.2 Time systems. Dynamic systems 52
2.3 Linear systems 53

3. Interconnections of systems 56
3.1 Formal definition of an interconnection of a family of systems . . 56
3.2 Input-output relations determined by an interconnection of a family of systems 61
3.3 Determinateness with respect to the empty set of realizability conditions 63
3.4 Determinateness with respect to a general set of realizability conditions 64
3.5 Illustrative example 67

Part II Differential Systems. The Module Structure

4. Generation of differential systems 75
4.1 Signal spaces and differential operators 75
4.2 Matrix differential equations 77

5. *The* **C**[*p*]*-module* $\mathscr{X}$ 80
5.1 Suitable signal spaces $\mathscr{X}$ 81
5.2 The ring **C**[*p*] of polynomial operators. The **C**[*p*]-module $\mathscr{X}$ . . 84
5.3 Relationship between polynomial matrix operators and matrices over **C**[*p*] 85
5.4 Fundamental properties of polynomial matrix operators . . . 86

6. *Differential input-output relations. Generators* 88
6.1 Introduction. Regular differential input-output relations and regular generators 89
6.2 Input-output equivalence. Complete invariants and canonical forms for input-output equivalence 91
6.3 The transfer matrix. Proper and strictly proper transfer matrices, generators, and differential input-output relations 94
6.4 Transfer equivalence. Complete invariants and canonical forms for transfer equivalence. Controllability 95
6.5 Proofs of theorems (6.2.1), (6.2.2) and (6.4.8) 100
6.6 Comments on canonical forms. Canonical row proper forms . . 104

7. *Analysis and synthesis problems* 106
7.1 An elimination procedure 107
7.2 Compositions and decompositions of regular differential input-output relations. Observability 110
7.3 A parallel composition 123
7.4 Parallel decompositions of regular differential input-output relations 126
7.5 Illustrative example 133
7.6 A series composition 137
7.7 Series and series-parallel decompositions of regular differential input-output relations 143
7.8 The Rosenbrock representation 145
7.9 The state-space representation 158
7.10 The Rosenbrock representation and the state-space representation as decompositions of regular differential input-output relations. Equivalence relations 160
7.11 Observer synthesis problem 176
7.12 Feedback compensator synthesis 184

Part III Differential Systems. The Vector Space Structure

8. *The projection method* 203
8.1 Reason for choosing a space of generalized functions as signal space 204
8.2 The basic signal space $\mathscr{D}$ of generalized functions. Projection mappings. Subspaces of $\mathscr{D}$. Generalized causality 205
8.3 The vector space $\mathscr{X}$ over **C**(*p*) 210
8.4 Compositions of projections and differential operators. Initial condition mappings 216
8.5 The projection method 226

9. *Interconnections of differential systems* 243
9.1 Two interconnections 243
9.2 Systems associated with compositions of input-output relations . . 250

9.3 The main results 253
9.4 Illustrative example 259

Part IV Difference Systems

10. Generation of difference systems 273
10.1 Signal spaces and shift operators 273
10.2 Matrix difference equations 274

11. The module structure 276
11.1 Suitable signal spaces 276
11.2 The rings **C**[z], **C**(z], **C**[1/z] and **C**(z) and modules over them . . 278
11.3 Polynomial and rational matrix operators and polynomial and rational matrices 279

12. Difference input-output relations. Generators 281
12.1 Regular difference input-output relations and regular generators . 282
12.2 Input-output equivalence. Canonical forms for input-output equivalence. Causality 283
12.3 The transfer matrix. Properness and causality 284
12.4 Transfer equivalence. Canonical forms for transfer equivalence Controllability. 285

13. Analysis and synthesis problems 288
13.1 Compositions and decompositions of regular difference input-output relations. Observability 288
13.2 The feedback composition 291

14. The vector space structure. The projection method 293
14.1 Signal space 293
14.2 The modules and vector spaces of quotients 295
14.3 Compositions of projections and delay operators. Initial condition mappings 296
14.4 The projection method 298

Appendices

A1 Fundamentals of abstract algebra 303

A2 Polynomials and polynomial matrices 321

A3 Polynomials and rational forms in an endomorphism . . . 340

A4 The space $\mathfrak{D}$ of generalized functions 342

References 349
Index 353

To Elseby (HB) and Maria (RY)

Symbols

The following list contains only symbols and notations which essentially have a fixed and unaltered meaning throughout the book. Symbols and notations of a more temporary nature are explained in the context where they appear and are not included in the list. Each symbol listed below is followed by a brief description of its meaning and, where motivated, by the number of the section or appendix where the symbol is introduced or where the underlying concept is discussed at some length.

$\triangleq$	equals by definition; denotes
$\propto$	is proportional to
$\in$	is an element of
$\notin$	does not belong to
$\cup$	union
$\cap$	intersection
$\supset$	contains
$\subset$	is a subset of; is contained in
$\{x \mid P\}$	set of all x having property P
$\Rightarrow$	implies
$\Leftarrow$	is implied by
$\Leftrightarrow$	implies and is implied by; if and only if
$\sim$	equivalence relation, appendix A1
$\times$	cartesian product
$\oplus$	direct sum

Symbols

$f: A \to B$	f is a mapping from the set A into the set B
$a \mapsto b$	b is the image of a under a certain mapping
A^B	set of mappings from the set B into the set A
A^n	set of n-lists ($n \in \{1, 2, 3, \ldots\}$) of elements belonging to the set A
$A^{m \times n}$	set of m by n-matrices ($m, n \in \{1, 2, 3, \ldots\}$) with entries belonging to the set (usually a ring) A
$A \circ B$	composition of the relations A and B (B followed by A)
$\mathbf{n}$	the set $\{1, 2, \ldots, n\}$
(t_0, t_1)	open interval $\{t \mid t \in \mathbf{R} \text{ and } t_0 < t < t_1\}$
$[t_0, t_1)$	semiclosed interval $\{t \mid t \in \mathbf{R} \text{ and } t_0 \leqslant t < t_1\}$
$[t_0, t_1]$	closed interval $\{t \mid t \in \mathbf{R} \text{ and } t_0 \leqslant t \leqslant t_1\}$
$(\cdot)_t^0$	initial condition mapping at t associated with an operator $(\cdot)$, sections 8.4, 8.5 and 14.3
$(\cdot)^*$	set of nonzero elements contained in $(\cdot)$; canonical form of $(\cdot)$
$\mathscr{A}$	family of system descriptions, section 3.1
$(\mathscr{A}, C, X)$	interconnection of a family of systems, sections 3.1, 9.1
$[A(p) \vdots -B(p)]$	generator for differential input-output relation, sections 1.1, 6.1
$[A(\mathscr{q}) \vdots -B(\mathscr{q})]$ $[A(\mathscr{z}) \vdots -B(\mathscr{z})]$	generators for difference input-output relations, sections 12.1 and 12.3
α	state, sections 3.1 to 3.4
$\mathscr{B}$	family of generators, section 7.2
$(\mathscr{B}, C)$	composition of a family of input-output relations, sections 7.2 and 9.2
C	interconnection matrix, sections 1.2, 3.1 and 7.2
$\mathbf{C}$	field of complex numbers
C^∞	space of complex-valued infinitely continuously differentiable functions

$\mathbf{C}[p]$	commutative ring of polynomials in p with complex coefficients, sections 1.3 and 5.2
$\mathbf{C}(p)$	the field of quotients of $\mathbf{C}[p]$, sections 1.3 and 8.3
$\mathbf{C}[z]$	commutative ring of polynomials in an indeterminate z with complex coefficients, appendix A2
$\mathbf{C}(z)$	field of quotients of $\mathbf{C}[z]$, appendix A2
$\mathbf{C}(z]$	ring of quotients of $\mathbf{C}[z]$ by the set of powers of z, appendix A2
$\mathbf{C}[z]_S$	ring of quotients of $\mathbf{C}[z]$ by the set S, appendix A2
	The above $\mathbf{C}[z]$, $\mathbf{C}(z)$, $\mathbf{C}(z]$ and $\mathbf{C}[z]_S$ occur with the indeterminate z replaced by the operator φ, $\imath$, $1/\imath$ or σ respectively, sections 11.2, 12.3, 12.4 and appendices A2, A3
CLT, CUT, CRP -form	Canonical lower triangular form, canonical upper triangular form, and canonical row proper form respectively, sections 1.3, 6.2, 6.6, 12.2, appendix A2
CPCUT-form	column permuted canonical upper triangular form of a polynomial matrix, appendix A2
$\lceil_t$, $\rceil_t$	projection mappings, sections 8.2 and 14.1
$\mathbf{C}^{\mathbf{N}_0}$	space of complex-valued functions on $\mathbf{N}_0$, section 11.1
$\mathbf{C}^{\mathbf{Z}}$	space of complex-valued functions on $\mathbf{Z}$, section 11.1
$\mathbf{C}^{\mathbf{Z}+}$, $\mathbf{C}_k^{\mathbf{Z}+}$, $\mathbf{C}^{\mathbf{Z}-}$, $\mathbf{C}_k^{\mathbf{Z}-}$	subspaces of $\mathbf{C}^{\mathbf{Z}}$, sections 11.1 and 14.1
$\mathcal{D}$	space of piecewise infinitely regularly differentiable functions, sections 8.1 to 8.5, appendix A4
$\mathcal{D}^+$, $\mathcal{D}_t^+$, $\mathcal{D}^-$, $\mathcal{D}_t^-$	subspaces of $\mathcal{D}$, section 8.2
$\mathsf{D}(\cdot)$	domain of $(\cdot)$
$\det(\cdot)$	determinant of $(\cdot)$

Symbols

$\delta_k(.)$	unit pulse at k, section 14.3
$\delta_t^{(k)}$	delta function of order k at t, appendix A4
$\partial(\cdot)$	degree of $(\cdot)$
$\mathcal{F}(\cdot)$	Fourier transform of $(\cdot)$, section 3.5
GCLD	greatest common left divisor
GCRD	greatest common right divisor
$\left.\begin{matrix}\mathcal{G}(p)\\ \mathcal{G}(q)\\ \mathcal{G}(z)\\ \tilde{\mathcal{G}}(1/z)\end{matrix}\right\}$	transfer matrices, rational matrices, sections 1.3, 6.3 and 12.3, appendix A2
$\left.\begin{matrix}h(.,.)\\ k(.,.)\end{matrix}\right\}$	parametric input-output mapping, sections 2.1, 8.5 and 9.1 to 9.3
$\mathcal{K}$	set of test functions, appendix A4
$\mathcal{K}'$	set of complex-valued generalized functions, appendix A4
ker(.)	kernel of (.)
$L(\mathcal{X}, \mathcal{X})$	algebra of linear mappings $\mathcal{X} \to \mathcal{X}$, section 5.2, appendix A3
$\mathcal{L}(\cdot)$	Laplace transform of $(\cdot)$, sections 3.5, 8.4 and 8.5
$\mathcal{M}$	set of regular generators, section 6.2
$\mathcal{M}^* \subset \mathcal{M}$	set of canonical forms for input-output equivalence, section 6.2
$\mathcal{M}_t^* \subset \mathcal{M}^*$	set of canonical forms for transfer equivalence, section 6.4
MR-system matrix	modified Rosenbrock system matrix, section 7.8
$\mathcal{MR}$	set of MR-system matrices, section 7.10
$\mathcal{MR}^* \subset \mathcal{MR}$	set of canonical forms for input-output equivalence, section 7.10
$\mathbf{N}$	set of natural numbers $\{1, 2, 3, \ldots\}$
$\mathbf{N}_0$	set of natural numbers including the zero $\{0, 1, 2, \ldots\}$
$p \triangleq d/dt$	differentiation operator considered as a linear mapping from a signal space into itself

PM	mapping determined by the projection method, section 8.5
$\mathcal{q}$	unit prediction operator, section 10.1
$\mathbf{R}$	field of real numbers
$\mathcal{R}$	set of regular differential input-output relations, section 6.2
$\mathcal{R}_m \subset \mathcal{R}$	set of minimal input-output relations, section 6.4
R(.)	range of (.)
R-representation	Rosenbrock representation, section 7.8
$\mathcal{r}$	unit delay operator, section 10.1
$R[z]$	ring of polynomials in z over R, appendix A2
S	input-output relation, sections 1.1 and 6.1
$S(0)$	zero input response set (space), section 2.3
S_i	internal input-output relation, sections 1.2, 3.2 and 7.2
S_{io}	internal-overall input-output relation, sections 1.2 and 7.2
S_o	overall input-output relation, sections 1.2, 3.2 and 7.2
$\mathcal{S}$	abstract input-output system, section 2.1
Sp(.)	vector space spanned by the (vector) elements of (.)
s	input-output mapping, sections 2.1, 2.2 and 2.3
Σ	state set, section 2.1
$(.)^T$	transpose of (.), appendix A2
$T \subset \mathbf{R}$	time set; (open) time interval, sections 1.1, 2.2, 4.1 and 10.1
$\mathcal{T}$	set of transfer matrices, section 6.4
$\mathcal{U}$	set of inputs
U_t	unit step at t, sections 8.2 and 9.4, appendix A4
u	input
X	state constraint, sections 3.1 and 9.1
$\mathcal{X}$	signal space, sections 1.1, 4.1, 5.1 and 10.1

Symbols

$\mathcal{Y}$	set of outputs
y	output
$\mathbf{Z}$	set of integers $\{\ldots, -2, -1, 0, 1, 2, \ldots\}$

1

Introduction

The early theory of control systems as developed and applied during and after World War II was based almost entirely on the transfer function technique. The mathematical framework needed for this purpose was provided by the mathematics of the Laplace and Fourier transforms and later also by the mathematics of the Z-transform. The transfer function technique was straightforward to apply and easy to grasp, at least as far as single input–single output systems were concerned. Its close relationship with experimental methods based on frequency response measurements made it particularly attractive to people working with practical control problems.

The early transfer function technique was a somewhat strange mixture of the rigorous mathematics of the transform methods used, of sound engineering skill and of more or less heuristically derived rules for how to deal with cancellations of common factors appearing in the numerators and denominators of the transfer functions involved. The very existence of such a cancellation problem was the direct consequence of the fact that the transfer function technique was based on the assumption that zero initial conditions prevailed. The eminent book by Truxal, 1955, probably represents the apex of what could be achieved with the aid of these tools in single input–single output cases.

During the time that followed several attempts were made to generalize the single input–single output transfer function technique to multiple input–multiple output cases. The books by Mesarovic, 1960, and Schwarz, 1967, are examples of this. The multivariable transfer function technique which emerged as a result of these attempts did not attract much attention at the time. There were many reasons for this. One was that the technique thus developed was not capable of coping with the multivariable version of the cancellation problem mentioned above, nor was it possible to give heuristically derived rules for this purpose. But the main reason for the lack of success of the multivariable transfer function technique was undoubtedly the appearance of a new and promising method for dealing with problems of the present kind—the famous "state-space approach" had been created.

The state-space approach uses a system description in the form of a differential (difference) equation in so-called normal form, and this form as such is well-known in the classical theory of differential equations. What was new was the discovery of the great inherent strength possessed by the normal form applied to problems concerning optimal control and (state-) feedback stabilization. The breakthrough for the state-space approach came with the early works by Kalman (e.g. Kalman, 1960) and the invention of Pontryagin's maximum principle (*cf.* Pontryagin *et al.*, 1962). Since then the state-space approach has been the universal tool for solving a great variety of control problems.

The state-space approach is based on sound and rigorous mathematics. There is no need to assume zero initial conditions and so no cancellation problem ever arises. Thus the ultimate tool seemed to have been created and the evolution in this respect to be complete.

But there are always some people who are not easily satisfied. Thus even during the era of the successful state-space approach, work has been going on in order to generalize the old transfer function technique and to eliminate its deficiencies. The aim has also been to create a common framework both for a new multivariable transfer function technique and the state-space approach. The work has been encouraged by a number of pertinent observations: The state-space approach has never been very popular among control people working in the field because of its lack of transparency—for instance the relationship between a state-space equation and a corresponding frequency response record is not apparent—and because it needs a rather heavy mathematical machinery not easily adopted by the people in the field. Further there are a number of problems to which the state-space approach is not so well suited—a typical problem of this kind being the dimensioning of a multivariable output feedback controller of restricted structure in order to improve the dynamics of a given system.

As a result of pertinacious research with the above-mentioned goal in mind a special "polynomial systems theory" has emerged. The pioneer work in this area has been done by Rosenbrock, Wolovich, and Blomberg and co-workers (*cf.* Rosenbrock, 1970, Wolovich, 1974, Blomberg *et al.*, 1969, and further references therein). The polynomial systems theory utilizes a polynomial operator algebra, which is a rather natural generalization of the usual operational calculus based on the Laplace and Z-transforms. The state-space representation has its natural place within this framework. The polynomial systems theory is particularly well suited to solving a number of typical problems relating to the analysis of linear time-invariant feedback systems and estimators. New results can be expected in this area.

The cornerstones of the polynomial systems theory are the properties possessed by polynomials and polynomial matrices. Here the crucial question is: What is the precise relationship between these polynomials and polynomial matrices on the one hand and various system aspects on the other? In this respect the pioneer works mentioned above show different interpretations. For instance, in the continuous-time case Rosenbrock and Wolovich form their polynomials and polynomial matrices formally by taking the Laplace transform of the system equations (ordinary linear time-invariant differential equations) under zero initial conditions, whereas Blomberg and his co-workers obtain their polynomials and polynomial matrices by interpreting the differentiation operation as defining a linear mapping between suitably chosen signal spaces. It turns out that the latter interpretation is in many respects more fruitful regarding the problems at hand—in fact it has in a very natural way led to a number of formulations and results which are not so easy to recognize within the framework of a Laplace transform study under zero initial conditions. This holds in particular for the significance and application of "equivalent generators" as explained later on.

The aim of this book is to give the reader rather detailed information on the structure and applicability of the polynomial systems theory as based on the mapping interpretation. We shall mainly be concerned with systems governed by ordinary linear time-invariant differential equations ("differential systems"). Systems governed by corresponding difference equations ("difference systems") are commented on only briefly at the end of the book. The methodology developed for differential systems can, under certain circumstances, be directly applied also to difference systems.

The mathematics of polynomials and polynomial matrices is as such not particularly difficult. It does, however, contain a great number of special concepts and features not known from the ordinary mathematics usually included in engineering courses. This means that people wishing to study the polynomial systems theory and its applications must devote some time to the mathematical preliminaries. The appendices at the end of the book should be helpful here.

The detailed presentation of the subject begins in chapter 2, but the reader has to wait until chapter 7 before s/he meets practical applications of the polynomial systems theory for differential systems thus developed. The reason for this is that the whole subject requires a painstakingly precise and detailed approach in order to build up the correct framework and to avoid mistakes. This point is emphasized by the material presented in chapters 8 and 9. This state of affairs could easily make the interested reader feel frustrated and lose interest before s/he ever gets to the point. The present authors therefore thought it wise to start with a quick and not so precise survey of the main material to be presented. The present introductory chapter is devoted to this task.

1.1 Differential input-output relations and systems

In forming a mathematical model of a real system we have first of all to decide what kind of signals the model should contain. The choice of signal space depends on one hand on what kind of signals appear in the real system, and on the other hand on the desired structure of the model. If we wish to construct a differential system model, say on some open time interval T, then the signals should be differentiable a sufficient number of times—preferably an infinite number of times. Then the differentiation operator $p \triangleq d/dt$ can be regarded as a mapping from the signal space into itself. Now suppose that the signal space $\mathscr{X}$ is, in addition, a vector space over $\mathbf{C}$ (or possibly over $\mathbf{R}$). This means that p is a linear mapping $\mathscr{X} \to \mathscr{X}$, and also that any polynomial in p of the form

1 $$a(p) \triangleq a_0 + a_1 p + \ldots + a_n p^n$$

with $a_0, a_1, \ldots, a_n \in \mathbf{C}$ and $n \in \{0, 1, 2, \ldots\}$ represents a linear mapping $\mathscr{X} \to \mathscr{X}$ given by the assignment ($p^0 x \triangleq 1x$, $p^i x \triangleq x^{(i)}$, the ith-order derivative of x)

2 $$x \mapsto a(p)x = a_0 x + a_1 px + \ldots + a_n p^n x.$$

More generally, any polynomial s by q-matrix $A(p)$ of the form ($s, q \in \{1, 2, 3, \ldots\}$)

3 $$A(p) \triangleq \begin{bmatrix} a_{11}(p) & a_{12}(p) \ldots a_{1q}(p) \\ a_{21}(p) & a_{22}(p) \ldots a_{2q}(p) \\ \vdots & \vdots \qquad \vdots \\ a_{s1}(p) & a_{s2}(p) \ldots a_{sq}(p) \end{bmatrix}$$

with the $a_{ij}(p)$ polynomials in p of the form (1) represents a linear mapping $\mathscr{X}^q \to \mathscr{X}^s$. For any $x = (x_1, x_2, \ldots, x_q) \in \mathscr{X}^q$ the corresponding $y = (y_1, y_2, \ldots, y_s) = A(p)x \in \mathscr{X}^s$ is evaluated by interpreting $y = A(p)x$ as a matrix equality with x and y regarded as column matrices.

Mappings of the form (1) and (3) are frequently called "differential operators". Differential operators of the form (1) and (3) will in this context be called "polynomial operators" and "polynomial matrix operators" respectively.

There are many signal spaces possessing the properties mentioned above. There are, however, also other conditions that could be imposed on the signal space. These conditions will be discussed in detail in chapter 5. For our present purpose it suffices to state that suitable signal spaces are the space C^∞ of all complex-valued infinitely continuously differentiable functions on T and the space $\mathscr{D}$ of all "piecewise infinitely regularly differentiable complex-valued generalized functions (distributions) on T" (*cf.* appendix A4). The latter space also allows differentiation of step functions. Both spaces are vector spaces over **C** with respect to the natural (pointwise) operations.

So it will, in the sequel, be assumed that the signal space $\mathscr{X}$ is either C^∞ or $\mathscr{D}$. It should be emphasized that not all the results given below necessarily hold if $\mathscr{X}$ is some other space.

The signal spaces chosen possess the property that for any $x \in \mathscr{X}$ and any polynomial operator $a(p) \neq 0$ there is always at least one $y \in \mathscr{X}$ such that $a(p)y = x$. This y is, on the other hand unique if and only if $a(p) = c = \text{constant} \neq 0$. This means that the polynomial operator $a(p) \neq 0$ has an inverse mapping $a(p)^{-1}:\mathscr{X} \rightarrow \mathscr{X}$ if and only if $a(p) = c = \text{constant} \neq 0$, in which case $a(p)^{-1} = c^{-1}$. $\mathscr{X}$ is moreover "rich enough" to contain all possible complex-valued solutions y to the differential equation represented by $a(p)y = 0$ for any nonzero polynomial operator $a(p)$. More generally a polynomial matrix operator of the form (3) has an inverse mapping $A(p)^{-1}:\mathscr{X}^s \rightarrow \mathscr{X}^q$ if and only if $s = q$ and $\det A(p) = c = \text{constant} \neq 0$. In this case $A(p)^{-1}$ is again a polynomial matrix operator of the form (3). It is obtained as the matrix inverse of $A(p)$. If $s = q$ and $\det A(p) \neq 0$ then there is, for any $x \in \mathscr{X}^s$, at least one $y \in \mathscr{X}^q$ such that $A(p)y = x$, and moreover, $\mathscr{X}^q$ is rich enough to contain all relevant solutions y to the system of differential equations represented by $A(p)y = 0$.

Now a system consisting of s linear time-invariant differential equations can be written as

4 $$A(p)y = B(p)u,$$

with $u \in \mathscr{X}^r$, $y \in \mathscr{X}^q$, and with $A(p)$ and $B(p)$ polynomial matrix operators of the form (3) of sizes s by q and s by r respectively. The *differential input-output relation* $S \subset \mathscr{X}^r \times \mathscr{X}^q$ determined by equation (4) is now defined as the set of all input-output pairs (u, y) that satisfy equation (4), i.e.

5 $$S \triangleq \{(u, y) \mid (u, y) \in \mathscr{X}^r \times \mathscr{X}^q \text{ and } A(p)y = B(p)u\}.$$

Equation (4) can also be written as

6 $$[A(p) \vdots -B(p)] \begin{bmatrix} y \\ \cdots \\ u \end{bmatrix} = 0.$$

Now interpreting $[A(p) \vdots -B(p)]$ as a polynomial s by $(q+r)$-matrix operator, the converse S^{-1} of S can be presented simply as

7 $$S^{-1} = \ker[A(p) \vdots -B(p)].$$

We shall say that S is *generated* by $[A(p) \vdots -B(p)]$, and that $[A(p) \vdots -B(p)]$ is a *generator* for S (this concept is closely related to the concept of a "system matrix" as introduced by Rosenbrock, 1970). S is clearly uniquely determined by its generator, but the same S can generally be generated by infinitely many generators.

S as given by equation (5) is a relation but generally not a mapping, i.e. to a given input signal u there may correspond more than one output signal y such that $(u, y) \in S$. If certain conditions are fulfilled, then it proves possible to construct input-output mappings from the relation S by fixing suitable sets of initial conditions. This is possible in particular if there is a generator $[A(p) \vdots -B(p)]$ for S with $s = q$ and $\det A(p) \neq 0$. In this case we shall say that S is *regular*, and we shall call $[A(p) \vdots -B(p)]$ a *regular* generator for S. Moreover, the degree of $\det A(p)$ is called the *order* of the generator and of the corresponding S. It can further be shown that if $s = q$, then S is regular if and only if $[A(p) \vdots -B(p)]$ is regular. A regular differential input-output relation S also possesses the very convenient property of having the whole of $\mathscr{X}^r$ as its domain. We shall, in what follows, mainly be interested in regular differential input-output relations, and unless otherwise stated, it is understood that regular generators are used for regular differential input-output relations.

Input-output relations as discussed above do not seem to be very useful as abstract (mathematical) models of real systems, because they cannot answer the question: Given the input signal under certain circumstances, what is the expected output signal? This is so because an input-output relation contains information only about the set of all possible output signals which can be expected with a given input signal—no information concerning the circumstances are contained in the relation. To answer the question posed we need a more detailed concept than just an input-output relation—we need what we call an *abstract input-output system*, that is a family of input-output mappings where every mapping corresponds to certain circumstances. Thus studies concerning the properties of real systems and their interconnections should generally be based on the use of abstract input-output systems as models of the real systems involved.

As was stated above it is, nevertheless, possible to form a family of input-output mappings, i.e. an abstract system, from a regular differential input-output relation by fixing a suitable set of initial conditions. This makes it possible to study the properties of real systems and their interactions on the basis of the corresponding input-output relations and their compositions. This is exactly what will be done in the main part of this book. For this purpose we have, of course, also to study the relationship between certain features of input-output relations and of corresponding abstract input-output systems.

In this introductory chapter we shall mainly be concerned with regular differential input-output relations and with compositions of such relations. Later chapters are devoted to more detailed studies.

1.2 Compositions of differential input-output relations

Models of complex real systems are usually formed by interconnecting separately constructed models of suitably formed subsystems of the real systems under consideration. In particular we shall here study interconnections of differential systems with the interconnections represented by compositions of corresponding differential input-output relations.

Suppose that the given real system is divided into subsystems represented by the regular differential input-output relations $S_1, S_2, \ldots, S_N$, and let $[A_1(p) \vdots -B_1(p)], [A_2(p) \vdots -B_2(p)], \ldots, [A_N(p) \vdots -B_N(p)]$ be corresponding regular generators. The interactions between the subsystems are described by means of an *interconnection matrix* C, which tells how the input and output signals of the subsystems are connected, and how the input and output signals of the overall system are chosen. Let $u_j \in \mathscr{X}^{r_j}, y_j \in \mathscr{X}^{q_j}. j = 1, 2, \ldots, N$, denote the input and output signals respectively of the jth subsystem, and let $u_0 \in \mathscr{X}^{r_0}$, $y_0 \in \mathscr{X}^{q_0}$ correspondingly denote the input and output signals of the overall system. The interconnection matrix C is then a $(q_0 + r)$ by $(r_0 + q)$-matrix with $r \triangleq \sum_{j=1}^{N} r_j$ and $q \triangleq \sum_{j=1}^{N} q_j$, which satisfies the conditions (for more details, see section 3.1):

(i) every entry is either a 1 or a 0.
(ii) every row contains one and only one 1.
(iii) each of the first r_0 columns contains at least one 1.
(iv) the first r_0 entries of the q_0 uppermost rows are zeros.

The given family $\{S_1, S_2, \ldots, S_N\}$ of regular differential input-output relations—or alternatively, the family of corresponding regular generators—together with the interconnection matrix C determine a composition of relations, and our aim is to find the resulting input-output relations determined by this composition. We are interested in particular in the resulting "overall" input-output relation S_o consisting of all possible pairs $(u_0, y_0) \in \mathscr{X}^{r_0} \times \mathscr{X}^{q_0}$ associated with the composition. We shall proceed as follows.

The interconnection constraints are expressed with the aid of the interconnection matrix according to

1 $$\begin{bmatrix} y_0 \\ u_1 \\ \vdots \\ u_N \end{bmatrix} = \underbrace{\begin{bmatrix} 0 & C_{01} & \cdots & C_{0N} \\ C_{10} & C_{11} & \cdots & C_{1N} \\ \vdots & \vdots & & \vdots \\ C_{N0} & C_{N1} & \cdots & C_{NN} \end{bmatrix}}_{C} \begin{bmatrix} u_0 \\ y_1 \\ \vdots \\ y_N \end{bmatrix},$$

or more briefly, using obvious notations,

2 $$\begin{bmatrix} y_0 \\ u \end{bmatrix} = \begin{bmatrix} 0 & C_2 \\ C_3 & C_4 \end{bmatrix} \begin{bmatrix} u_0 \\ y \end{bmatrix}.$$

The system of differential equations describing the subsystems are written in the form

3 $$\underbrace{\begin{bmatrix} A_1(p) & 0 & \cdots & 0 \\ 0 & A_2(p) & \cdots & 0 \\ \vdots & \vdots & & \vdots \\ 0 & 0 & \cdots & A_N(p) \end{bmatrix}}_{A(p)} \begin{bmatrix} y_1 \\ y_2 \\ \vdots \\ y_N \end{bmatrix} = \underbrace{\begin{bmatrix} B_1(p) & 0 & \cdots & 0 \\ 0 & B_2(p) & \cdots & 0 \\ \vdots & \vdots & & \vdots \\ 0 & 0 & \cdots & B_N(p) \end{bmatrix}}_{B(p)} \begin{bmatrix} u_1 \\ u_2 \\ \vdots \\ u_N \end{bmatrix},$$

or more briefly, again using obvious notations,

4 $$A(p)y = B(p)u$$

with $\det A(p) \neq 0$. Thus (4) determines a regular differential input-output relation $S \subset \mathscr{X}^r \times \mathscr{X}^q$, and a regular generator for S is

5 $$[A(p) \vdots -B(p)].$$

From (2) and (4) we get the equation

6 $$\begin{bmatrix} A(p) - B(p)C_4 & 0 \\ -C_2 & I \end{bmatrix} \begin{bmatrix} y \\ y_0 \end{bmatrix} = \begin{bmatrix} B(p)C_3 \\ 0 \end{bmatrix} u_0,$$

which determines a differential input-output relation S_{io}—called the "internal-overall" input-output relation—determined by the composition. S_{io} consists of all signal triples of the form $(u_0, (y, y_0)) \in \mathscr{X}^{r_0} \times (\mathscr{X}^q \times$

$\mathscr{X}^{q_0}$) satisfying equation (6), i.e. a generator for S_{io} is given by

7
$$\left[\begin{array}{c:c:c} A(p) - B(p)C_4 & 0 & -B(p)C_3 \\ \hdashline -C_2 & I & 0 \end{array}\right].$$

The determinant of the left-hand part of (7) is det $(A(p) - B(p)C_4)$ (recall that the determinant of a square block triangular matrix with square diagonal blocks is equal to the product of the determinants of the diagonal blocks). Thus S_{io} is regular if and only if det $(A(p) - B(p)C_4) \neq 0$. If S_{io} is regular, then the composition under consideration is also said to be regular. Only regular compositions are generally useful as representations of interconnections of real systems.

From S_{io} we can obtain two further input-output relations called the "internal" input-output relation S_i and the "overall" input-output relation S_o. S_i is found as the set of all pairs of the form $(u_0, y) \in \mathscr{X}^{r_0} \times \mathscr{X}^{q}$, and S_o as the set of all pairs of the form $(u_0, y_0) \in \mathscr{X}^{r_0} \times \mathscr{X}^{q_0}$ that can be formed from the elements $(u_0, (y, y_0)) \in S_{io}$.

From (6) it is immediately concluded that S_i is a differential input-output relation generated by

8
$$[A(p) - B(p)C_4 \vdots -B(p)C_3].$$

S_i is regular if and only if S_{io} is regular.

The situation is not so clear with respect to S_o—although it should be reasonably clear that S_o is also a differential input-output relation, no generator for S_o can be read directly from (6). To obtain a generator for S_o, equation (6) must first be transformed to a suitable form. We shall say that the elements (u_0, y_0) of S_o are obtained through "elimination" of the variable y from the elements $(u_0, (y, y_0))$ of S_{io}. The elimination procedure is discussed below in section 1.4 and in more detail in section 7.1.

The block diagram in Fig. 1.1 illustrates equation (6) and the various input-output relations mentioned above. The interpretation of the diagram should be quite obvious.

Below we shall illustrate the points presented above by means of a few examples.

9 **Example.** Fig. 1.2 shows a series composition of two regular differential input-output relations S_1 and S_2 generated by the regular generators $[A_1(p) \vdots -B_1(p)]$ and $[A_2(p) \vdots -B_2(p)]$ respectively. The diagram is formed in accordance with Fig. 1.1, except that the interconnection constraints are directly indicated by means of connecting lines without the use of special connecting elements as was done in Fig. 1.1.

In this example the relevant quantities are as follows.

S is generated by the regular generator (to help the reader in interpreting various matrices appearing in the context, we occasionally label columns and/or rows with symbols indicating with which variables the columns and rows in question are associated)

10
$$\underbrace{\begin{array}{cc} y_1 & y_2 \\ \left[\begin{array}{c:c} A_1(p) & 0 \\ \hdashline 0 & A_2(p) \end{array}\right. \end{array}}_{A(p)} \begin{array}{c} \\ \vdots \end{array} \underbrace{\begin{array}{cc} u_1 & u_2 \\ \left.\begin{array}{c:c} -B_1(p) & 0 \\ \hdashline 0 & -B_2(p) \end{array}\right] \end{array}}_{-B(p)}.$$

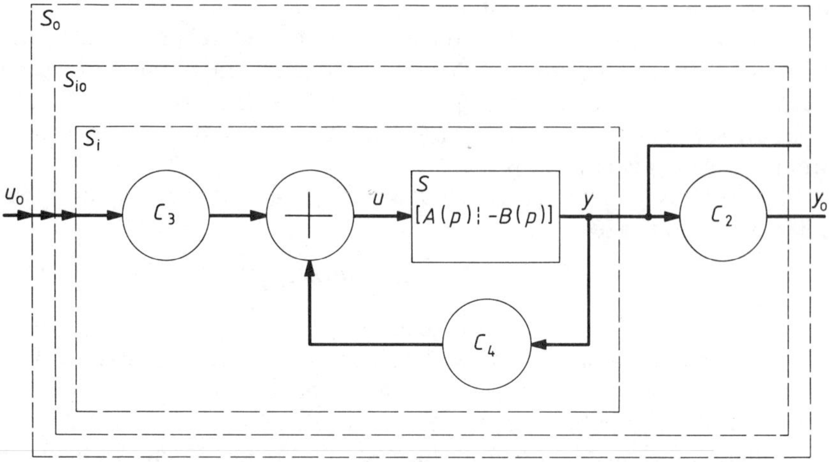

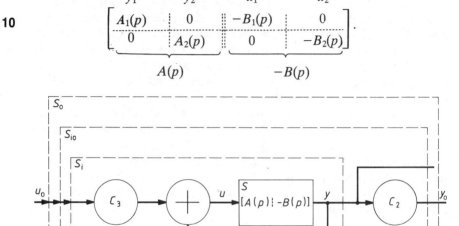

Fig. 1.1. Block diagram illustrating the composition represented by equation (1.2.6).

The interconnection matrix C is

11
$$C = \begin{array}{c} \\ y_0 \\ u_1 \\ u_2 \end{array} \begin{array}{c} \begin{array}{ccc} u_0 & y_1 & y_2 \end{array} \\ \left[\begin{array}{c:c:c} 0 & 0 & I \\ \hdashline I & 0 & 0 \\ \hdashline 0 & I & 0 \end{array}\right] \end{array}.$$

From (10) and (11) the generator for S_{io} as given by (7) is obtained as

12
$$\begin{array}{cccc} y_1 & y_2 & y_0 & u_0 \end{array}$$
$$\left[\begin{array}{c:c:c:c} A_1(p) & 0 & 0 & -B_1(p) \\ \hdashline -B_2(p) & A_2(p) & 0 & 0 \\ \hdashline 0 & -I & I & 0 \end{array}\right],$$

and the corresponding generator (8) for S_i is thus

13
$$\begin{array}{ccc} y_1 & y_2 & u_0 \\ \left[\begin{array}{c|c|c} A_1(p) & 0 & -B_1(p) \\ \hline -B_2(p) & A_2(p) & 0 \end{array}\right]. \end{array}$$

The regularity of S_{io} and S_i follows from the regularity of S, i.e. from $\det A_1(p) \det A_2(p) \neq 0$.

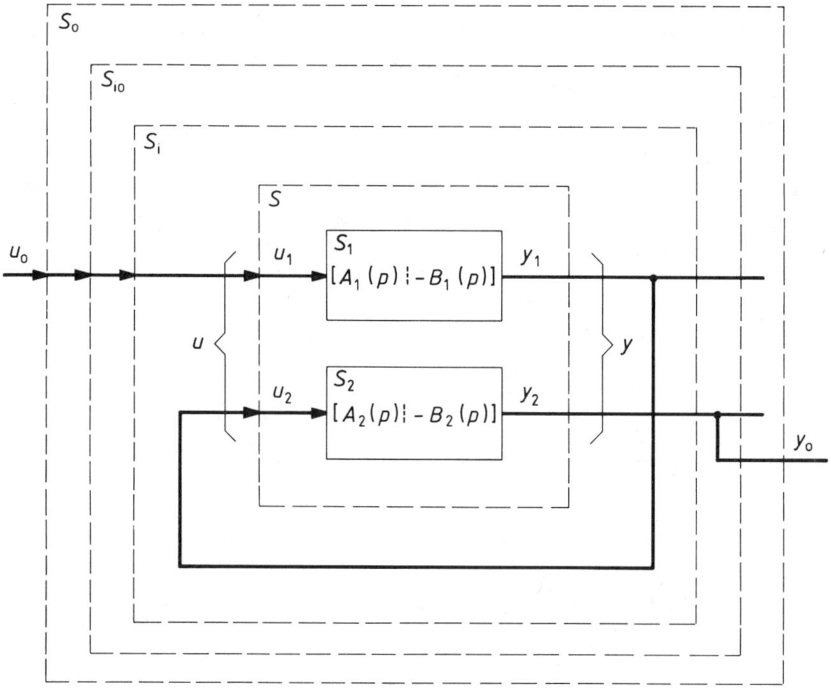

Fig. 1.2. Series composition of two regular differential input-output relations S_1 and S_2.

On the basis of the results obtained so far nothing definite can be said about the resulting overall input-output relation S_o and its generator. The final results concerning the present series composition are derived in section 7.6 with the aid of the elimination procedure mentioned above. A numerical example is considered in example (1.4.9) below. □

14 **Note.** Example (9) above was treated in accordance with the general

formalism presented earlier in this paragraph. Now it can be seen that in most cases a simplified procedure can be applied. Consider, for instance, the bottom block row in (7) and (12). The structure of C_2 means simply a renaming of some of the components of y, for instance in (12) it means that y_2 is renamed y_0 in forming the overall input-output relation S_o. This renaming can, however, equally well be performed already in forming S_i and its generator (8). If this is done, then S_{io} can be replaced by S_i without loss of information. In example (9) we can thus replace the label y_2 in (13) by the label y_0, which makes (12) superfluous. For more details concerning this matter, see section 7.2.

This simplification possibility will be applied in the sequel, i.e. we shall generally not explicitly indicate S_{io} and its generator. Other obvious simplifications are also possible—thus Fig. 7.6 shows in a simplified and more customary form the same composition as Fig. 1.2. □

15 **Example.** As a second example consider a feedback composition, as shown in Fig. 1.3, around a regular differential input-output relation S generated by the regular generator

16
$$\begin{array}{ccccc} y & & u_1 & & u_2 \\ [A(p) & \vdots & -B_1(p) & \vdots & -B_2(p)]. \end{array}$$

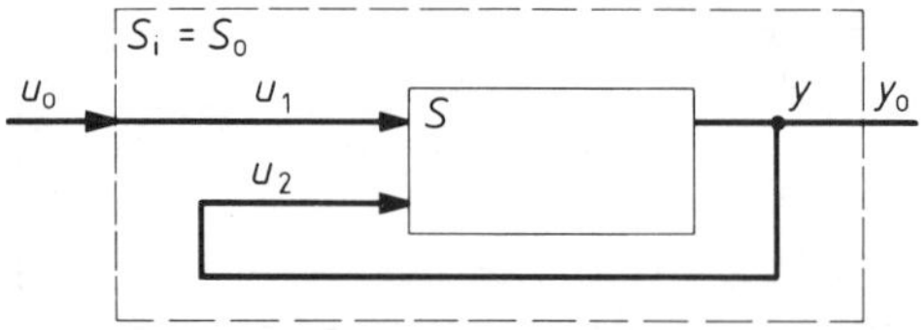

Fig. 1.3. Feedback composition around a regular differential input-output relation S.

The interconnection matrix C is found to be

17
$$C = \begin{array}{c} \\ y_0 \\ u_1 \\ u_2 \end{array} \begin{array}{c} \begin{array}{cc} u_0 & y \end{array} \\ \left[\begin{array}{c|c} 0 & I \\ \hline I & 0 \\ \hline 0 & I \end{array}\right] \end{array}.$$

Here S_i is equal to S_o, and a generator for $S_i = S_o$ is obtained from (8) as

18
$$\begin{array}{ccc} y_0 & & u_0 \\ [A(p) - B_2(p) & \vdots & -B_1(p)]. \end{array}$$

$S_i = S_o$ is regular if and only if $\det(A(p) - B_2(p)) \neq 0$. In this case we obtained a generator for S_o without any elimination procedure.

We shall return to feedback compositions below in section 1.5 and in more detail in section 7.12. □

1.3 Algebraic foundations

The introductory discussion presented above showed that in general not all the interesting results concerning a composition of differential input-output relations can be obtained just by forming the relevant equations. In addition we need a machinery which allows us to transform the equations obtained in a suitable way. As it happens such a machinery is provided by the abstract algebra.

The signal space as a module

Let $\mathbf{C}[p]$ denote the set of all polynomial operators of the form (1.1.1), i.e.

1
$$\mathbf{C}[p] \triangleq \{a(p) | a(p) = a_0 + a_1 p + \ldots + a_n p^n, \quad a_0, a_1, \ldots, a_n \in \mathbf{C}, \quad n \in \{0, 1, 2, \ldots\}\}.$$

With the signal space $\mathscr{X}$ chosen as described above it turns out that p is an indeterminate over $\mathbf{C}$, and that the elements of $\mathbf{C}[p]$ are thus ordinary polynomials in this indeterminate. This means that all the usual concepts and methods concerning ordinary polynomials in an indeterminate apply to the elements of $\mathbf{C}[p]$. Hence sums and products of elements of $\mathbf{C}[p]$ can be formed according to the usual rules for polynomials—the sum of two polynomials representing the pointwise sum of the corresponding mappings, and the product representing the composition of those mappings. It is easily seen that $\mathbf{C}[p]$ is a commutative ring and an integral domain with unity with respect to these operations. $\mathbf{C}[p]$ is, however, not a field, because the only invertible elements of $\mathbf{C}[p]$ are the nonzero constants.

The image $a(p)x$ of $x \in \mathscr{X}$ under $a(p) \in \mathbf{C}[p]$ can be interpreted as the product of $a(p)$ and x. With this interpretation $\mathscr{X}$ becomes a (left) $\mathbf{C}[p]$-module.

Then consider a set of module equations represented by a matrix equation of the form $A(p)y = x$ with $A(p)$ a polynomial matrix operator (*cf.* (1.1.3)). Such a matrix equation cannot generally be solved with respect to y for a given x in the usual vector space sense, because we cannot perform division by arbitrary nonzero scalars. Thus the question arises: Would it be possible to strengthen the $\mathbf{C}[p]$-module $\mathscr{X}$ to a vector space $\mathscr{X}$ over some suitable field which contains $\mathbf{C}[p]$ as a subring?

Now it is known that we can form the field of quotients of $\mathbf{C}[p]$, denoted by $\mathbf{C}(p)$, in much the same way as the field of rational numbers is formed from the ring of integers. Formally the field of quotients of $\mathbf{C}[p]$ is given by

2 $$\mathbf{C}(p) \triangleq \{b(p)/a(p) \,|\, a(p), \quad b(p) \in \mathbf{C}[p], \quad a(p) \neq 0\},$$

and in this field every nonzero polynomial is invertible. Thus we may expect it to be possible to extend the set of scalars $\mathbf{C}[p]$ to $\mathbf{C}(p)$ and so to make $\mathscr{X}$ a vector space over $\mathbf{C}(p)$. It can, however, be shown that this is not possible in this case, because no multiplication between the elements of $\mathbf{C}(p)$ and $\mathscr{X}$ satisfying the condition $(b(p)/1)x = b(p)x$ exists. A signal space that is a vector space over $\mathbf{C}(p)$ can be constructed only if the present signal space $\mathscr{X}$ is replaced by another suitably chosen signal space—but this would in turn lead to loss of important structural properties. At this stage we shall therefore be content with the module structure obtained—the vector space structure and its implications are studied in detail in part III of the book.

It remains for us to find out what else can be done with module equations if they cannot be solved in the usual vector space sense. In this context various forms—canonical and others—of polynomial matrices play an important role.

Polynomial matrices

Thus let us next study the properties of polynomial matrices in some detail.

Let $\mathbf{C}[p]^{s\times q}$ denote the set of all s by q-matrices with entries belonging to $\mathbf{C}[p]$, and let $A(p)$, $B(p) \in \mathbf{C}[p]^{s\times q}$. Then $A(p)$ is invertible as a polynomial matrix if and only if $s = q$ and $\det A(p) = c = \text{constant} \neq 0$. Such a matrix is said to be *unimodular*. Further, $A(p)$ and $B(p)$ are said to be *row equivalent* if there is a unimodular matrix $P(p) \in \mathbf{C}[p]^{s\times s}$ such that $A(p) = P(p)B(p)$, or equivalently, $B(p) = P(p)^{-1}A(p)$, where $P(p)^{-1}$ is the inverse of $P(p)$ as a polynomial matrix. Correspondingly, $A(p)$ and $B(p)$ are *column equivalent* if $A(p) = B(p)Q(p)$ with $Q(p) \in \mathbf{C}[p]^{q\times q}$ unimodular. Every square polynomial matrix is row equivalent to an upper right (or lower left, as desired) triangular matrix where every entry above a nonzero diagonal entry is of lower degree than the diagonal entry itself. This matrix can moreover be found by means of a *triangularization procedure* (Rosenbrock, 1970; Polak, 1969). The triangularization can be performed using *elementary row operations*. These are (*cf.* Rosenbrock, 1970, p. 7; appendix A1):

(i) interchange any two rows.
(ii) add to any row any other row multiplied by any element of $\mathbf{C}[p]$.
(iii) multiply any row by a nonzero constant.

Note that these operations do not include multiplication of a row by a nonconstant polynomial.

The elementary row operations can also be expressed as premultiplication of the matrix under consideration by *elementary matrices* of proper sizes, i.e. by matrices that are obtained by applying the elementary row operations to the identity matrix I of the proper size. The elementary matrices and their products are unimodular.

Now the determinant of a triangular matrix is equal to the product of the diagonal entries. A unimodular polynomial matrix must therefore be row equivalent to the identity matrix. It then follows that every unimodular matrix can be expressed as a product of elementary matrices and these matrices can be found by triangularization of the matrix under consideration.

The triangularization procedure is described in detail for instance in Polak, 1969. Here we shall be content with giving an illustrative example. The reader should have no difficulties in interpreting and generalizing the different operations applied.

3 **Example.** Consider a 3 by 3 polynomial matrix $A(p)$ with $\det A(p) \neq 0$ as given by

4
$$A(p) = \begin{bmatrix} 6p^2 + 3p & 4p^3 + 9p & 8p^2 \\ 6p^2 + 5p + 1 & 4p^3 + p^2 + 9p + 3 & 8p^2 + 2p \\ 4p^3 + 2p^2 & 2p^4 + 6p^2 & 4p^3 + 2 \end{bmatrix}.$$

We shall apply the elementary row operations mentioned above to the augmented matrix $[A(p) \,\vdots\, I]$, where I is the 3 by 3 identity matrix. If the final result is $[\tilde{A}(p) \,\vdots\, P(p)]$, then we know that the operations applied to $A(p)$ to obtain $\tilde{A}(p)$ correspond to premultiplication of $A(p)$ by $P(p)$.

Step 1. The rows of $[A(p) \,\vdots\, I]$ are reordered, if necessary, so that the entry in the upper left corner, i.e. the entry in the position (1, 1), has the lowest degree among all the nonzero entries in the first column. It is seen that (4) fulfils this condition without any reordering.

Step 2. The degree of the entries in the positions (2, 1) and (3, 1) are now lowered in the following way.

The first row multiplied by -1 is added to the second row.

The third row is multiplied by 3 (this multiplication is performed only to avoid fractions in the next operation).

The first row multiplied by $-2p$ is added to the third row.
Through these operations $[A(p) \vdots I]$ is brought to the form

5 $$\begin{bmatrix} 6p^2+3p & 4p^3+9p & 8p^2 & \vdots & 1 & 0 & 0 \\ 2p+1 & p^2+3 & 2p & \vdots & -1 & 1 & 0 \\ 0 & -2p^4 & -4p^3+6 & \vdots & -2p & 0 & 3 \end{bmatrix}.$$

Step 3. The rows of (5) are reordered so that the entry in the position (1, 1) has the lowest degree among all the nonzero entries in the first column. This is achieved by moving the second row to the uppermost position.

Step 4. The degree of the entry in the position (2, 1) in the matrix thus obtained is now lowered in the following way.

The first row multiplied by $-3p$ is added to the second row. The resulting matrix is:

6 $$\begin{bmatrix} 2p+1 & p^2+3 & 2p & \vdots & -1 & 1 & 0 \\ 0 & p^3 & 2p^2 & \vdots & 3p+1 & -3p & 0 \\ 0 & -2p^4 & -4p^3+6 & \vdots & -2p & 0 & 3 \end{bmatrix}.$$

Step 5. The first column has now attained a satisfactory form. The first column and the first row are therefore momentarily deleted to yield

7 $$\begin{bmatrix} p^3 & 2p^2 & \vdots & 3p+1 & -3p & 0 \\ -2p^4 & -4p^3+6 & \vdots & -2p & 0 & 3 \end{bmatrix}.$$

The procedure above is now repeated and applied to (7).

Step 6. The entry in the position (1, 1) in (7) is of lower degree than the other nonzero entry in the first column. No reordering of the rows is therefore performed.

The first row multiplied by $2p$ is added to the second row. This results in

8 $$\begin{bmatrix} p^3 & 2p^2 & \vdots & 3p+1 & -3p & 0 \\ 0 & 6 & \vdots & 6p^2 & -6p^2 & 3 \end{bmatrix}.$$

Step 7. The first column in (8) has attained a satisfactory form, and so has the entry in the position (2, 2). At this stage we therefore augment (8) with the previously deleted row and column to get

9 $$\begin{bmatrix} 2p+1 & p^2+3 & 2p & \vdots & -1 & 1 & 0 \\ 0 & p^3 & 2p^2 & \vdots & 3p+1 & -3p & 0 \\ 0 & 0 & 6 & \vdots & 6p^2 & -6p^2 & 3 \end{bmatrix}.$$

Step 8. The first column in (9) is satisfactory.

The second column is also satisfactory, because the entry in the position (1, 2) is of lower degree than the corresponding diagonal entry in the position (2, 2).

The third column is not yet in a satisfactory form, because the entries in the positions (1, 3) and (2, 3) are of higher degree than the corresponding diagonal entry in the position (3, 3). Therefore we proceed in the following way.

The third row multiplied by $-1/3\,p^2$ is added to the second row.

The third row multiplied by $-1/3\,p$ is added to the first row.

The result reads

10 $$\left[\begin{array}{ccc:ccc} 2p+1 & p^2+3 & 0 & -2p^3-1 & 2p^3+1 & -p \\ 0 & p^3 & 0 & -2p^4+3p+1 & 2p^4-3p & -p^2 \\ 0 & 0 & 6 & 6p^2 & -6p^2 & 3 \end{array}\right].$$

The left-hand part is seen to be of the particular upper right triangular form mentioned above, and the right-hand part represents the elementary row operations applied to $A(p)$. Note that the determinant of $A(p)$ is $\propto 6\,p^3(2p+1) \neq 0$. □

11 **Note. (i)** The triangularization operations performed above can, of course, be carried out without the repeated reordering of the rows as suggested in steps 1, 3, and 6. The final reordering to an upper triangular form can be made at the end of the procedure.

(ii) The above operations were chosen solely on the basis of the leading terms appearing in the entries under consideration. This leads to a simple and easily performed procedure. The number of elementary row operations required to obtain the final result could in some cases be reduced by choosing the operations on the basis of a complete division algorithm as outlined in Polak, 1969, rather than on the basis of a partial division algorithm as applied above.

(iii) Note that the first entry in the first column of the final matrix (10), i.e. $2p + 1$, is a greatest common divisor of the entries in the first column of the original matrix $A(p)$ as given by (4). In the same way the first entry in the first column of (8), i.e. p^3, is a greatest common divisor of the entries in the first column of (7). Quite generally the triangularization procedure amounts to finding greatest common divisors of certain polynomials, and the final form depends on the appearance of such divisors.

If the polynomial matrix under consideration represents a differential equation model of a real physical system, then the coefficients appearing in the matrix depend on the actual values of the parameters of the real system. There may then be difficulties in identifying "true" common divisors, i.e. divisors, which remain common divisors for every possible set of parameter values. This problem imposes special requirements on the computational procedures applied for the present purpose. □

Canonical forms

An important question concerns the uniqueness of the triangular equivalent form of a given polynomial matrix. Suppose that $A(p)$, $B(p) \in \mathbf{C}[p]^{q \times q}$, $\det A(p) \neq 0$, $\det B(p) \neq 0$, are row equivalent and of the particular upper right triangular form mentioned above. Then the transformation matrix $P(p) \in \mathbf{C}[p]^{q \times q}$ satisfying $A(p) = P(p)B(p)$ proves to be a diagonal matrix whose diagonal entries are nonzero constants. That is to say that the rows of $A(p)$ and $B(p)$ coincide up to constant multiples. If the matrices are normalized so that the diagonal entries are made monic (i.e. the leading coefficients of the diagonal entries are equal to one), then $A(p) = B(p)$. Such a form can accordingly be called a *canonical upper triangular form*, or a CUT-form for short (*cf.* appendix A2).

Correspondingly, if we choose a lower left triangular form, then the normalization leads to a *canonical lower triangular form*, or CLT-form.

12 **Example.** The CUT-form of $A(p)$ as given by (4) is (*cf.* (10))

13
$$\begin{bmatrix} p + 1/2 & 1/2 p^2 + 3/2 & 0 \\ 0 & p^3 & 0 \\ 0 & 0 & 1 \end{bmatrix}.$$
□

More generally also row equivalent matrices of the form

14
$$A(p) = [A_1(p) \vdots A_2(p)], \text{ or}$$

15
$$A(p) = \begin{bmatrix} A_1(p) \\ \cdots \\ 0 \end{bmatrix},$$

where $A_1(p)$ is of CUT-form ($A_1(p)$ square, $\det A_1(p) \neq 0$) coincide. Hence such matrices can also be said to be of CUT-form.

Let $P(p)$ be a unimodular matrix and let $B(p)$ be a suitable polynomial matrix so that $P(p)B(p)$ is of CUT-form. Then $P(p)$ is unique if $P(p)B(p)$ is of the form (14), but nonunique if $P(p)B(p)$ is of the form (15) with one or more zero rows.

With the aid of elementary column operations corresponding to the elementary row operations mentioned above, it is also possible to transform a polynomial matrix to a canonical lower left (or upper right, as desired) triangular form. Any unimodular matrix is column equivalent to the identity matrix.

To avoid ambiguity, we shall use the CUT- and CLT-notations only for canonical forms with respect to row equivalence. If canonical forms with respect to column equivalence are meant, then we shall say so.

Later on, in the main part of the text, we shall meet other important canonical forms.

Common divisors

Let $L(p)$, $A(p)$, $A_1(p) \in \mathbf{C}[p]^{q\times q}$, $B(p)$, $B_1(p) \in \mathbf{C}[p]^{q\times r}$ be such that $A(p) = L(p)A_1(p)$ and $B(p) = L(p)B_1(p)$. $L(p)$ is then said to be a *common left divisor* of $A(p)$ and $B(p)$. $L(p)$ is a *greatest common left divisor* (GCLD) of $A(p)$ and $B(p)$ if every other common left divisor $M(p)$ of $A(p)$ and $B(p)$ is such that $L(p) = M(p)N(p)$ for some $N(p) \in \mathbf{C}[p]^{q\times q}$. One GCLD of $A(p)$ and $B(p)$ can be found by transforming $[A(p) \vdots B(p)]$ to a column equivalent matrix of the form $[L(p) \vdots 0]$—for instance with $L(p)$ of lower triangular form. $L(p)$ is then a GCLD of $A(p)$ and $B(p)$. In particular, if the GCLD of $A(p)$ and $B(p)$ are unimodular, then $A(p)$ and $B(p)$ are said to be *left coprime*. In this case any common left divisor of $A(p)$ and $B(p)$ is unimodular, and $[A(p) \vdots B(p)]$ is column equivalent to $[I \vdots 0]$. For $A(p)$ and $B(p)$ left coprime there is thus a unimodular matrix $Q(p)$ such that

16 $$[A(p) \vdots B(p)] = [I \vdots 0] \underbrace{\left[\begin{array}{c:c} Q_1(p) & Q_2(p) \\ \hdashline Q_3(p) & Q_4(p) \end{array}\right]}_{Q(p)} = [Q_1(p) \vdots Q_2(p)],$$

that is, there are matrices $Q_3(p)$ and $Q_4(p)$ such that

17 $$Q(p) \triangleq \left[\begin{array}{c:c} A(p) & B(p) \\ \hdashline Q_3(p) & Q_4(p) \end{array}\right]$$

is unimodular. Note that a $Q(p)$ as given by (17) can be found in the following way. Determine a unimodular $P(p)$ such that $[A(p) \vdots -B(p)]P(p) = [I \vdots 0]$. Then $Q(p) = P(p)^{-1}$.

Further, let $P(p)$ be a unimodular matrix of the form

18 $$P(p) \triangleq \left[\begin{array}{c:c} P_1(p) & P_2(p) \\ \hdashline P_3(p) & P_4(p) \end{array}\right]$$

with $P_1(p)$ or $P_2(p)$ square. Then $P_1(p)$ and $P_2(p)$ are left coprime, and so are $P_3(p)$ and $P_4(p)$.

Common right divisors and *greatest common right divisors* (GCRD) of

$A(p) \in \mathbf{C}[p]^{q \times q}$ and $B(p) \in \mathbf{C}[p]^{r \times q}$ can be studied analogously. The above considerations can, of course, be extended to any number of matrices.

19 **Example.** Consider the polynomial matrices $A(p)$ and $B(p)$ contained in a generator (*cf.* (1.1.7)) $[A(p) \vdots -B(p)]$ given by

20
$$\begin{bmatrix} 2p^3 + p^2 - 8p - 4 & 2p + 1 & \vdots & -4p^2 - 2p \\ p^2 - 4 & p^3 + 5p^2 + 8p + 5 & \vdots & -3p - 2 \end{bmatrix}.$$

By means of elementary column operations the augmented matrix (*cf.* example (3))

21
$$\begin{bmatrix} A(p) & \vdots & -B(p) \\ \hline & I & \end{bmatrix}$$

can be transformed to

22
$$\begin{bmatrix} L(p) & \vdots & 0 \\ \hline & P(p) & \end{bmatrix}$$

$$= \begin{bmatrix} 2p+1 & 0 & \vdots & 0 \\ 1 & p+2 & \vdots & 0 \\ \hline 1/47(-24p^2 - 26p - 6) & 1/47(-92p^2 - 84p + 24) & & 2p^3 + 6p^2 + 4p - 1 \\ 1/47(-12p + 23) & 1/47(-46p + 96) & & p^2 - 4 \\ 1/47(-12p^3 - 13p^2 + 45p + 46) & 1/47(-46p^3 - 42p^2 + 196p + 145) & & p^4 + 3p^3 - 2p^2 - 12p - 8 \end{bmatrix}$$

with

23
$$[L(p) \vdots 0] = [A(p) \vdots -B(p)]\, P(p),$$

i.e. $P(p)$ represents the elementary column operations applied to $[A(p) \vdots -B(p)]$ to obtain $[L(p) \vdots 0]$. Note that here $\det P(p) = 1$.

It follows that $L(p)$ given by

24
$$\begin{bmatrix} 2p + 1 & 0 \\ 1 & p + 2 \end{bmatrix}$$

is a GCLD of $A(p)$ and $(-)$ $B(p)$. Further we can write

25
$$[A(p) \vdots -B(p)] = L(p)[A_1(p) \vdots -B_1(p)]$$

with $A_1(p)$, $B_1(p)$ left coprime. Interpreting (25) as an equation in rational matrices (*cf.* below), we can compute $[A_1(p) \vdots -B_1(p)]$ as $L(p)^{-1}$ $[A(p) \vdots -B(p)]$ yielding

26
$$\begin{bmatrix} p^2 - 4 & 1 & \vdots & -2p \\ 0 & p^2 + 3p + 2 & \vdots & -1 \end{bmatrix}.$$

Thus

$$\det A(p) = \det L(p) \det A_1(p) = (2p+1)(p+2)(p^2-4)(p^2+3p+2) \neq 0,$$

and (20) is consequently a regular generator.

Note the following properties possessed by the unimodular transformation matrix $P(p)$ as given by (22) and satisfying (23).

The last column is unique up to a constant multiple.

The first and second columns are not unique. In fact, let $\tilde{P}(p)$ be formed from $P(p)$ so that the first and second columns of $\tilde{P}(p)$ are formed from the corresponding columns of $P(p)$ by adding arbitrary polynomial multiples of the last column to the first and second columns. Evidently then $[L(p) \vdots 0] = [A(p) \vdots -B(p)]P(p) = [A(p) \vdots -B(p)]\tilde{P}(p)$, i.e. $\tilde{P}(p)$ would do the same job as $P(p)$. Here this fact has been used to reduce the degrees of the entries in the first and second columns of $P(p)$ to their lowest possible values. □

Rational matrices

The existence of the field of quotients $\mathbf{C}(p)$ (*cf.* (2)) also allows us to introduce rational matrices, i.e. matrices with entries belonging to $\mathbf{C}(p)$. For instance, consider a regular differential input-output relation S generated by the regular generator

27 $$[A(p) \vdots -B(p)].$$

Now if $A(p)$ and $B(p)$ are interpreted as rational matrices, then $A(p)$ is invertible as a rational matrix and we can form the rational matrix $\mathcal{G}(p)$ given by

28 $$\mathcal{G}(p) \triangleq A(p)^{-1} B(p).$$

$\mathcal{G}(p)$ is called the *transfer matrix* determined by $[A(p) \vdots -B(p)]$, or equivalently (*cf.* sections 6.3, 6.4), by S. In single input—single output cases the term "transfer ratio" is occasionally used for (28).

The transfer matrix plays an important role in what follows (*cf.* chapter 6). Note however, that no mapping involving signal spaces is assigned to the transfer matrix at this stage.

Properness and strict properness

We shall also use the features of the transfer matrix to define the concepts of "properness" and "strict properness". For every $s \in \mathbf{C}$, let $\mathcal{G}(s)$ denote the complex matrix obtainable from (28) by substituting s for p in (28). We shall say that S, $[A(p) \vdots -B(p)]$, as well as $\mathcal{G}(p)$ above are *proper* if

$\lim_{|s|\to\infty} \mathscr{G}(s) = K =$ a constant matrix. If $K = 0$, then these quantities are said to be *strictly proper.*

29 **Example.** Let a regular generator $[A(p) \vdots -B(p)]$ be given according to (20) above. The corresponding transfer matrix $\mathscr{G}(p) = A(p)^{-1} B(p)$ can be computed as $A_1(p)^{-1} B_1(p)$ with $[A_1(p) \vdots -B_1(p)]$ given by (26). The result is

30
$$\mathscr{G}(p) = \begin{bmatrix} \dfrac{2p^3 + 6p^2 + 4p - 1}{(p^2 - 4)(p^2 + 3p + 2)} \\ \dfrac{1}{p^2 + 3p + 2} \end{bmatrix},$$

a strictly proper transfer matrix. □

1.4 Properties of differential input-output relations

It has already been mentioned that a differential input-output relation does not uniquely determine a corresponding generator. It is therefore possible to choose different generators for a given input-output relation depending on what properties we are looking for. We shall therefore next discuss the problem of how a given generator can be transformed without changing the underlying input-output relation. We have the following important result.

A fundamental result

Let S_1 and S_2 be two regular differential input-output relations generated by the regular generators $[A_1(p) \vdots -B_1(p)]$ and $[A_2(p) \vdots -B_2(p)]$ respectively. It is found (*cf.* theorems (6.2.1) and (6.2.2)) that $S_2 \subset S_1$ if and only if there exists a square polynomial matrix $L(p)$ such that

1
$$[A_1(p) \vdots -B_1(p)] = L(p)[A_2(p) \vdots -B_2(p)],$$

which also implies that $\det L(p) \neq 0$. Moreover $S_2 = S_1$ if and only if $L(p)$ is unimodular.

Canonical representations

It follows from the above fundamental result that all possible regular generators for a given regular differential input-output relation are mutually row equivalent—they in fact constitute an equivalence class. This in turn implies that there is, for every regular differential input-output relation, a unique corresponding generator of CUT-form. This is a nice feature from an identification point of view because the CUT-form contains a relatively small number of parameters.

To emphasize the system aspect of row equivalence we shall also say that two row equivalent regular generators are *input-output equivalent.*

Stability

A regular differential input-output relation S generated by a regular generator $[A(p) \vdots -B(p)]$ is said to be *asymptotically stable* if all the roots of $\det A(p)$ have negative real parts (*cf.* Zadeh and Desoer, 1963, sections 5.7 and 7.5). If so, then all the solutions y to $A(p)y = 0$ asymptotically approach zero with increasing time. The monic polynomial corresponding to $\det A(p)$ is called the *characteristic polynomial* of S as well as of $[A(p) \vdots -B(p)]$.

Because the determinant of a triangular matrix is simply the product of the diagonal entries, it is often advantageous to transform a given generator to a triangular form before computing its characteristic polynomial.

An elimination procedure

In section 1.2 above we discussed compositions of differential input-output relations. The discussion led to the following problem. In the case of a regular composition we obtained regular generators (1.2.7) and (1.2.8) for the input-output relations S_{io} and S_i determined by the composition, but a corresponding generator for the overall input-output relation S_o could not be found at that stage. Now we are in a position to give a solution to this problem. Referring to note (1.2.14) we shall proceed in the following way (for more details, see sections 7.1 and 7.2).

Suppose that the internal input-output relation S_i determined by the composition in question is regular and generated by the regular generator $[A_i(p) \vdots -B_i(p)]$. Suppose furthermore that the components of the internal output y are ordered so that y can be written as $y = (y_1, y_0)$, where y_0 represents those components of y which appear as outputs of the overall input-output relation S_o. $[A_i(p) \vdots -B_i(p)]$ can then be written in partitioned form as

2

$$\begin{array}{ccc} y_1 & y_0 & u_0 \\ \end{array}$$
$$\left[\begin{array}{c:c:c} A_{i1}(p) & A_{i2}(p) & -B_{i1}(p) \\ \hdashline A_{i3}(p) & A_{i4}(p) & -B_{i2}(p) \end{array}\right]$$

with $A_{i1}(p)$, $A_{i4}(p)$ square. Now our task is to generate the set S_o as the set of all pairs (u_0, y_0) that can be formed from the elements $(u_0, (y_1, y_0))$ contained in S_i, i.e. the variable y_1 shall be "eliminated" from the elements $(u_0, (y_1, y_0))$ of S_i.

To this end the generator (2) is transformed to an upper right block triangular form by means of elementary row operations, for instance to CUT-form. The new generator is then input-output equivalent to (2), i.e. it is a regular generator for S_i, and it has the form

3
$$\begin{array}{c} \\ \\ \end{array}\begin{bmatrix} \overset{y_1}{\tilde{A}_{i1}(p)} & \overset{y_0}{\tilde{A}_{i2}(p)} & \overset{u_0}{-\tilde{B}_{i1}(p)} \\ 0 & A_0(p) & -B_0(p) \end{bmatrix}$$

with $\det \tilde{A}_{i1}(p) \neq 0$ and $\det A_0(p) \neq 0$. Note that the $\tilde{A}_{i1}(p)$ appearing above is a GCRD of $A_{i1}(p)$ and $A_{i3}(p)$ in (2).

Now it is asserted that S_o is a regular differential input-output relation generated by the regular generator

4
$$\overset{y_0}{[A_0(p)} \;\vdots\; \overset{u_0}{-B_0(p)]}$$

obtained from (3). The validity of this assertion is a direct consequence of the fact that, according to (3), every (u_0, y_0) that can be formed from an element $(u_0, (y_1, y_0))$ belonging to S_i must satisfy the equation $A_0(p)y_0 = B_0(p)u_0$, and conversely, for every (u_0, y_0) satisfying $A_0(p)y_0 = B_0(p)u_0$ there is always a y_1 such that $(u_0, (y_1, y_0))$ belongs to S_i (*cf.* section 1.1). In fact, any y_1 satisfying

5
$$\tilde{A}_{i1}(p)y_1 = -\tilde{A}_{i2}(p)y_0 + \tilde{B}_{i1}(p)u_0$$

would do.

Observability

A special case is obtained if $\tilde{A}_{i1}(p)$ above is unimodular, i.e. if $A_{i1}(p)$ and $A_{i3}(p)$ in (2) are right coprime. If so then y_1 in (5) is uniquely determined by (u_0, y_0) according to

6
$$y_1 = \tilde{A}_{i1}(p)^{-1}(-\tilde{A}_{i2}(p)y_0 + \tilde{B}_{i1}(p)u_0).$$

Conversely, if we require that the y_1 corresponding to any $(u_0, y_0) \in S_o$ be unique, then $\tilde{A}_{i1}(p)$ must be unimodular. A regular composition possessing this remarkable property is said to be *observable*. This concept turns out to be a generalization of the usual observability concept for state-space representations (*cf.* section 7.9).

Algebraic equations

Consider again a regular differential input-output relation S generated by the regular generator $[A(p) \vdots -B(p)]$, i.e. by the equation $A(p)y = B(p)u$. It may happen that this equation contains purely algebraic relationships between the various components of u and y. The situation is closely related to the observability concept considered above, and it is always possible to extract, from $A(p)y = B(p)u$, relevant equations representing such algebraic relationships. This question is discussed in detail in section 7.1.

Controllability

Now suppose that the generator $[A(p) \vdots -B(p)]$ for S above is such that

7 $$[A(p) \vdots -B(p)] = L(p)[A_1(p) \vdots -B_1(p)],$$

where $L(p)$ is a GCLD of $A(p)$ and $B(p)$. What does a relationship of this kind signify from a system point of view beyond what was contained in the fundamental result cited at the beginning of this paragraph? This question proves to be rather complicated, and here we shall therefore be content with making the following brief observation. If $L(p)$ above is not unimodular, i.e. if $A(p)$ and $B(p)$ are not left coprime, then S can be represented as a parallel composition of the form shown in Fig. 7.5 (note that our present notations deviate from those used in Fig. 7.5). The essential thing is that the upper part in Fig. 7.5 represents an input-output relation whose output does not depend on the input, and which exhibits zero input responses not present in the zero input response space of the lower part in the diagram. It is found that this leads to the result that—in a certain sense—the overall output cannot be efficiently "controlled" by the overall input. On the other hand, if $L(p)$ is unimodular and $A(p)$ and $B(p)$ consequently left coprime, then no such deficiency appears. This point will be further illuminated in note (8.5.35), (iii).

In view of the above brief observation we shall say that the S generated by (7) is *controllable* if $A(p)$ and $B(p)$ are left coprime.

We shall encounter this controllability concept on many occasions in the sequel. In particular it is found in section 7.9 that our controllability concept—suitably interpreted—is in agreement with the usual controllability concept for state-space representations.

8 **Example.** The regular differential input-output relation S generated by the regular generator $[A(p) \vdots -B(p)]$ as given by (1.3.20) is not controllable, because $A(p)$ and $B(p)$ have $L(p)$, $\det L(p) \neq$ constant, given by (1.3.24) as GCLD. Moreover, S is not asymptotically stable, because $\det A(p)$ has a root at $+2$. □

9 **Example.** Consider the series composition previously discussed in example (1.2.9) and depicted in Fig. 1.2 (*cf.* also Fig. 7.6), and let the regular differential input-output relations S_1 and S_2 involved in the composition be generated by the regular generators $[A_1(p) \vdots -B_1(p)]$ and $[A_2(p) \vdots -B_2(p)]$ given by

10 $$[(p-2)(p+3) \vdots -(p+1)(p+3)]$$

and

11 $$[(p+1)(p+4) \vdots -(p-2)(p+4)]$$

respectively.

We note the following points.

(i) S_1 is not asymptotically stable ($\det A_1(p)$ has a root at $+2$), whereas S_2 is asymptotically stable.

(ii) Neither S_1, nor S_2 is controllable ($(p+3)$ is a greatest common left and right divisor of $A_1(p)$ and $B_1(p)$, and $(p+4)$ is a greatest common left and right divisor of $A_2(p)$ and $B_2(p)$).

(iii) The transfer ratios determined by S_1 and S_2 are

12 $$\mathscr{G}_1(p) \triangleq A_1(p)^{-1}B_1(p) = \frac{p+1}{p-2}$$

and

13 $$\mathscr{G}_2(p) \triangleq A_2(p)^{-1}B_2(p) = \frac{p-2}{p+1}$$

respectively. Thus S_1 and S_2 are both proper. Further, it is evidently to be expected that the transfer ratio $\mathscr{G}_o(p)$ determined by the resulting overall input-output relation S_o will in this case be $\mathscr{G}_o(p) = \mathscr{G}_2(p)\,\mathscr{G}_1(p) = 1$.

Now we can apply the general results obtained above as well as the more special results obtained in example (1.2.9) for the series composition.

Thus a regular generator corresponding to (2) above for the internal input-output relation S_i is in this case obtained from (1.2.13) as

14 $$\begin{array}{ccc} y_1 & y_0 & u_0 \end{array}$$
$$\left[\begin{array}{cc:c} (p-2)(p+3) & 0 & -(p+1)(p+3) \\ -(p-2)(p+4) & (p+1)(p+4) & 0 \end{array}\right].$$

Next we form the generator corresponding to (3) by transforming (14) to an upper right triangular form, say to CUT-form, by means of elementary row operations. The desired result is obtained by premultiplication of (14)

by the unimodular matrix

15 $$\begin{bmatrix} -1 & -1 \\ p+4 & p+3 \end{bmatrix}$$

and the resulting CUT-form of (14) is

16 $$\begin{bmatrix} \overset{y_1}{p-2} & \overset{y_0}{-(p+1)(p+4)} & \vdots & \overset{u_0}{(p+1)(p+3)} \\ 0 & (p+1)(p+3)(p+4) & \vdots & -(p+1)(p+3)(p+4) \end{bmatrix}.$$

Consequently, a regular generator for the resulting overall input-output relation S_o is obtained as (*cf.* (3), (4) above)

17 $$[(p+1)(p+3)(p+4) \vdots -(p+1)(p+3)(p+4)].$$

Thus $\mathscr{G}_0(p) = 1$ as expected. Moreover,
(i) the series composition considered is not observable (the entry in the upper left hand corner of (16) is not equal to 1),
(ii) S_i is not asymptotically stable (*cf.* (16)), but S_o is (*cf.* (17)),
(iii) S_i and S_o are not controllable.

Note that the generator $[A_0(p) \vdots -B_0(p)]$ as given by (17) can in this case be written as

18 $$[A_0(p) \vdots -B_0(p)] = (p-2)^{-1}[A_1(p)A_2(p) \vdots -B_1(p)B_2(p)],$$

i.e. although it here holds that $A_1(p)A_2(p) = B_1(p)B_2(p)$, only $(p-2)$ is cancelled when $[A_0(p) \vdots -B_0(p)]$ is formed! A more thorough discussion of this and other general features of the series composition will follow in section 7.6. □

1.5 Feedback compensation and control

In this paragraph we shall give a brief exposé of one of the most important areas of application of the present polynomial systems theory, namely various problems concerning feedback compensation and control. A more detailed discussion will follow later in section 7.12.

We shall start with a feedback composition of the kind depicted in Fig. 1.4.

In Fig. 1.4 S_1 is a regular differential input-output relation representing a given system whose dynamic properties we are trying to modify with the aid of a feedback compensator represented by the regular differential input-output relation S_2. If nothing else is explicitly stated, it will be assumed that all the signals indicated in Fig. 1.4 have one or more components. We shall in this context consider only certain aspects concerning the internal input-output relation S_i determined by the composition. The particular results obtained (e.g. stability) then carry over to any overall input-output relation S_o—with the output signal y_0 determined by $y_0 = C_2 y$ for some C_2 (*cf.* section 1.2)—that can be assigned to the composition.

Without loss of generality we can assume that a regular generator for

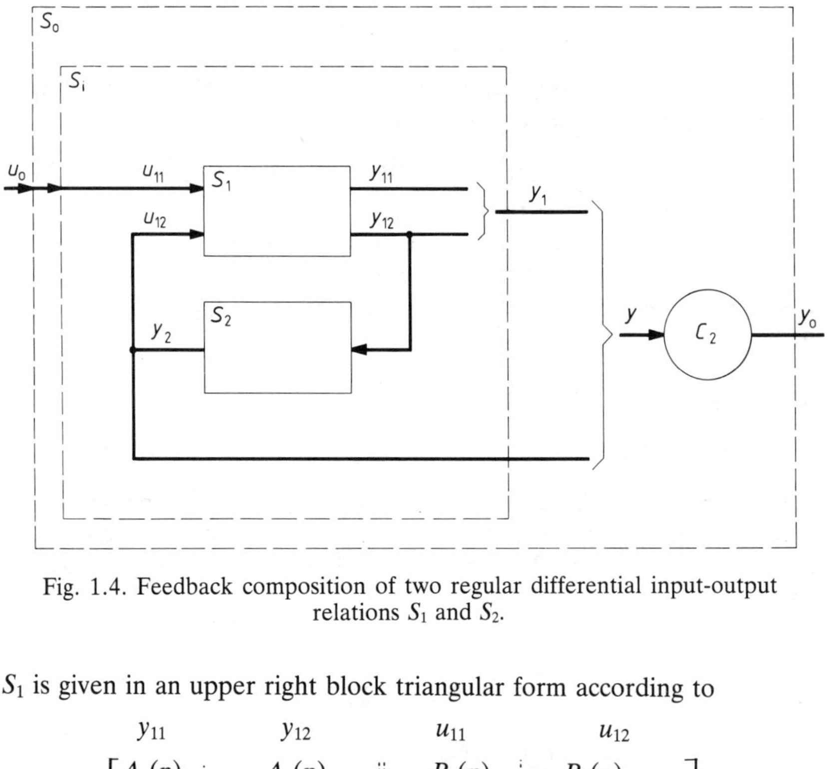

Fig. 1.4. Feedback composition of two regular differential input-output relations S_1 and S_2.

S_1 is given in an upper right block triangular form according to

1 $$\begin{array}{cccc} y_{11} & y_{12} & u_{11} & u_{12} \end{array}$$
$$\left[\begin{array}{c:c::c:c} A_1(p) & A_2(p) & -B_1(p) & -B_2(p) \\ \hdashline 0 & L(p)A_{41}(p) & -B_3(p) & -L(p)B_{41}(p) \end{array}\right],$$

where $L(p)$ is a GCLD of $L(p)A_{41}(p)$ and $L(p)B_{41}(p)$. This means that the regular differential input-output relation generated by $[L(p)A_{41}(p) \vdots -L(p)B_{41}(p)]$ is controllable if and only if $\det L(p) =$ constant. Note that if (1) is proper (strictly proper), then so is $[A_{41}(p) \vdots -B_{41}(p)]$.

The characteristic polynomial of S_1 is thus obtained from (1) as the monic polynomial corresponding to

2 $$\det A_1(p) \det L(p) \det A_{41}(p) \neq 0.$$

If there is no y_{11}, i.e. if the whole output y_1 of S_1 is used as input u_2 to the compensator S_2, then the upper block row along with the first block column on the left are deleted from (1), and $\det A_1(p)$ in (2) is interpreted as being equal to 1.

The problem is to choose S_2 generated by a regular generator of the form

3
$$\begin{array}{c} \quad y_2 \qquad\quad u_2 \\ [C(p) \;\vdots\; -D(p)] \end{array}$$

so that the resulting composition behaves in a satisfactory way.

Now the first step is to form a generator for the internal input-output relation S_i given as the set of all $(u_0, (y_{11}, y_{12}, y_2))$ generated by the composition. Such a generator can be constructed along the general lines discussed in section 1.2. The resulting generator for S_i then reads

4
$$\begin{array}{cccc} y_{11} & y_{12} & y_2 & u_0 \end{array}$$
$$\left[\begin{array}{c:c:c::c} A_1(p) & A_2(p) & -B_2(p) & -B_1(p) \\ \hdashline 0 & L(p)A_{41}(p) & -L(p)B_{41}(p) & -B_3(p) \\ \hdashline 0 & -D(p) & C(p) & 0 \end{array}\right].$$

On inspection it is easily confirmed that (4) represents all the relevant equations associated with the present composition.

Again, if there is no y_{11}, then (4) is reduced correspondingly, and $\det A_1(p)$ below (*cf.* (7)) is interpreted as being equal to 1.

We shall in this context be content with studying a particular stabilization and pole assignment problem associated with S_i. The problem reads as follows.

5 **Problem.** Consider the feedback composition shown in Fig. 1.4 and suppose that S_1 is given, regular and strictly proper. The problem is to find a regular and proper (alternatively strictly proper) S_2 such that S_i is regular and the roots of the characteristic polynomial of S_i, i.e. the roots of the characteristic polynomial of (4), are located in a satisfactory way with regard to the resulting dynamic properties of the feedback composition under consideration—in particular it is required that all these roots have negative real parts, implying an asymptotically stable S_i.

In order to make the problem nontrivial let it be assumed that the characteristic polynomial of S_1 as given by (2) has at least one root that is *not* satisfactorily located. □

6 **Remark.** Problem (5) above is a rather idealized problem formulation. In

more realistic formulations there are additional constraints concerning structural and other properties of the compensator. Consider, for instance, a situation where the given system (S_1 in Fig. 1.4) contains various controllers (say P-controllers), which are designed to guarantee a certain steady-state (static) control accuracy. In order not to disturb this accuracy it may in such a case be required that the feedback compensator S_2 be chosen so that the steady-state value of its output signal y_2 is zero irrespective of the steady-state value of its input signal $u_2 = y_{12}$.

Additional requirements appear if higher order control accuracy (control of the derivative of a signal) is also demanded, and in decentralized control. In most cases an asymptotically stable compensator would be desirable. □

Thus our main subject of interest is the location of the roots of the characteristic polynomial of (4), i.e. the roots of

7 $$\det A_1(p) \det L(p) \det \left[\begin{array}{c:c} A_{41}(p) & -B_{41}(p) \\ \hdashline -D(p) & C(p) \end{array}\right].$$

Now comparison of (2) and (7) shows that $\det A_1(p) \det L(p)$ is a common divisor of the characteristic polynomials of S_1 and S_i irrespective of how S_2 is chosen. Further it is found (*cf.* below) that there always exists a proper (or strictly proper as desired) S_2 such that the roots of the last factor in (7) are located within an arbitrarily chosen region of the complex plane. Moreover, S_2 can always be chosen controllable (a GCLD of $D(p)$ and $C(p)$ in (7) can always be deleted without deteriorating the resulting dynamic properties). This leads immediately to the following

8 **Conclusion.** Problem (5) has a solution if and only if the roots of $\det A_1(p) \det L(p)$ are satisfactorily located with regard to the resulting dynamic properties of the feedback composition under consideration. □

Hence in trying to solve a problem of the present kind the first thing to do would be to determine a generator (1) for different orderings and different groupings of the components of $y_1 = (y_{11}, y_{12})$, and then to choose an ordering and a grouping—if it exists—yielding a satisfactory location of the roots of $\det A_1(p) \det L(p)$. Of course, a corresponding analysis could also be performed with respect to the choice of u_{12}—it may happen that a satisfactory result can be obtained using only some of the control inputs available as components of u_{12}.

9 **Example.** Consider problem (5) specified as follows.

Suppose that there is no y_{11}, and let the corresponding generator (1) for S_1 be given as (S_1 can be supposed to consist of an adder followed by a single-input relation)

10
$$
\begin{array}{c}
\quad y_{12} \qquad\qquad u_{11} \qquad\qquad u_{12} \\
[L(p)A_{41}(p) \;\vdots\vdots\; -B_3(p) \;\vdots\; -L(p)B_{41}(p)]
\end{array}
$$

11
$$
= \left[\begin{array}{cc:c:c}
\multicolumn{2}{c}{y_{12}} & u_{11} & u_{12} \\
2p^3 + p^2 - 8p - 4 & 2p + 1 & -4p^2 - 2p & -4p^2 - 2p \\
p^2 - 4 & p^3 + 5p^2 + 8p + 5 & -3p - 2 & -3p - 2
\end{array}\right].
$$

Note that (11) is essentially the same strictly proper generator $[A(p) \,\vdots\, -B(p)]$ that was studied in examples (1.3.19), (1.3.29) (*cf.* (1.3.20), (1.3.30)). Note also that the present formulation implies that we are allowed to use feedback from both the output components of the given system. There is only one control input available.

According to the results obtained in example (1.3.19) we have (*cf.* (1.3.24))

12
$$
L(p) = \begin{bmatrix} 2p + 1 & 0 \\ 1 & p + 2 \end{bmatrix}
$$

and (*cf.* (1.3.26))

13
$$
[A_{41}(p) \,\vdots\, -B_{41}(p)]
$$

14
$$
= \left[\begin{array}{cc:c} p^2 - 4 & 1 & -2p \\ 0 & p^2 + 3p + 2 & -1 \end{array}\right].
$$

The characteristic polynomial of S_1 is thus found to be (*cf.* (2))

15
$$
\tfrac{1}{2}(2p + 1)(p + 2)(p^2 - 4)(p^2 + 3p + 2),
$$

and $\det A_1(p) \det L(p)$ is equal to

16
$$
(2p + 1)(p + 2).
$$

S_1 is accordingly not asymptotically stable because (15) has a root at $+2$. This root is consequently not satisfactorily located. Problem (5) has nevertheless a solution (*cf.* conclusion (8)) if we consider the roots at $-1/2$ and -2 of (16) as satisfactorily located. □

17 **Example.** Let example (9) studied above be altered in such a way that we are allowed to use feedback only from the second component of the output of the given system. This means that the equality (10), (11) now takes the

form

$$
\begin{array}{cccc}
y_{11} & y_{12} & u_{11} & u_{12}
\end{array}
$$

18
$$
\left[
\begin{array}{cc|cc}
A_1(p) & A_2(p) & -B_1(p) & -B_2(p) \\
\hline
0 & L(p)A_{41}(p) & -B_3(p) & -L(p)B_{41}(p)
\end{array}
\right]
$$

$$
\begin{array}{cccc}
y_{11} & y_{12} & u_{11} & u_{12}
\end{array}
$$

19
$$
= \left[
\begin{array}{cc|cc}
p^2 - 4 & p^3 + 5p^2 + 8p + 5 & -3p - 2 & -3p - 2 \\
\hline
0 & (2p^2 + 5p + 2)(p^2 + 3p + 2) & -2p^2 - 5p - 2 & -2p^2 - 5p - 2
\end{array}
\right],
$$

where (19) has been obtained from (11) by means of elementary row operations.

It follows that in this case

20
$$
A_1(p) = p^2 - 4,
$$
$$
L(p) = 2p^2 + 5p + 2 = (2p + 1)(p + 2),
$$
$$
[A_{41}(p) : -B_{41}(p)] = [p^2 + 3p + 2 : -1].
$$

The characteristic polynomial of S_1 is as before given by (15), and det $A_1(p)$ det $L(p)$ is equal to

21
$$
(p^2 - 4)(2p + 1)(p + 2).
$$

(21) has a root at $+2$, and hence the roots of det $A_1(p)$ det $L(p)$ are certainly not satisfactorily located, i.e. problem (5) has no solution in this case, and stability can therefore not be achieved.   □

22 **Example.** Let us finally study the above case when we are allowed to use feedback only from the first component of the output of the given system. The generator (1) for this case is obtained from (11) or (19) by interchanging the first two columns and then transforming the generator thus obtained to an upper right triangular form by means of elementary row operations. The significant part of the result obtained in this way reads:

23
$$
A_1(p) = 1,
$$
$$
L(p) = 2p^2 + 5p + 2 = (2p + 1)(p + 2),
$$
$$
[A_{41}(p) : -B_{41}(p)]
$$
$$
= [(p^2 - 4)(p^2 + 3p + 2) : -2p^3 - 6p^2 - 4p + 1].
$$

The characteristic polynomial of S_1 is as before given by (15), and det $A_1(p)$ det $L(p)$ is equal to

24 $$(2p+1)(p+2),$$

which coincides with (16). Thus problem (5) has a solution in this case if we consider the roots of (24) as satisfactorily located. □

Now suppose that we are faced with a problem (5) such that the roots of the corresponding det $A_1(p)$ det $L(p)$ (*cf.* (7)) are satisfactorily located. Consider then the last factor in (7) and let it be renamed as

25 $$\det\left[\begin{array}{c|c} A(p) & -B(p) \\ \hline -D(p) & C(p) \end{array}\right],$$

where $A(p)$ and $B(p)$ are now left coprime and $[A(p) \vdots -B(p)]$ strictly proper. Further, $A(p)$ and $C(p)$ are square with $\det A(p) \neq 0$ and $\det C(p) \neq 0$. The solvability of problem (5) relies directly on the fundamental result presented below.

A fundamental result

To begin with recall the result (1.3.17) which implies the following. Given $[A(p) \vdots -B(p)]$ as in (25), there are always left coprime $Q_3(p)$ and $Q_4(p)$ of appropriate sizes ($Q_4(p)$ square) such that

26 $$\det\left[\begin{array}{c|c} A(p) & -B(p) \\ \hline Q_3(p) & Q_4(p) \end{array}\right] = \text{constant} \neq 0,$$

moreover, such $Q_3(p)$ and $Q_4(p)$ can be constructed as explained in section 1.3.

Now the fundamental result referred to above reads (for a proof and a detailed discussion, see section 7.12):

Consider (25) and (26) as given above. Let $[A(p) \vdots -B(p)]$ be given, and let the corresponding $Q_3(p)$, $Q_4(p)$ be constructed. Then it holds that: **(i)** For any regular and proper (or strictly proper as desired) generator $[C(p) \vdots -D(p)]$ the degree of (25) is equal to the sum of the degrees of det $A(p)$ and det $C(p)$ implying that (25) is nonzero. Further there exist corresponding polynomial matrices $T_3(p)$ and $T_4(p)$ of appropriate sizes ($T_4(p)$ square, det $T_4(p) \neq 0$) such that

27 $$\left[\begin{array}{c|c} A(p) & -B(p) \\ \hline -D(p) & C(p) \end{array}\right] = \underbrace{\left[\begin{array}{c|c} I & 0 \\ \hline T_3(p) & T_4(p) \end{array}\right]}_{T(p)} \left[\begin{array}{c|c} A(p) & -B(p) \\ \hline Q_3(p) & Q_4(p) \end{array}\right],$$

with

28 $$\det\begin{bmatrix} A(p) & -B(p) \\ -D(p) & C(p) \end{bmatrix} \propto \det T_4(p).$$

(ii) For any (nonempty) set C of complex numbers there is always a regular and proper (or strictly proper as desired) generator $[C(p) \vdots -D(p)]$ of appropriate size such that (25) is nonzero and the roots of (25) belong to C. Moreover, $C(p)$ and $D(p)$ can always be chosen left coprime.

Generally there are, in fact, infinitely many generators $[C(p) \vdots -D(p)]$ of various orders possessing the properties mentioned in item (ii) above. It should also be reasonably clear that there are, among these generators, minimal order generators—the actual minimal order depending both on $[A(p) \vdots -B(p)]$ and on the set C. Some results concerning this matter are given later in section 7.12.

According to the above results, problem (5) has many solutions under the present assumptions. The next step would be to actually construct a suitable solution, i.e. a regular and proper (strictly proper) generator $[C(p) \vdots -D(p)]$ such that the roots of (25) are satisfactorily located. We would also like to find a $[C(p) \vdots -D(p)]$ of fairly low order, implying that $C(p)$ and $D(p)$ be left coprime (a controllable S_2).

Constructing candidates for $[C(p) \vdots -D(p)]$

Here the construction will be based on the equality (27). To this end note first that if $T_1(p), T_2(p), \ldots, T_k(p)$ are transformations of the type $T(p)$ in (27), then the product $T_k(p) \ldots T_2(p)T_1(p)$ is again a transformation of the same type. Hence we can proceed according to the following scheme.

Let an arbitrary candidate for

29 $$\begin{bmatrix} A(p) & -B(p) \\ -D(p) & C(p) \end{bmatrix}$$

be denoted by

30 $$Z(p) \triangleq \begin{bmatrix} A(p) & -B(p) \\ X(p) & Y(p) \end{bmatrix},$$

$\det Z(p) \neq 0$, yielding

31 $$[Y(p) \vdots X(p)]$$

as a candidate for

32 $$[C(p) \vdots -D(p)].$$

We shall require of a candidate $Z(p)$ as given by (30) that the roots of $\det Z(p)$ are satisfactorily located, but we shall not require that $X(p)$ and

$Y(p)$ be left coprime. For various reasons, which will gradually become apparent, we shall further require the upper part $[A(p) \vdots -B(p)]$ as well as the lower part $[X(p) \vdots Y(p)]$ of a candidate (30) to be of row proper form, i.e. the matrices having as their rows the leading coefficients of the rows of $[A(p) \vdots -B(p)]$ and $[X(p) \vdots Y(p)]$ respectively are required to be of full rank (*cf.* appendix A2). The row proper forms required can always be obtained by means of elementary row operations (*cf.* appendix A2; note here that $\det Z(p) \neq 0$ guarantees that $[X(p) \vdots Y(p)]$ is of full rank).

Next let us form the matrix having as its rows the leading coefficients of the rows of (30), and let this matrix be denoted by (we use an obvious notational convention)

33 $$\left[\begin{array}{c:c} A_1 & -B_1 \\ \hdashline X_1 & Y_1 \end{array}\right]$$

with A_1 and Y_1 square. $[A(p) \vdots -B(p)]$ was assumed regular, row proper, and strictly proper, which is equivalent to $\det A_1 \neq 0$ and $B_1 = 0$.

Now we shall regard (30) and (31) as a pair of *satisfactory* candidates for (29) and (32) respectively if, in (33), $\det Y_1 \neq 0$ ($X_1 = 0$ if a strictly proper S_2 is required). This condition is necessary and sufficient for (31) to represent a generator that is regular as well as proper (strictly proper), *cf.* appendix A2. If the condition is fulfilled, then a feasible solution to the given problem, i.e. a feasible generator $[C(p) \vdots -D(p)]$ for S_2 is obtained from $[Y(p) \vdots X(p)]$ on cancelling any (nonunimodular) GCLD from $Y(p)$ and $X(p)$ that may appear.

34 **Note.** If (30) happens to be a satisfactory candidate for (29), then the coefficient of the leading term of the determinant of (30) is just equal to the determinant of (33), i.e. equal to $\det A_1 \det Y_1$.

More important still, a satisfactory candidate (30) also possesses a rudimentary degree of insensitivity with respect to parameter variations in the following sense. Let C be any open set of complex numbers containing the roots of the determinant of a satisfactory candidate (30). Let a be any coefficient appearing in some entry of (30) such that either $a \neq 0$ or else $a = 0$, but the degree of the determinant of the candidate does not change if a is given any sufficiently small nonzero value. Then the roots mentioned above remain in C for sufficiently small variations of a. This means that the resulting composition obtained by using a satisfactory candidate (30) is asymptotically stable by construction and that it also remains asymptotically stable for sufficiently small variations of a.

This kind of insensitivity with respect to parameter variations may, nevertheless, not be enough from a practical point of view. To see this recall first that no finite dimensional mathematical model of a real system can account for all the time constants appearing in the system. Thus, in order for a proposed solution to be practically sensible we must at least require that the stability properties of the solution are not destroyed if sufficiently small positive time constants are added to the original models. Some further comments concerning this matter are given below in example (49). □

If (30) is not a satisfactory candidate for (29), then a new candidate is constructed from (30) according to

35
$$T(p)Z(p) = \begin{bmatrix} I & 0 \\ \hline T_3(p) & T_4(p) \end{bmatrix} \begin{bmatrix} A(p) & -B(p) \\ \hline X(p) & Y(p) \end{bmatrix}$$
$$= \begin{bmatrix} A(p) & -B(p) \\ \hline T_3(p)A(p) + T_4(p)X(p) & -T_3(p)B(p) + T_4(p)Y(p) \end{bmatrix}$$

with $\det T(p)Z(p) = \det T_4(p) \det Z(p)$, where $T(p)$ is a suitable transformation of the type appearing in (27) so chosen that the roots of $\det T_4(p)$ are satisfactorily located, and further, so that the lower part of the resulting matrix $T(p)Z(p)$ above is of row proper form. This lower part is then taken as a new candidate for (32). The candidates thus constructed are finally renamed according to (30) and (31) respectively.

In order to start the above procedure we first construct a suitable unimodular (*cf.* (26))

36
$$\begin{bmatrix} A(p) & -B(p) \\ \hline Q_3(p) & Q_4(p) \end{bmatrix}$$

having the upper and lower parts in an appropriate row proper form. The determinant of (36) is a nonzero constant, that is, it has no unsatisfactorily located roots. (36) can thus be taken as the first candidate $Z(p)$ (*cf.* (30)) for (29) yielding $[Q_4(p) \vdots Q_3(p)]$ as the first candidate for (32).

Quite generally it is, however, found that this first pair of candidates cannot be satisfactory (why?). Therefore we apply the above scheme repeatedly and construct new candidates until (hopefully) a satisfactory pair of candidates emerges. Of course, the crucial problem here is how to construct new candidates from old ones in order to guarantee that we eventually end up with a satisfactory result. One difficulty is that consecutive transformations of the type (35) form an irreversible process, i.e. if $\tilde{Z}(p) = T(p)Z(p)$ according to (35), then it is generally not possible to reconstruct $Z(p)$ as $T_1(p)\tilde{Z}(p)$ with $T_1(p)$ some transformation of the same type as $T(p)$. This means that if the transformation process "goes wrong" at some point, then the mistake cannot generally be corrected later.

Choosing suitable transformations $T(p)$

Here the choice of suitable transformations $T(p)$ will be based on the following reasoning.

Suppose that we have obtained a candidate (30) with the corresponding coefficient matrix given by (33). Suppose further that (30) is not a satisfactory candidate, i.e. that $\det Y_1 = 0 (X_1 \neq 0$ if a strictly proper S_2 is required). Loosely speaking this means that the leading terms of the rows of $[X(p) \vdots Y(p)]$ are concentrated too much on $X(p)$. A new candidate (30) should therefore be constructed from an old one according to (35) so that, in the new candidate, the leading terms of the rows of $[X(p) \vdots Y(p)]$ are concentrated more on $Y(p)$ than before. To this end the transformation $T(p)$ is chosen so that in the sum (*cf.* (35)) $T_3(p)A(p) + T_4(p)X(p)$ suitable leading terms of $T_4(p)X(p)$ are cancelled by corresponding leading terms of $T_3(p)A(p)$.

In view of this the passing from an old and unsatisfactory candidate (30) with the corresponding matrix of leading coefficients given by (33) to a new and possibly satisfactory candidate could be performed, for instance, on a row-by-row basis according to the following scheme. In this scheme we have assumed that (30) is a $(q + r)$ by $(q + r)$-matrix.

Step 1. In this first step we choose a suitable row of $[X(p) \vdots Y(p)]$ as a basis for the treatment. The row chosen should be a "most unsatisfactory" one. A reasonable way to make the choice would be the following. Consider the rows of $[X_1 \vdots Y_1]$ and let them be denoted by $[X_{i1} \vdots Y_{i1}]$ for $i = 1, 2, \ldots, r$. If $r = 1$, then the choice is trivial. Hence suppose that $r > 1$. If there are rows with $X_{i1} \neq 0$ and $Y_{i1} = 0$, then choose such a row, otherwise choose a row with $X_{i1} \neq 0$, $Y_{i1} \neq 0$. In the latter case a row with Y_{i1} a linear combination of some other rows of Y_1 would be preferred. Let the row thus chosen be the kth row of $[X(p) \vdots Y(p)]$. Go to Step 2.

Step 2. Denote the kth row of $[X(p) \vdots Y(p)]$ by $[X_k(p) \vdots Y_k(p)]$, and the corresponding row of the coefficient matrix (33) by $[X_{k1} \vdots Y_{k1}]$. Compute the constant row matrix

37 $$T_{3k1} \triangleq [t_1 \quad t_2 \ldots t_q]$$

from (note that A_1 in (33) is invertible)

38 $$T_{3k1} = -X_{k1} A_1^{-1}.$$

Go to Step 3.

Step 3. Let the row degrees of $A(p)$ be $r_1, r_2, \ldots, r_q$. Let the degree of the row $[X_k(p) \vdots Y_k(p)]$ be s_k. Determine the integers $\tau_1, \tau_2, \ldots, \tau_q$ according to the rule, for $i = 1, 2, \ldots, q$:

39
$$\tau_i = \begin{cases} 0 \text{ if } t_i = 0, \\ s_k - r_i \text{ otherwise.} \end{cases}$$

Denote

40
$$\tau_m \triangleq \min_{i \in \{1,2,\ldots,q\}} \tau_i.$$

If $\tau_m \geqslant 0$, go to Step 4. Otherwise, go to Step 5.

Step 4. Form the transformation matrix $T(p)$ according to

41
$$T(p) = \left[\begin{array}{c|c} I & 0 \\ \hline T_3(p) & I \end{array}\right]$$

with $T_3(p)$ given by

42
$$T_3(p) = \begin{array}{c} 1 \\ 2 \\ \vdots \\ k \\ \vdots \\ r \end{array} \left[\begin{array}{c} 0 \\ \hline 0 \\ \hline \vdots \\ \hline T_{3k1} \\ \hline \vdots \\ \hline 0 \end{array}\right] \begin{bmatrix} p^{\tau_1} & 0 \ldots 0 \\ 0 & p^{\tau_2} \ldots 0 \\ \vdots & \vdots \quad \vdots \\ 0 & 0 \ldots p^{\tau_q} \end{bmatrix}.$$

Compute

43
$$\left[\begin{array}{c|c} I & 0 \\ \hline T_3(p) & I \end{array}\right] \left[\begin{array}{c|c} A(p) & -B(p) \\ \hline X(p) & Y(p) \end{array}\right] = \left[\begin{array}{c|c} A(p) & -B(p) \\ \hline T_3(p)A(p) + X(p) & -T_3(p)B(p) + Y(p) \end{array}\right].$$

It is easily found that the above choice of $T_3(p)$ leads indeed to cancellations of leading terms in the kth row of $T_3(p)A(p) + X(p)$.

Transform, if necessary, the lower part of (43) to row proper form by means of elementary row operations. The resulting matrix thus obtained from (43) is renamed as (30) and taken as a new and possibly satisfactory candidate for (29). This candidate is then judged as described previously.

Step 5. If $\tau_m < 0$ (*cf.* (39), (40)), this means that s_k in (39) is too small—it should be increased at least by $-\tau_m$. Therefore, form the transformation matrix $T(p)$ according to

44
$$T(p) = \left[\begin{array}{c|c} I & 0 \\ \hline 0 & T_4(p) \end{array}\right]$$

with $T_4(p)$ having the form

45 $$T_4(p) = \begin{array}{c} \\ 1 \\ 2 \\ \vdots \\ k \\ \vdots \\ r \end{array} \begin{array}{c} \begin{array}{cccc} 1 & 2\ldots & k & \ldots r \end{array} \\ \begin{bmatrix} 1 & 0\ldots & 0 & \ldots 0 \\ 0 & 1\ldots & 0 & \ldots 0 \\ \vdots & \vdots & \vdots & \vdots \\ 0 & 0\ldots & t(p) & \ldots 0 \\ \vdots & \vdots & \vdots & \vdots \\ 0 & 0\ldots & 0 & \ldots 1 \end{bmatrix} \end{array},$$

where $t(p)$ is a suitable polynomial in p of degree $\geqslant -\tau_m$ and such that the roots of $t(p)$ are satisfactorily located.

In real parameter systems $t(p)$ can conveniently be formed as a product of factors of the form

46 $$(p + \alpha)$$

and

47 $$((p + \alpha)^2 + \beta^2) = (p^2 + 2\alpha p + \alpha^2 + \beta^2)$$

with α and β real, $\alpha > 0$. A factor of the form (46) adds a root at $-\alpha$, and a factor of the form (47) a pair of complex conjugate roots at $-\alpha \pm i\beta$, to the roots of the determinant of the candidate under consideration.

Next compute

48 $$\left[\begin{array}{c:c} I & 0 \\ \hdashline 0 & T_4(p) \end{array}\right] \left[\begin{array}{c:c} A(p) & -B(p) \\ \hdashline X(p) & Y(p) \end{array}\right].$$

It is easily seen that (48) formally qualifies as a candidate for (29). It is, however, no more satisfactory than the original candidate. Therefore rename the resulting matrix (48) as (30) and take it as a new unsatisfactory candidate for (29). Form the corresponding new coefficient matrix (33). After this start the procedure again from step 1 noting that some of the old results are still relevant.

Now it should be reasonably clear that if we apply consecutive transformations chosen according to the above general scheme, then we must, sooner or later, arrive at a satisfactory candidate (30) yielding a corresponding feasible solution to our problem. The details of the scheme can be implemented in many different ways, and the order and structure of the final solution $[C(p) \vdots -D(p)]$ may well depend on these details. Because the solution is constructed on a step-by-step basis, there will be ample opportunities for the designer to use his skill and intuition.

We shall return to the feedback problem later in section 7.12, where we shall take a slightly more direct course of attack.

Let us conclude the present paragraph with an illuminating numerical example.

49 **Example.** Let us return to example (9) above, and let problem (5) be formulated according to this example. It was already concluded that this problem has a solution.

In the present case $[A(p) \vdots -B(p)]$ is thus given by (14), i.e. by

50 $$[A(p) \vdots -B(p)] = \left[\begin{array}{cc:c} p^2-4 & 1 & -2p \\ 0 & p^2+3p+2 & -1 \end{array}\right],$$

and the corresponding matrix having as its rows the leading coefficients of the rows of (50) is consequently (*cf.* (33))

51 $$[A_1 \vdots -B_1] = \left[\begin{array}{cc:c} 1 & 0 & 0 \\ 0 & 1 & 0 \end{array}\right].$$

Thus $\det A_1 \neq 0$ and $B_1 = 0$, i.e. (50) is indeed strictly proper and row proper.

The next step would be to construct a matrix (36) corresponding to (50) as explained in section 1.3 (*cf.* (1.3.16), (1.3.17)). But in example (1.3.19) we already constructed a unimodular matrix $P(p)$ such that $[A(p) \vdots -B(p)]P(p) = [I \vdots 0]$ with $[A(p) \vdots -B(p)]$ as in (50) (*cf.* (1.3.20), . . . , (1.3.26)), and this $P(p)$ is given by the lower part of (1.3.22). To get a suitable matrix (36) as a first candidate $Z(p)$ for (29) we have accordingly first of all to compute $P(p)^{-1}$. The lower part of $P(p)^{-1}$ consists in this case of one single nonzero row—this part is therefore necessarily always row proper. $P(p)^{-1}$ would thus as such qualify as the first candidate. For the sake of convenience, we shall here take $P(p)^{-1}$ with the bottom row multiplied by 47 as our first candidate $Z(p)$ for (29). This gives

52 $$Z(p) \triangleq \left[\begin{array}{c:c} A(p) & -B(p) \\ \hdashline X(p) & Y(p) \end{array}\right] = \left[\begin{array}{cc:c} p^2-4 & 1 & -2p \\ 0 & p^2+3p+2 & -1 \\ \hdashline 12p-23 & 46p+42 & -24 \end{array}\right]$$

with $\det Z(p) = 47$. The coefficient matrix (33) for (52) is accordingly

53 $$\left[\begin{array}{c:c} A_1 & -B_1 \\ \hdashline X_1 & Y_1 \end{array}\right] = \left[\begin{array}{cc:c} 1 & 0 & 0 \\ 0 & 1 & 0 \\ \hdashline 12 & 46 & 0 \end{array}\right].$$

Hence we have $\det Y_1 = Y_1 = 0 (X_1 \neq 0)$, and (52) is, as expected, not a satisfactory candidate for (29). Therefore we shall, starting from (52), construct new candidates according to the general scheme outlined above.

Step 1, . . . , step 5 below refer to the corresponding steps as described in the scheme.
Step 1. In this case the lower part of (52) has one single row. Therefore $k = 1$. Go to Step 2.
Step 2. From (52), (53) we obtain

54 $$A_1^{-1} = \begin{bmatrix} 1 & 0 \\ 0 & 1 \end{bmatrix}$$

and

55 $$X_{11} = X_1 = [12 \quad 46].$$

This gives, according to (38),

56 $$T_{311} = -X_{11}A_1^{-1} = [t_1 \quad t_2] = [-12 \quad -46].$$

Go to Step 3.
Step 3. The row degrees of $A(p)$ are $r_1 = r_2 = 2$, and the row degree of $[X_1(p) \vdots Y_1(p)] = [X(p) \vdots Y(p)]$ is $s_1 = 1$. From (39) and (40) we get

57 $$\tau_1 = \tau_2 = \tau_{\mathrm{m}} = 1 - 2 = -1.$$

Because $\tau_{\mathrm{m}} < 0$, we next apply Step 5.
Step 5. Let us choose

58 $$t(p) = p + 1,$$

implying that we regard a root at -1 as satisfactorily located. This gives us from (52) according to (45) and (48) the following new and unsatisfactory candidate (30)

59 $$\left[\begin{array}{cc:c} p^2 - 4 & 1 & -2p \\ 0 & p^2 + 3p + 2 & -1 \\ \hdashline 12p^2 - 11p - 23 & 46p^2 + 88p + 42 & -24p - 24 \end{array}\right]$$

with the corresponding coefficient matrix (33) given as before by (53).
Using these new matrices we start again from Step 1.
Step 1. $k = 1$ as before. Go to Step 2.
Step 2. The result (56) for T_{311} still holds. Go to Step 3.
Step 3. We have $\tau_1 = \tau_2 = \tau_{\mathrm{m}} = 1 - 1 = 0$. Next we therefore apply Step 4.
Step 4. According to (42) we get for $T_3(p)$

60 $$T_3(p) = T_{311} = [-12 \qquad -46],$$

then we apply (43). This yields the new candidate (30) given by

61 $$\left[\begin{array}{cc:c} p^2-4 & 1 & -2p \\ 0 & p^2+3p+2 & -1 \\ \hdashline -11p+25 & -50p-62 & 22 \end{array}\right]$$

and the new coefficient matrix (33) given by

62 $$\left[\begin{array}{cc:c} 1 & 0 & 0 \\ 0 & 1 & 0 \\ \hdashline -11 & -50 & 0 \end{array}\right].$$

Still $\det Y_1 = Y_1 = 0$ ($X_1 \neq 0$), and (61) is consequently not a satisfactory candidate for (29). Therefore we start again from Step 1.

The reader is advised to perform the various steps as outlined above over and over again, always choosing

63 $$t(p) = p + 1$$

in Step 5.

Note that the product of all the $t(p)$'s chosen in step 5 determines in a unique way the details of the whole process in this single-row case.

After a number of iterations, a candidate (30) as given by

64 $$\left[\begin{array}{cc:c} p^2-4 & 1 & -2p \\ 0 & p^2+3p+2 & -1 \\ \hdashline 49p+145 & -8p-40 & 47p-4 \end{array}\right]$$

and having a determinant equal to $47\,(p+1)^5$ emerges. The corresponding coefficient matrix (33) is

65 $$\left[\begin{array}{cc:c} 1 & 0 & 0 \\ 0 & 1 & 0 \\ \hdashline 49 & -8 & 47 \end{array}\right],$$

i.e. $\det Y_1 = Y_1 = 47 \neq 0$ ($X_1 \neq 0$). It follows that (64) is a satisfactory candidate for (29) if S_2 is required to be only proper, but it is not yet satisfactory if S_2 is required to be strictly proper. In the latter case it suffices to apply the above scheme once more (exercise for the reader).

Thus, for S_2 proper, (64) gives the following feasible solution $[C(p) \vdots -D(p)]$ to our problem (there is no GCLD to be cancelled)

66 $$[47p-4 \vdots 49p+145 \quad -8p-40],$$

corresponding to a transfer matrix $\mathcal{G}_2(p) = C(p)^{-1} D(p)$ given by

67 $$\begin{bmatrix} \dfrac{-49p - 145}{47p - 4} & \dfrac{8p + 40}{47p - 4} \end{bmatrix}.$$

The characteristic polynomial of the resulting internal input-output relation S_i will be (*cf.* (16))

68 $$\tfrac{1}{2}(2p + 1)(p + 2)(p + 1)^5.$$

All the modes are thus aperiodically damped.

The S_2 thus obtained is, however, not asymptotically stable (the characteristic polynomial of (66) has a root at $+4/47$). Therefore let us apply the above scheme all over again starting from (52) and this time using in Step 5 $t(p)$'s such that the product of all these $t(p)$'s amounts to

69 $$(p + 1)^3((p + 1)^2 + 1),$$

where the last factor is of the form (47) with $\alpha = \beta = 1$.

As a result we obtain a candidate (30) of the form

70 $$\left[\begin{array}{cc:c} p^2 - 4 & 1 & -2p \\ 0 & p^2 + 3p + 2 & -1 \\ \hdashline 44p + 182 & -35p - 81 & 47p + 6 \end{array}\right]$$

having a determinant equal to $47(p + 1)^3((p + 1)^2 + 1)$. The corresponding coefficient matrix (33) is

71 $$\left[\begin{array}{cc:c} 1 & 0 & 0 \\ 0 & 1 & 0 \\ \hdashline 44 & -35 & 47 \end{array}\right],$$

i.e. $\det Y_1 = Y_1 = 47 \neq 0$ ($X_1 \neq 0$). It follows again that (70) is a satisfactory candidate for (29) if S_2 is required to be proper, but it is not satisfactory if S_2 is required to be strictly proper.

Accordingly, for S_2 proper, (70) gives the following feasible solution to our problem (no cancellations)

72 $$[47p + 6 \vdots 44p + 182 \quad -35p - 81],$$

the corresponding transfer matrix $\mathcal{G}_2(p) = C(p)^{-1} D(p)$ being equal to

73 $$\begin{bmatrix} \dfrac{-44p - 182}{47p + 6} & \dfrac{35p + 81}{47p + 6} \end{bmatrix}.$$

The characteristic polynomial of the internal input-output relation S_i is now (*cf.* (16))

74 $$\tfrac{1}{2}(2p+1)(p+2)(p+1)^3((p+1)^2+1).$$

There are now modes which are not aperiodically damped, but we have indeed found an asymptotically stable S_2 (the characteristic polynomial of (72) has a single root at $-6/47$).

The solutions (66) and (72) are both of first order. The reader is advised to compare this number with the order estimated on the basis of the results in section 7.12.

The reader is further advised to discuss the sensitivity aspects of the above solutions (*cf.* note (34) above). In particular it should be shown that additional sufficiently small positive time constants can be added to the candidates (64) and (70) without destroying the asymptotical stability of the resulting composition. *Hint*: let the above candidates be denoted by $Z(p)$ according to (30), multiply every entry of $A(p)$ and $Y(p)$ in turn by $(Tp+1)$, where T represents a time constant, let $Z'(p)$ denote the new $Z(p)$'s obtained in this way, compare in every case $\det Z(p)$ and $\det Z'(p)$ for sufficiently small positive values of T.

Finally note that the sensitivity properties of the candidates are generally not invariant with respect to row equivalence—they depend on the particular form used. □

1.6 Concluding remarks

The authors do hope that the material covered by the previous paragraphs has given the reader a fair idea and a working knowledge of the aim and means of the polynomial systems theory as developed and presented in this book.

The mathematical machinery and the main system concepts needed in this context were introduced above in a rather informal manner without many details or proofs. Essential points were clarified and illustrated by means of numerical examples.

As to practical applications of the present methodology, only one significant and typical problem was considered. This problem concerned the design of a feedback compensator, and the actual problem formulation was rather elementary. A numerical example concluded the presentation. The main text, in particular chapter 7, contains a few additional examples of problems to which the polynomial systems theory is particularly well suited. It should, however, be recognized that the practical applications considered here are no more than just a scratch on the surface—numerous other applications can be thought of, for instance, in the area of identification and diagnostics, decoupling theory, sensitivity analysis, and hierarchical and decentralized control. It is left to the interested reader to explore the applicability of the theory in various situations.

Of course, the reason for the relatively sparse appearance of practical applications is that it has been necessary to lay the main emphasis on creating the correct mathematical framework, and on interpreting various systems theoretical concepts in terms of this framework. The aim of the applications presented has merely been to show that the methodology is indeed applicable and not just a kind of fruitless logical gymnastics.

Part I

Basic Concepts of Systems Theory

The formation of models of large-scale real systems is a difficult task of paramount importance to applied systems theory. This modelling is usually effected by interconnecting known models of different interacting parts of the real system under consideration in such a way that the resulting overall model appropriately describes the relevant behaviour of the system as a whole.

It may happen that an interconnected model fails to describe the behaviour of the system adequately even though the models of the interconnected parts as such are perfectly satisfactory. From a practical point of view it is, of course, important to understand the reason for such a failure and to know what to do in order to correct the situation.

We shall, in this part of the book, be discussing this problem in some detail. For this purpose we shall first present some basic facts concerning mathematical models in general and dynamic models in particular. After some introductory remarks we shall then give an exact definition of an interconnection of a set of models, and introduce the concept of determinateness of such an interconnection. The results obtained will finally be illustrated by means of examples.

The term "system" will generally be used to mean a model of the kind mentioned above. When reference is made to the underlying physical reality, the term "real system" is employed. The presentation is based on a great number of books, papers and reports. The works by Arbib, 1966, Arbib and Zeiger, 1969, Blomberg, 1972a and b, 1975, Blomberg *et al.*, 1969, Kalman, 1960, Kalman, Falb and Arbib, 1969, Klir, 1969, Mesarovic, 1969, Mesarovic, Macko and Takahara, 1970, Mesarovic and Takahara, 1975, Orava, 1973 and 1974, Rosenbrock, 1970, Salovaara, 1967, Sinervo and Blomberg, 1971, Willems, 1971, Windeknecht, 1967 and 1971, Wymore, 1967, Ylinen, 1975 and 1980 and Zadeh and Desoer, 1963 have in particular contributed in a significant way to the content of this part of the book.

2

Systems and system descriptions

In order to be able to predict events occurring as a result of the occurrence of other events, or to be able to choose suitable actions for the purpose of causing certain desired events to occur, scientists and practicians working in various fields utilize *mathematical* or *abstract models* describing relevant *cause-effect relationships* observed among phenomena appearing in the surrounding world. These models reflect basic properties associated with such cause-effect relationships, for instance the causality property, meaning briefly that the effect cannot occur prior in time to the cause. Under different circumstances the same cause may, furthermore, result in different effects.

We shall devote this chapter to the presentation of a number of basic concepts relating to the general theory of abstract models—generally called "systems" in the sequel. The concepts introduced are mainly formulated in pure set theoretical terms using as little structure as possible. They are, moreover, chosen so as to make them particularly suitable for the purpose just mentioned.

A model of a cause-effect relationship should first of all be able to give a definite answer to the following question: Given the circumstances and the cause, what effect will be expected? Accordingly, we shall basically regard a system as a collection of mappings from sets of possible causes to sets of possible effects so that for given circumstances there is always a corresponding unique member of this collection, and this member assigns the appropriate expected effect to the given cause.

Now it has been customary in the relevant literature to choose the relation consisting of all possible cause-effect pairs that can be associated with a particular cause-effect relationship as the basic quantity (system) describing this relationship (see e.g. Mesarovic and Takahara, 1975) rather than the corresponding collection of mappings mentioned above. Generally such a relation is, however, clearly not satisfactory as a model, because it is not capable of answering the above question—it does not contain any information concerning the circumstances under which different cause-effect pairs are obtained.

There are, nevertheless, cases in which a given set of cause-effect pairs determines a natural corresponding collection of mappings of the kind just mentioned. This happens, for instance, with cause-effect relationships described by certain classes of linear differential and difference equations. In these cases it is therefore possible to develop analytical tools for treating various problems concerning such relationships using either collections of mappings or simple relations as a starting point. We shall discuss these special cases in detail in parts II to IV of the book. For the present we shall keep to our original model interpretation.

2.1 Parametric input-output mappings. Abstract input-output systems. System descriptions

The basic concept in describing cause-effect relationships will in this context be the concept of a *parametric input-output mapping*. Such a mapping is of the following general form.

There is a set $\mathcal{U}$ of (possible) *inputs*, a set $\mathcal{Y}$ of (possible) *outputs*, and a parameter, or *state* set Σ. A parametric input-output mapping h is then a mapping according to

1 $$h : \mathrm{D}h \to \mathcal{Y},\ \mathrm{D}h \subset \Sigma \times \mathcal{U}, \quad \text{with } \mathrm{D}(\mathrm{D}h) = \Sigma.$$

The domain of h is thus a relation from Σ to $\mathcal{U}$ such that the domain of this relation coincides with Σ.

In using such a mapping as a model of a cause-effect relationship, the input set $\mathcal{U}$ represents possible causes, the output set $\mathcal{Y}$ possible effects, and the state set Σ possible circumstances. A cause $u \in \mathrm{D}h(\alpha,.) \subset \mathcal{U}$ occurring under the circumstances $\alpha \in \Sigma$ is thus expected to result in the effect $y \in \mathcal{Y}$ given by $y = h(\alpha, u)$.

Two states α_1, $\alpha_2 \in \Sigma$ are said to be *equivalent* if $h(\alpha_1,.) = h(\alpha_2,.)$. Moreover, a state set Σ is called *minimal* if it does not contain any distinct equivalent states.

Abstract input-output systems

The mapping h as given by (1) defines a *family* $\mathcal{S}$ of *input-output mappings* according to

2 $$\mathcal{S} = \{h(\alpha,.)\,|\,\alpha \in \Sigma\}.$$

Such a family of input-output mappings will be called an *abstract input-output system*, or just a *system* for short. It should be emphasized that a given system $\mathcal{S}$ according to this definition can generally be generated by an infinite number of parametric input-output mappings. If $h_1 : \mathsf{D}h_1 \to \mathcal{Y}_1$, $\mathsf{D}(\mathsf{D}h_1) = \Sigma_1$ and $h_2 : \mathsf{D}h_2 \to \mathcal{Y}_2$, $\mathsf{D}(\mathsf{D}h_2) = \Sigma_2$, are such that $\mathcal{S} = \{h_1(\alpha_1,.)\,|\,\alpha_1 \in \Sigma_1\} = \{h_2(\alpha_2,.)\,|\,\alpha_2 \in \Sigma_2\}$, then h_1 and h_2 are said to be *equivalent* parametric input-output mappings for the system $\mathcal{S}$.

Associated with $\mathcal{S}$ there is also a corresponding *input-output relation S* consisting of the set of all input-output pairs determined by $\mathcal{S}$, i.e.

3 $$S = \cup\, \mathcal{S} = \{(u, y)\,|\,(u, y) \in s \text{ for some } s \in \mathcal{S}\} \subset \mathcal{U} \times \mathcal{Y}.$$

Of course, $\cup\, \mathcal{S}_1 = \cup\, \mathcal{S}_2$ does not generally imply $\mathcal{S}_1 = \mathcal{S}_2$.

System descriptions

From a practical point of view it is convenient to introduce a suitable *description* of a given system. Such a description can be formed as a list of symbols representing relevant quantities associated with the system. A description of a system $\mathcal{S}$ as given by (1), . . . , (3) above could thus be formed, for instance, as the list

4 $$(\mathcal{U}, \mathcal{Y}, \Sigma, S, \mathcal{S}, h),$$

or the list

5 $$(\Sigma, \mathcal{S}, h),$$

or just as the pair

6 $$(\mathcal{S}, h).$$

Because Σ, S, and $\mathcal{S}$ are uniquely determined by h, it will in general suffice to specify only h explicitly. Different descriptions of a given system $\mathcal{S}$ are said to be *equivalent.*

Schematically, a system $\mathcal{S}$ of the kind mentioned above can be presented as shown in Fig. 2.1.

Fig. 2.1. Alternative block diagram representations of a system $\mathcal{S}$ determined by a parametric input-output mapping h.

To summarize, we shall in this context regard a system just as a family of mappings between suitable sets. Note that if $\mathcal{S}$ is *any* family of mappings, then $\mathcal{S}$ can always be generated trivially by a parametric mapping h of the form (1) with Σ taken as equal to $\mathcal{S}$ itself.

2.2 Time systems. Dynamic systems

Realizability conditions for time systems

If a parametric input-output mapping h as given by (2.1.1) is used to describe a cause-effect relationship concerning sequences of consecutive events occurring in the course of time, then the input and output sets $\mathsf{D}h(\alpha,.)$ and $\mathsf{R}h(\alpha,.)$ are, for all $\alpha \in \Sigma$, sets of suitable *time functions* ("signals"). The corresponding system $\mathcal{S}$ as given by (2.1.2) is then said to be a *time system.* In such a case it is generally required that $\mathcal{S}$ and its elements fulfil additional conditions dictated by physical realities. These conditions are called *realizability conditions.* They may vary from case to case. Usually they comprise at least a causality condition, but often requirements concerning, for instance, the "richness" of the input and output sets of the mappings $s \in \mathcal{S}$ are also included. The last requirement is in conformity with the idea that it generally makes sense from a physical point of view to speak of a cause-effect relationship only if it is possible for the cause and effect to vary or to be varied in some way.

Dynamic systems. Causality

In this context particular attention will be paid to a special class of time systems called "dynamic systems".

Consider a time system $\mathcal{S}$ and let $S = \cup \mathcal{S}$. $\mathcal{S}$ is called a *uniform time system* on T if $\mathsf{D}S$ and $\mathsf{R}S$ are sets of time functions (signals) defined on a naturally ordered (nontrivial) common time set $T \subset \mathbf{R}$. In this case it is assumed that there are natural numbers r and q, and sets $U_1, U_2, \ldots, U_r$ and $Y_1, Y_2, \ldots, Y_q$ such that $\mathsf{D}S \subset U_1^T \times U_2^T \times \ldots \times U_r^T$ and $\mathsf{R}S \subset Y_1^T \times Y_2^T \times \ldots \times Y_q^T$.

Now let T be a time set and let

1 $$s : \mathsf{D}s \to \mathsf{R}s,\ \mathsf{D}s \subset U_1^T \times U_2^T \times \ldots \times U_r^T,\ \mathsf{R}s \subset Y_1^T \times Y_2^T \times \ldots \times Y_q^T$$

be an input-output mapping. We shall use the following causality definition.

2 **Definition.** An input-output mapping s as given by (1) is said to be *causal* if it fulfils the following condition: for every $t \in T$ and $u_1, u_2 \in \mathsf{D}s$ it holds that $u_1|T \cap (-\infty, t] = u_2|T \cap (-\infty, t]$ implies $s(u_1)|T \cap (-\infty, t] =$

$s(u_2)|T \cap (-\infty, t]$. □

Note that the restriction of, say, $u_1 = (u_{11}, u_{12}, \ldots, u_{1r}) \in \mathcal{U}$ to $T \cap (-\infty, t]$, denoted by $u_1|T \cap (-\infty, t]$, should be interpreted as $(u_{11}|T \cap (-\infty, t], u_{12}|T \cap (-\infty, t], \ldots, u_{1r}|T \cap (-\infty, t])$.

Let $\mathcal{S}$ be a uniform time system on T, and let every $s \in \mathcal{S}$ fulfil a given set of realizability conditions comprising at least the causality property: Every $s \in \mathcal{S}$ is causal according to definition (2); further conditions concerning, for instance, the richness and structure of the sets $\mathsf{D}s$ and $\mathsf{R}s$ of the mappings $s \in \mathcal{S}$ may also be included. A system $\mathcal{S}$ of this kind will henceforth be called a *dynamic system* (or even simply a "causal system") for short.

The formal definition of causality given above corresponds to the causality property mentioned previously, i.e. to the property that the effect cannot occur prior in time to the cause. Note also that a mapping consisting of one single pair of input-output time functions is always causal in a trivial way.

The time set T can, for instance, be an interval of $\mathbf{R}$, or a set of discrete points. It will often be assumed that T contains a first element, i.e. that $t_0 \triangleq \min T$ exists. If so, then t_0 is termed the *initial time* of the system. In the case of a dynamic system it is then also natural to assign the elements of Σ to the initial time t_0, and to call them possible *initial states* of the system.

3 **Remark.** Stochastic dynamic systems can also be treated within the framework sketched above. How this can be done has been outlined in Blomberg, 1975, chapter 1. □

2.3 Linear systems

Linearity is a strong property which brings on a wealth of structure. In this context we are particularly interested in the following kind of linearity. For details, see Ylinen, 1975, section 3.2.

1 **Definition.** Let $\mathcal{S}$ be a system according to the definition given in section 2.1 and let $S = \cup \mathcal{S}$. $\mathcal{S}$ is said to be a (vector space) *linear system* if the following conditions are fulfilled.

(i) $\mathsf{D}S = \mathsf{D}s$ for every $s \in \mathcal{S}$, and $\mathsf{D}S$ is a vector space over a field F.

(ii) There exists a vector space $\mathcal{Y}$ over F such that $\mathsf{R}(S) \subset \mathcal{Y}$.

(iii) There exists a linear mapping (i.e. a morphism of vector spaces over a common field) $\gamma: \mathsf{D}S \to \mathcal{Y}$ possessing the properties:

For every $s \in \mathcal{S}$ and every $u \in \mathsf{D}S$ it holds that

2 $$s(u) = s(0) + \gamma(u).$$

(iv) The set

$$S(0) \triangleq \{y \mid (0, y) \in S\} = \{y \mid y = s(0) \text{ for some } s \in \mathcal{S}\}$$

is a subspace of $\mathcal{Y}$.

If, in addition, $S(0)$ is finite dimensional, then $\mathcal{S}$ is said to be a *finite dimensional* linear system, and the dimension of $\mathcal{S}$ is equal to the dimension of $S(0)$. □

A number of relevant properties possessed by linear systems are collected in the following note. The proofs of most of the statements are either trivial or else they can be found in Ylinen, 1975, section 3.2 in a slightly more general setting.

3 **Note.** Let $\mathcal{S}$ throughout items (i), . . . , (vii) below be a linear system according to definition (1), and let S, $S(0)$, $\mathcal{Y}$, and γ denote the corresponding quantities mentioned in the definition. Then the following statements are true.

(i) $\mathsf{R}S$ is a subspace of $\mathcal{Y}$.

(ii) S and γ are subspaces of $\mathsf{D}S \times \mathcal{Y}$. ($\mathsf{D}S \times \mathcal{Y}$ is made a vector space over F in a natural way).

(iii) There are the following relationships between $\mathcal{S}$, S, $S(0)$, and γ:
S determines uniquely $S(0)$.
$\mathcal{S}$ determines uniquely S, $S(0)$, and γ.
$S(0)$ and γ together determine uniquely $\mathcal{S}$ and S.

Note also that S can be expressed as the direct sum

$$S = (\{0\} \times S(0)) \oplus \gamma.$$

In particular it should be emphasized that S alone does not generally determine a unique corresponding linear system $\mathcal{S}$ such that $\cup\, \mathcal{S} = S$. Item (vii) below gives a result relating to this matter.

(iv) By a *linear description* of $\mathcal{S}$ we mean a description of the form ($\mathcal{U}$, $\mathcal{Y}$, Σ, S, $\mathcal{S}$, h) with $\mathcal{U} = \mathsf{D}S$, $\mathcal{Y}$, and Σ vector spaces over F, and with $h: \Sigma \times \mathcal{U} \to \mathcal{Y}$ a linear mapping ($\Sigma \times \mathcal{U}$ is made a vector space over F in a natural way). Note that $\mathcal{U} = \mathsf{D}S$ and $\mathcal{Y}$ according to our above agreement immediately fulfil the required conditions. For every $s \in \mathcal{S}$ and every $u \in \mathsf{D}S$ there is then a corresponding $\alpha \in \Sigma$ such that

4 $$s(u) = h(\alpha, u) = h(\alpha, 0) + h(0, u).$$

Comparison with (2) shows that $h(\alpha, 0)$ represents $s(0)$, whereas $h(0, u)$ represents $\gamma(u)$. In (4), $h(\alpha, 0)$ is called the *zero input response* corresponding to h and $\alpha \in \Sigma$, and $h(0, u)$ the *zero state response* of $\mathcal{S}$ corresponding to $u \in \mathrm{D}S$. Note that $h(\alpha, 0)$ depends on the actual description, whereas $h(0, u)$ is uniquely determined by $\mathcal{S}$ itself.

Finally it is noted that any linear system $\mathcal{S}$ has a linear description of the form mentioned above. Such a description is obtained by choosing for instance Σ equal to $S(0)$, and h as given by the assignment

$$(\alpha, u) \mapsto \alpha + \gamma(u).$$

(v) $\mathcal{S}$ constitutes a partition of S, i.e. given a pair $(u, y) \in S$, there is a unique $s \in \mathcal{S}$ such that $y = s(u) = s(0) + \gamma(u)$. Note that (u, y) determines $s(0)$ according to $s(0) = y - \gamma(u)$. The elements of $\mathcal{S}$ can accordingly be said to be "distinguishable by a single input-output pair".
(vi) If $\mathcal{S}$ is a uniform time system, then the elements of $\mathcal{S}$ are causal if and only if γ is causal.
(vii) It was pointed out above under item (iii) that S alone does not generally determine a unique corresponding linear system $\mathcal{S}$ with $\cup\mathcal{S} = S$. Thus let us, in addition to the original linear system $\mathcal{S}$, construct another linear system $\mathcal{S}_1$ by means of a linear mapping $\gamma_1 : \mathrm{D}S \rightarrow \mathcal{Y}$ as follows:

5
$$\mathcal{S}_1 \triangleq \{s | s : \mathrm{D}S \rightarrow \mathcal{Y}, s(.) = y_0 + \gamma_1(\cdot) \text{ for some } y_0 \in S(0)\}.$$

It can then be shown that $S = \cup\mathcal{S}_1$ if and only if $\gamma - \gamma_1$ is a linear mapping $\mathrm{D}S \rightarrow S(0)$. □

Unless otherwise explicitly stated, it will be assumed in what follows that a description of a linear system is always chosen as linear in the sense of note (3), (iv).

Special concepts

Observability, reconstructibility, controllability, and reachability are well-known concepts in the theory of linear systems. These concepts could be given quite general interpretations in simple set theoretical terms. We shall not, however dwell on this question here—a number of comments relating to these concepts are given later in a more special context.

3

Interconnections of systems

The basic ideas relating to the concept of a system as a model of a cause-effect relationship were presented in the previous chapter. In this chapter we shall present and discuss some problems concerning families of systems. We shall in particular be interested in how new composed systems can be formed from given families of systems by means of "interconnections".

We shall start with a precise definition of an interconnection of a family of systems. We shall then study sets of input-output pairs generated by such an interconnection. One crucial question arises: Are the sets of input-output pairs generated in this way satisfactory with regard to the original purpose of the interconnection? This question leads naturally to the "determinateness" concept. Briefly, an interconnection of a family of systems can be satisfactory as a model of a composite cause-effect relationship only if it fulfils certain conditions, i.e. only if it is determinate.

The presentation is largely along the lines of some earlier works by the authors (Blomberg, 1972 a and b; Blomberg, 1975; Ylinen, 1975) and a thesis (Timonen, 1974).

3.1 Formal definition of an interconnection of a family of systems

Formally, we define an *interconnection of a family of systems* as a triple

1 $$(\mathscr{A}, C, X)$$

which satisfies certain conditions. These conditions as well as the meaning of the three elements appearing in (1) will now be explained.

A family of system descriptions

There is a given (indexed) family $\{\mathcal{S}_1, \mathcal{S}_2, \ldots, \mathcal{S}_N\}$ of systems involved in the interconnection. This family is represented in (1) by the family $\mathcal{A}$ of corresponding suitable system descriptions of the form (2.1.4). Thus

2 $$\mathcal{A} = \{(\mathcal{U}_j, \mathcal{Y}_j, \Sigma_j, S_j, \mathcal{S}_j, h_j) \,|\, j = 1, 2, \ldots, N\}.$$

For every j, $\mathcal{U}_j$ and $\mathcal{Y}_j$ in (2) are assumed to be of the form

3 $$\mathcal{U}_j \subset \mathcal{U}_{j1} \times \mathcal{U}_{j2} \times \ldots \times \mathcal{U}_{jr_j},$$
$$\mathcal{Y}_j \subset \mathcal{Y}_{j1} \times \mathcal{Y}_{j2} \times \ldots \times \mathcal{Y}_{jq_j}.$$

System $\mathcal{S}_j$ is said to have r_j distinct input components and q_j distinct output components.

System $\mathcal{S}_j$ is graphically represented by a block diagram of the form shown in Fig. 3.1, where α_j represents some element of Σ_j and (u_j, y_j) some input-output pair with $y_j = h_j(\alpha_j, u_j)$.

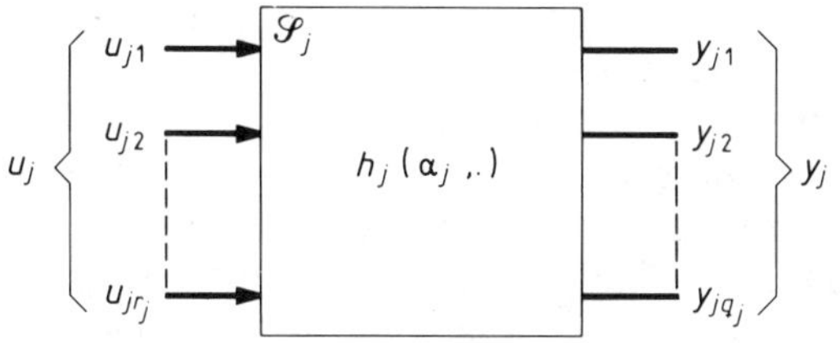

Fig. 3.1. Block diagram representing system $\mathcal{S}_j$. $u_j = (u_{j1}, u_{j2}, \ldots, u_{jr_j})$, $y_j = (y_{j1}, y_{j2}, \ldots, y_{jq_j})$.

An interconnection matrix

C is an *interconnection matrix* which describes a set of interconnection constraints. These constraints restrict the set of simultaneously permissible input-output pairs in a certain way.

C has in partitioned form the following general structure.

4 $$C = \begin{array}{c} \\ y_0 \\ u_1 \\ u_2 \\ \vdots \\ u_N \end{array} \begin{array}{c} \begin{array}{ccccc} u_0 & y_1 & y_2 & \ldots & y_N \end{array} \\ \left[\begin{array}{c:c:c:c:c} C_{00} & C_{01} & C_{02} & \ldots & C_{0N} \\ \hdashline C_{10} & C_{11} & C_{12} & \ldots & C_{1N} \\ \hdashline C_{20} & C_{21} & C_{22} & \ldots & C_{2N} \\ \hdashline \vdots & \vdots & \vdots & & \vdots \\ \hdashline C_{N0} & C_{N1} & C_{N2} & \ldots & C_{NN} \end{array}\right] \end{array},$$

where the submatrices C_{ij} are Boolean matrices with entries 1 or 0.

The rows of the lower part of C are assigned to the inputs, denoted by $u_1, u_2, \ldots, u_N$, of the systems $\mathcal{S}_1, \mathcal{S}_2, \ldots, \mathcal{S}_N$ so that every input component has its own row and every system input its own group of rows. There is, in addition, an upper group y_0 of rows containing the rows $y_{01}, y_{02}, \ldots, y_{0q_0}$. These rows are assigned to the components of the "overall output" y_0 of an "overall system" $\mathcal{S}_o$ which we shall associate with the interconnection (1) provided that certain conditions are fulfilled.

The columns of C are analogously assigned to the components of the outputs, denoted by $y_1, y_2, \ldots, y_N$, of the systems $\mathcal{S}_1, \mathcal{S}_2, \ldots, \mathcal{S}_N$. The additional group u_0 of columns containing the columns $u_{01}, u_{02}, \ldots, u_{0r_0}$ is assigned to the overall input u_0 of the overall system $\mathcal{S}_o$.

C is qualified to be called an interconnection matrix if it fulfils the following conditions.

5 **Conditions on** C. **(i)** Every entry of C is either a 1 or a 0.
(ii) Every row contains one and only one 1.
(iii) Each of the columns $u_{01}, u_{02}, \ldots, u_{0r_0}$ belonging to the group u_0 of columns contains at least one 1.
(iv) The submatrix C_{00} is a zero matrix. □

C describes the interconnection constraints in the following way.

Consider, for instance, a row u_{ik} of C and suppose that this row contains its only 1 in column y_{jl}. This then implies that the input component u_{ik} of system $\mathcal{S}_i$ is constrained to be equal to the output component y_{jl} of system $\mathcal{S}_j$, i.e. we have the equality constraint

$$u_{ik} = y_{jl}.$$

We shall also express such a constraint by saying that u_{ik} is "connected" to y_{jl}.

A particular component of $y_0, u_1, u_2, \ldots, u_N$ cannot generally be equal to more than one of the distinct components of $u_0, y_1, y_2, \ldots, y_N$ at a time. For the interconnection to make sense it is, on the other hand, necessary for every component of $y_0, u_1, u_2, \ldots, u_N$ to equal some component of $u_0, y_1, y_2, \ldots, y_N$. Item (ii) above takes care of these requirements.

The purpose of items (iii) and (iv) of (5) is only to avoid indefiniteness owing to undefined input and output sets. Indefiniteness may arise if any one of the components of u_0 is connected directly to any one of the components of y_0, or if u_0 contains components which are not connected anywhere.

Suitable additional systems may, if necessary, be introduced in order to satisfy conditions (5).

As a result, there is for every $\alpha = (\alpha_1, \alpha_2, \ldots, \alpha_N) \in \Sigma_1 \times \Sigma_2 \times \ldots \times \Sigma_N$ a corresponding set of inputs consisting of all the possible inputs u_0 of system $\mathscr{S}_o$ and determined by the input sets associated with those components of $u_1, u_2, \ldots, u_N$ to which the components of u_0 are connected.

Finally, there is no need for particular restrictions concerning the numbers of 1 occurring in the columns assigned to the components of $y_1, y_2, \ldots, y_N$ because any of these components may be connected simultaneously to any number of the components of $u_1, u_2, \ldots, u_N$.

It is now possible to express the interconnection constraints in a compact matrix form as

6 $$\underbrace{\begin{bmatrix} y_0 \\ \hdashline u_1 \\ u_2 \\ \vdots \\ u_N \end{bmatrix}}_{\mu} = \begin{matrix} \\ y_0 \\ u_1 \\ \vdots \\ u_N \end{matrix} \underbrace{\overset{\begin{matrix} u_0 & y_1 \cdots y_N \end{matrix}}{\left[\begin{array}{c:c} 0 & C_2 \\ \hdashline C_3 & C_4 \end{array}\right]}}_{C} \underbrace{\begin{bmatrix} u_0 \\ \hdashline y_1 \\ y_2 \\ \vdots \\ y_N \end{bmatrix}}_{\eta},$$

where C has been partitioned further in an obvious way.

The product $C\eta$ in (6) is evaluated formally by applying the usual rules for matrix multiplication with the interpretation that multiplication of η by a row of C as result yields that entry of η which corresponds to the position of the 1 in the considered row of C. Products of this kind also appear in the sequel and they are interpreted analogously.

The interconnection matrix introduced above is, of course, closely related to the "adjacency matrix" used in the theory of graphs (*cf.* Klir, 1969).

State constraint

The third element of the triple (1) stands for a *state constraint*

7 $$X \subset \Sigma_1 \times \Sigma_2 \times \ldots \times \Sigma_N$$

and it implies that the interconnection is defined only for N-lists $(h_1(\alpha_1, .), h_2(\alpha_2, .), \ldots, h_N(\alpha_N, .))$ of input-output mappings with $(\alpha_1, \alpha_2, \ldots, \alpha_N) \in X$.

It is easily seen that there are cases in which it makes sense to take X as a proper subset of $\Sigma_1 \times \Sigma_2 \times \ldots \times \Sigma_N$. Recall to this end that an N-list $(\alpha_1, \alpha_2, \ldots, \alpha_N) \in \Sigma_1 \times \Sigma_2 \times \ldots \times \Sigma_N$ represents specific circumstances under which the different cause-effect relationships involved in the interconnection are considered. It may thus clearly happen that some of these circumstances simultaneously influence several of the cause-effect relationships considered—that is to say—there may be $(\alpha_1, \alpha_2, \ldots, \alpha_N) \in \Sigma_1 \times \Sigma_2 \times \ldots \times \Sigma_N$ which cannot occur in reality. Such elements are therefore not included in X. This point will be further discussed later on in part III of the book in a rather special setting.

8 **Note.** It should already be evident at this point that the given family $\{\mathcal{S}_1, \mathcal{S}_2, \ldots, \mathcal{S}_N\}$ of systems involved in the interconnection and represented by the family $\mathcal{A}$ of system descriptions (*cf.* (2) above) could equally well be replaced by a singleton family comprising the single system $\mathcal{S}$ represented by the system description

$$(\mathcal{U}, \mathcal{Y}, \Sigma, S, \mathcal{S}, h) \text{ with}$$

$$\mathcal{U} = \mathcal{U}_1 \times \mathcal{U}_2 \times \ldots \times \mathcal{U}_N,$$

$$\mathcal{Y} = \mathcal{Y}_1 \times \mathcal{Y}_2 \times \ldots \times \mathcal{Y}_N,$$

9 $$\Sigma = \Sigma_1 \times \Sigma_2 \times \ldots \times \Sigma_N, \text{ and}$$

$$h\colon \mathsf{D}h \to \mathcal{Y}, \quad \mathsf{D}h \subset \Sigma \times \mathcal{U}, \quad \mathsf{D}(\mathsf{D}h) = \Sigma,$$

$$(\alpha, u) = ((\alpha_1, \alpha_2, \ldots, \alpha_N), (u_1, u_2, \ldots, u_N)) \mapsto$$

$$(h_1(\alpha_1, u_1), h_2(\alpha_2, u_2), \ldots, h_N(\alpha_N, u_N)).$$

We shall feel free in the sequel to replace the given family $\mathcal{A}$ of system descriptions by the singleton family $\{(\mathcal{U}, \mathcal{Y}, \Sigma, S, \mathcal{S}, h)\}$ as given by (9) above whenever this seems convenient. □

10 **Remark. (i)** We have here chosen to define an interconnection of a family of systems with the aid of a corresponding family of system descriptions. A more direct definition of such an interconnection could now be given also on the basis of the given family of systems itself (Ylinen, 1975, section 3.3). This is seen as follows.

Consider an interconnection $(\mathcal{A}, C, X)$ as given above. Every $\alpha = (\alpha_1, \alpha_2, \ldots, \alpha_N) \in X$ then determines a corresponding unique element of $\mathcal{S}_1 \times \mathcal{S}_2 \times \ldots \times \mathcal{S}_N$ given by $(h_1(\alpha_1, .), h_2(\alpha_2, .), \ldots, h_N(\alpha_N, .))$, and thus the whole X determines a corresponding unique subset $X_1 \subset \mathcal{S}_1 \times \mathcal{S}_2 \times \ldots \times \mathcal{S}_N$. Hence it is clear that $(\mathcal{A}, C, X)$ could equally well be represented by the triple $(\{\mathcal{S}_1, \mathcal{S}_2, \ldots, \mathcal{S}_N\}, C, X_1)$ with C as before.

(ii) There are some formal difficulties concerning the precise interpretation of matrix representations of lists and of partitioned matrix equations of the form (6). Take, for instance, μ in (6). This column matrix can be interpreted as the matrix representation of either $(y_0, (u_1, u_2, \ldots, u_N))$, or

$(y_0, u_1, u_2, \ldots, u_N)$, or $(y_{01}, y_{02}, \ldots, y_{0q_0}, u_{11}, u_{12}, \ldots, u_{1r_1}, \ldots, u_{N1}, u_{N2}, \ldots, u_{Nr_N})$, etc. For the sake of notational simplicity, and following common usage, we shall in what follows feel free to "identify" different interpretations of this kind with each other. This also implies that we shall identify a Cartesian product of the form, say, $\mathcal{Y}_0 \times (\mathcal{U}_1 \times \mathcal{U}_2 \times \ldots \times \mathcal{U}_N)$ where $\mathcal{Y}_0 = \mathcal{Y}_{01} \times \mathcal{Y}_{02} \times \ldots \times \mathcal{Y}_{0_{q0}}$, $\mathcal{U}_1 = \mathcal{U}_{11} \times \mathcal{U}_{12} \times \ldots \times \mathcal{U}_{1r_1}$, $\mathcal{U}_2 = \mathcal{U}_{21} \times \mathcal{U}_{22} \times \ldots \times \mathcal{U}_{2r_2}, \ldots, \mathcal{U}_N = \mathcal{U}_{N1} \times \mathcal{U}_{N2} \times \ldots \times \mathcal{U}_{Nr_N}$, with $\mathcal{Y}_0 \times \mathcal{U}_1 \times \mathcal{U}_2 \times \ldots \times \mathcal{U}_N$ and with $\mathcal{Y}_{01} \times \mathcal{Y}_{02} \times \ldots \times \mathcal{Y}_{0_{q0}} \times \mathcal{U}_{11} \times \mathcal{U}_{12} \times \ldots \times \mathcal{U}_{1r_1} \times \ldots \times \mathcal{U}_{N1} \times \mathcal{U}_{N2} \times \ldots \times \mathcal{U}_{Nr_N}$ etc. This should cause no ambiguity.

(iii) We have here used a rather unusual way of labeling the rows and columns of the interconnection matrix C by using symbols corresponding to various signals appearing in the context. This has been done merely for the sake of clarity and in order to make it easier for the reader to grasp the significance of different parts of C. We shall also find this notational convention very convenient elsewhere in this book. □

3.2 Input-output relations determined by an interconnection of a family of systems

Let us consider an interconnection of a family of systems given by (*cf.*, (3.1.1))

1 $$(\mathcal{A}, C, X),$$

where $\mathcal{A}$ is a singleton family of system descriptions as given by

2 $$\mathcal{A} = \{(\mathcal{U}, \mathcal{Y}, \Sigma, S, \mathcal{S}, h)\}$$

with $\mathcal{U}$, $\mathcal{Y}$ of the form

3 $$\mathcal{U} \subset \mathcal{U}_1 \times \mathcal{U}_2 \times \ldots \times \mathcal{U}_r,$$
$$\mathcal{Y} \subset \mathcal{Y}_1 \times \mathcal{Y}_2 \times \ldots \times \mathcal{Y}_q.$$

In view of note (3.1.8) it is clear that the choice of $\mathcal{A}$ equal to a singleton family of system descriptions does not mean any loss of generality.

The interconnection matrix C has the partitioned form (*cf.* (3.1.6); u and y below correspond to $(u_1, u_2, \ldots, u_N)$ and $(y_1, y_2, \ldots, y_N)$ in (3.1.6))

4 $$C = \begin{matrix} & u_0 & y \\ y_0 & 0 & C_2 \\ u & C_3 & C_4 \end{matrix}$$

with q_0 rows assigned to the overall output y_0, r rows assigned to the input

u of $\mathcal{S}$, r_0 columns assigned to the overall input u_0, and q columns assigned to the output y of $\mathcal{S}$.

The interconnection constraints (3.1.6) in this case thus read

5 $$\begin{bmatrix} y_0 \\ \hdashline u \end{bmatrix} = \begin{matrix} \\ y_0 \\ u \end{matrix} \begin{matrix} \begin{matrix} u_0 & \quad y \end{matrix} \\ \left[\begin{array}{c:c} 0 & C_2 \\ \hdashline C_3 & C_4 \end{array}\right] \end{matrix} \begin{bmatrix} u_0 \\ \hdashline y \end{bmatrix}.$$

Fundamental relations associated with the interconnection

Let $\alpha \in X \subset \Sigma$, and let u_0, y_0, u, and y be such that they simultaneously satisfy the interconnection constraints (5) as well as the system equations, i.e. such that

6 $$\begin{aligned} u &= [C_3 \,\vdots\, C_4] \begin{bmatrix} u_0 \\ \hdashline y \end{bmatrix}, \\ y_0 &= C_2 y, \\ u &\in \mathrm{D}h(\alpha,.), \quad \text{and} \\ y &= h(\alpha, u). \end{aligned}$$

These are now the fundamental relations associated with the interconnection under consideration. They are schematically represented by the block diagram shown in Fig. 3.2.

The interesting part of the diagram in Fig. 3.2 is the part denoted by $\mathcal{S}_i$, which represents an "internal system" associated with the given interconnection (1). If there is a well-defined $\mathcal{S}_i$, then the "overall system" $\mathcal{S}_o$ is obtained in a trivial way just by picking appropriate outputs of $\mathcal{S}_i$ as outputs of $\mathcal{S}_o$. $\mathcal{S}_i$ contains further information about all the input and output variables associated with $\mathcal{S}$. Because of this, our attention will subsequently be concentrated mainly on $\mathcal{S}_i$ rather than on $\mathcal{S}_o$.

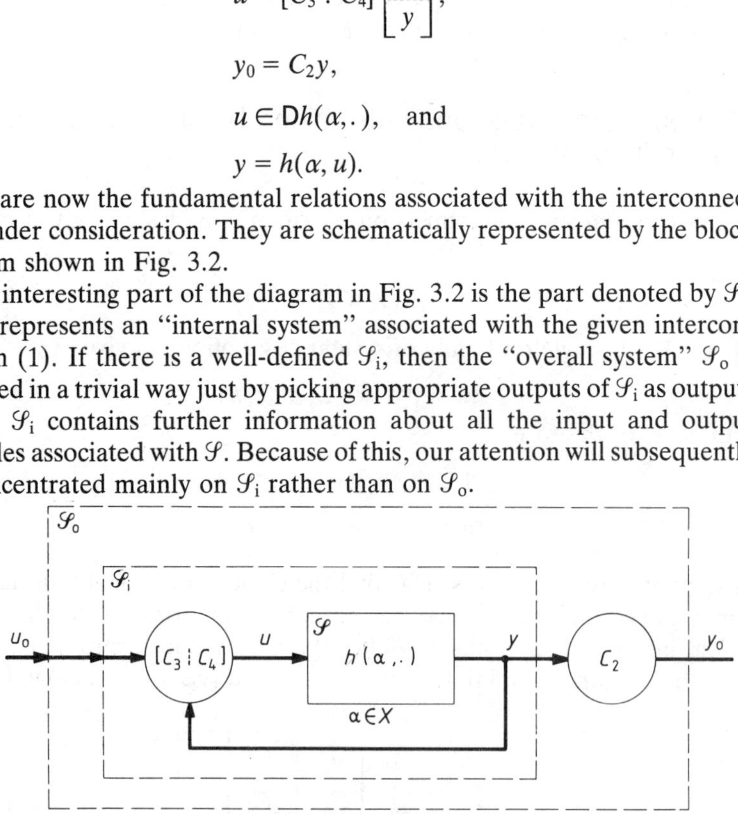

Fig. 3.2. Block diagram representation of the interconnection (3.2.1) and the relations (3.2.6).

Input-output relations determined by the interconnection

Now let $S_{i\alpha}$ and $S_{o\alpha}$ denote the sets of all (u_0, y) and (u_0, y_0) respectively corresponding to $\alpha \in X$ and determined by the interconnection in accordance with (6), i.e.

7

$$S_{i\alpha} \triangleq \{(u_0, y) | u_0, y \quad \text{are such that}$$
$$[C_3 \vdots C_4] \begin{bmatrix} u_0 \\ \hdashline y \end{bmatrix} \in \mathrm{D}h(\alpha, .) \quad \text{and}$$
$$y = h(\alpha, [C_3 \vdots C_4] \begin{bmatrix} u_0 \\ \hdashline y \end{bmatrix})\}, \quad \text{and}$$
$$S_{o\alpha} \triangleq \{(u_0, y_0) | (u_0, y_0) = (u_0, C_2 y) \text{ for some } (u_0, y) \in S_{i\alpha}\}.$$

The set of all possible outputs y corresponding to a given pair (α, u_0) is thus obtained as the set of all fixed points of the mapping

$$h(\alpha, [C_3 \vdots C_4] \begin{bmatrix} u_0 \\ \hdashline \cdot \end{bmatrix}).$$

Let us next take a closer look at the relation $S_{i\alpha}$ as given above, and let us to begin with note that $S_{i\alpha}$ may obviously be either the empty set, a nonempty mapping, or a nonempty relation but not a mapping.

3.3 Determinateness with respect to the empty set of realizability conditions

Consider again an interconnection $(\mathscr{A}, C, X)$ as given by (3.2.1). If no particular realizability conditions are specified, i.e. if the systems involved are regarded just as families of input-output mappings, then, in order to get a well-defined internal system $\mathscr{S}_i$ and an associated overall system $\mathscr{S}_o$, both of them determined uniquely by the given interconnection $(\mathscr{A}, C, X)$, we have evidently to demand that $S_{i\alpha}$ be a mapping for every $\alpha \in X$. Here we shall formally also regard an empty $S_{i\alpha}$ as a mapping, i.e. as the empty mapping.

If the above condition is fulfilled, then we shall say that the interconnection $(\mathscr{A}, C, X)$ is *determinate* with respect to the empty set of realizability conditions. This definition of determinateness corresponds to the terminology used in Zadeh and Desoer, 1963, section 2.6 (see in particular remark (2.6.7)).

Suppose now that $(\mathscr{A}, C, X)$ is determinate with respect to the empty set of realizability conditions. To avoid trivialities, suppose also that there is a nonempty subset $\Sigma_i \subset X$ such that $S_{i\alpha}$—and consequently also $S_{o\alpha}$, *cf.* (3.2.7)—is a nonempty mapping for every $\alpha \in \Sigma_i$. Parametric input-output mappings h_i and h_o defining the internal system $\mathscr{S}_i$ and the overall system $\mathscr{S}_o$ are for every $\alpha \in \Sigma_i$ then clearly given by

1
$$h_i(\alpha,.) = S_{i\alpha},$$
$$h_o(\alpha,.) = S_{o\alpha}.$$

Determinateness thus guarantees in this special case the existence of well-defined systems $\mathscr{S}_i$ and $\mathscr{S}_o$ associated with the given interconnection. The general case is discussed in the next section.

2 **Remark.** The block diagram shown in Fig. 3.2 illustrates only one possible representation of the interconnection $(\mathscr{A}, C, X)$. In particular it is observed that the whole output y of $\mathscr{S}_i$ is "fed back" to the input side whether all the components of y are needed for the forming of u or not. Other representations where the feedback contains only those components of y which are really needed can readily be given (see e.g. Blomberg, 1972 a, and b). □

3.4 Determinateness with respect to a general set of realizability conditions

The considerations of the previous paragraphs led to the concept of determinateness of an interconnection $(\mathscr{A}, C, X)$ with respect to the empty set of realizability conditions. In this paragraph the concept of determinateness of an interconnection will be generalized to also cover cases with nonempty sets of realizability conditions.

A nonempty set of realizability conditions

So consider again an interconnection $(\mathscr{A}, C, X)$ as given by (3.2.1), where $\mathscr{A}$ comprises the description of a time system $\mathscr{S}$. Suppose further that there is a specified set of realizability conditions (*cf.* section 2.2), and that the input-output mappings contained in $\mathscr{S}$ satisfy these conditions. $\mathscr{S}$ may, for instance, be a dynamic system in the sense of section 2.2.

The important question is: Are there any well-defined systems $\mathscr{S}_i$ and $\mathscr{S}_o$ (*cf.* section 3.2) determined by $(\mathscr{A}, C, X)$ such that the input-output mappings contained in these systems also satisfy the given set of realizability conditions? In the case of dynamic systems we thus also require $\mathscr{S}_i$ and $\mathscr{S}_o$ to be dynamic systems.

Referring to the discussion above (*cf.* section 3.2) concerning the simple relationship between the internal system $\mathcal{S}_i$ and the corresponding overall system $\mathcal{S}_o$, we shall quite generally assume that the existence of a system $\mathcal{S}_i$ whose elements satisfy the given set of realizability conditions also implies the existence of a corresponding system $\mathcal{S}_o$ whose elements satisfy the same set of conditions.

Accordingly, let us consider the set $S_{i\alpha}$ of input-output pairs determined by the interconnection and corresponding to $\alpha \in X$ as given by (3.2.7). If no realizability conditions are specified, we found in the previous paragraph that a well-defined internal system $\mathcal{S}_i$ can be associated with the interconnection if and only if $S_{i\alpha}$ is a mapping for every $\alpha \in X$. With a nonempty set of realizability conditions it is not necessary to require that $S_{i\alpha}$ be a mapping for every $\alpha \in X$—evidently it suffices to require that a unique mapping satisfying the given set of realizability conditions can be *extracted* from $S_{i\alpha}$ for every $\alpha \in X$. This requirement can now be formalized in the following way.

Families of mappings satisfying the given set of realizability conditions and determined by $S_{i\alpha}$

Starting with $S_{i\alpha}$ according to (3.2.7), a family $\mathfrak{Q}_{i\alpha}$ of mappings is constructed as follows:

1 $$\mathfrak{Q}_{i\alpha} \triangleq \{q \mid q \subset S_{i\alpha} \text{ and } q \text{ is a mapping satisfying the given set of realizability conditions}\}.$$

Note that the empty mapping is considered as satisfying any set of realizability conditions and that it may thus happen that $\mathfrak{Q}_{i\alpha} = \{\phi\}$.

$\mathfrak{Q}_{i\alpha}$ above thus consists of all input-output mappings that can be formed from the elements of $S_{i\alpha}$ and which satisfy the realizability conditions.

Formal definition of the concept of determinateness

We are now ready to define the concept of determinateness with respect to a general set of realizability conditions.

2 **Definition.** Let $(\mathcal{A}, C, X)$ be an interconnection as given by (3.2.1) and let the system $\mathcal{S}$ represented by $\mathcal{A}$ be a time system such that every element of $\mathcal{S}$ satisfies a given set (empty or nonempty) of realizability conditions. Further let $S_{i\alpha}$ and $\mathfrak{Q}_{i\alpha}$ be formed for every $\alpha \in X$ according to (3.2.7) and (1) above. $(\mathcal{A}, C, X)$ is then said to be *determinate* with respect to the given set of realizability conditions if

3 $$\cup \mathfrak{Q}_{i\alpha} \in \mathfrak{Q}_{i\alpha}$$

for every $\alpha \in X$, otherwise $(\mathscr{A}, C, X)$ is termed *indeterminate.* □

Condition (3) means that the union of all the members of $\mathfrak{Q}_{i\alpha}$ shall also be a member of $\mathfrak{Q}_{i\alpha}$, i.e. $\cup \mathfrak{Q}_{i\alpha}$ shall be a mapping satisfying the given set of realizability conditions.

4 **Note.** The following observations concerning definition (2) can be made.

(3) is always fulfilled if $S_{i\alpha}$ is a mapping satisfying the given set of realizability conditions, because then it holds that $\cup \mathfrak{Q}_{i\alpha} = S_{i\alpha} \in \mathfrak{Q}_{i\alpha}$.

Definition (2) also makes sense if the given set of realizability conditions is empty. If so then $\mathfrak{Q}_{i\alpha}$ according to (1) is the family of *all* mappings that can be formed from $S_{i\alpha}$ and (3) is satisfied if and only if $S_{i\alpha}$ is a mapping.

Suppose that the set of realizability conditions contains the causality property only. Then any mapping consisting of a single input-output pair satisfies these conditions. This leads to the result that $\cup \mathfrak{Q}_{i\alpha} = S_{i\alpha}$ for every $\alpha \in X$. Determinateness then follows if and only if $S_{i\alpha}$ is a causal mapping for every $\alpha \in X$. If the set of realizability conditions includes conditions concerning, for instance, the richness of the input sets (*cf.* section 3.5 below), then determinateness can follow even if $S_{i\alpha}$ for some $\alpha \in X$ is a relation but not a mapping.

If $\mathfrak{Q}_{i\alpha} = \{\phi\}$, then $\cup \mathfrak{Q}_{i\alpha} = \phi \in \mathfrak{Q}_{i\alpha}$ and condition (3) is formally regarded as fulfilled. □

Significance of determinateness

The significance of determinateness is most easily understood by considering the consequences of indeterminateness. Hence suppose that the interconnection $(\mathscr{A}, C, X)$ is indeterminate. There then exists an $\alpha \in X$ such that $\cup \mathfrak{Q}_{i\alpha} \notin \mathfrak{Q}_{i\alpha}$. This in turn implies that $\cup \mathfrak{Q}_{i\alpha} \subset S_{i\alpha}$ is a relation but not a mapping, or else a mapping which does not satisfy the given set of realizability conditions. All the input-output mappings contained in $\mathfrak{Q}_{i\alpha}$ thus individually satisfy the set of realizability conditions but they cannot be joined together, nor do we know which one to choose. The situation is indeed indeterminate.

An indeterminate interconnection is clearly useless as a model of a real system.

Systems determined by a determinate interconnection

Hence suppose that the interconnection $(\mathscr{A}, C, X)$ is determinate (with respect to a given set of realizability conditions), i.e. that (3) holds for every $\alpha \in X$. Parametric input-output mappings h_i and h_o describing the internal system $\mathscr{S}_i$ and the overall system $\mathscr{S}_o$ respectively associated with

$(\mathcal{A}, C, X)$ according to the considerations of section 3.2 can now be uniquely determined as follows (recall that the members of $\mathcal{S}_i$ and $\mathcal{S}_o$ were required to satisfy the set of realizability conditions).

As state sets Σ_i and Σ_o for h_i and h_o we take the set

5 $$\Sigma_i = \Sigma_o = \{\alpha | \alpha \in X \quad \text{and } \mathcal{Q}_{i\alpha}$$
$$(cf.\ (1)) \text{ contains at least one nonempty mapping}\}.$$

Suppose that $\Sigma_i = \Sigma_o$ is nonempty. In this case we define for every $\alpha \in \Sigma_i = \Sigma_o$,

6 $$h_i(\alpha, .) = \cup\, \mathcal{Q}_{i\alpha},$$
$$h_o(\alpha, .) = C_2 h_i(\alpha, .).$$

If $\Sigma_i = \Sigma_o = \phi$, then $\mathcal{S}_o$ and $\mathcal{S}_i$ consist of the empty mapping only.

3.5 Illustrative example

By means of an example we shall here illlustrate the meaning and applicability of various concepts introduced previously in this chapter. Other examples serving the same purpose can be found in the literature (*cf.* in particular Willems, 1971, section 4.3, and further references therein). The example considered below will, in addition, explain certain particular features which characterize operational methods based on the Laplace and Fourier transforms. These features have in the past caused much trouble and confusion.

It is assumed that the reader is familiar with the fundamental theory of the Fourier and Laplace transforms as presented in a great number of excellent textbooks.

The following notations are introduced:

$\mathcal{X}_F$ denotes the space of all (real-valued) Fourier transformable functions defined on **R** such that for every $u \in \mathcal{X}_F$ it holds that $\mathcal{F}^{-1}(\mathcal{F}(u)) = u$, where $\mathcal{F}$ denotes the Fourier transform formula and $\mathcal{F}^{-1}$ the inverse Fourier transform formula. With suitable interpretations, the usual L_2-space qualifies as $\mathcal{X}_F$ (*cf.* Willems, 1971, section 1.3).

$\mathcal{X}_{L_0}$ denotes the space of all (real-valued) Laplace transformable functions defined on **R** such that for every $u \in \mathcal{X}_{L_0}$ it holds that $u(t) = 0$ for every $t \in (-\infty, 0)$ and $\mathcal{L}^{-1}(\mathcal{L}(u)) = u$, where $\mathcal{L}$ denotes the Laplace transform formula and $\mathcal{L}^{-1}$ the inverse Laplace transform formula.

$\mathcal{X}_L$ denotes the space of all functions defined on **R** obtainable by finite translations of the members of $\mathcal{X}_{L_0}$.

$\mathcal{W}$ and $\mathcal{Z}$ denote the following sets of functions:

$$\mathcal{W} = \mathcal{X}_F \cup \mathcal{X}_L, \quad \text{and}$$

$$\mathcal{Z} = \mathcal{X}_F \times \mathcal{X}_F \cup \mathcal{X}_L \times \mathcal{X}_L.$$

$\mathcal{X}_F$, $\mathcal{X}_{L_0}$, and $\mathcal{X}_L$ are vector spaces over **R**, but $\mathcal{W}$ and $\mathcal{Z}$ are not. $\mathcal{X}_F \cap \mathcal{X}_L$ contains elements other than the zero function.

Now let **R** be regarded as a time set T and let us introduce three time systems $\mathcal{S}$, $\mathcal{S}'$, and $\mathcal{S}''$, each consisting of one single input-output mapping s, s', and s'' respectively. These mappings are defined as follows (p denotes d/dt, c is a real constant):

1
$$\begin{aligned}
&s:\mathcal{Z} \to \mathcal{W},\\
&s = \{((u_1, u_2), y)\,|\,u_1, u_2, y \in \mathcal{X}_F \quad \text{or}\\
&u_1, u_2, y \in \mathcal{X}_L, \quad \text{and}\,(p+1)y = c(u_1 + u_2)\},\\
&s' = s|\mathcal{X}_F \times \mathcal{X}_F : \mathcal{X}_F \times \mathcal{X}_F \to \mathcal{X}_F\\
&(s' \text{ is the restriction of } s \text{ to } \mathcal{X}_F \times \mathcal{X}_F),\\
&s'' = s|\mathcal{X}_L \times \mathcal{X}_L : \mathcal{X}_L \times \mathcal{X}_L \to \mathcal{X}_L\\
&(s'' \text{ is the restriction of } s \text{ to } \mathcal{X}_L \times \mathcal{X}_L).
\end{aligned}$$

The three systems are thus generated by the same ordinary linear time-invariant differential equation $(p + 1)y = c(u_1 + u_2)$. They differ from each other with respect to the input and output signal sets.

On the basis of the well-known theory of the Fourier and Laplace transforms and their use in solving differential equations it should now be clear that s, s', and s'' really are mappings, and that the output value $y(t)$ for any input $u = (u_1, u_2) \in \mathrm{D}s$, $\mathrm{D}s'$, or $\mathrm{D}s''$ and any $t \in \mathbf{R}$ in all three cases is given by

2
$$y(t) = \int_{-\infty}^{t} ce^{-(t-\tau)}(u_1(\tau) + u_2(\tau))\,d\tau.$$

s, s', and s'' are evidently causal and $s = s' \cup s''$. $\mathcal{S}'$ and $\mathcal{S}''$ are linear systems, but $\mathcal{S}$ is not. Note also that $s' \cap s''$ contains pairs other than the zero pair.

It should be emphasized that we have above referred to the Fourier and Laplace transforms only as tools by means of which expression (2) can be found. The expression itself depends only on the differential equation and the function spaces involved.

Set of realizability conditions

As a suitable set of realizability conditions we shall now choose the causality property along with the requirement that if u belongs to the domain of a particular input-output mapping, then $u + \varepsilon$ should also belong to this domain, where ε is any suitable pulse of finite duration and amplitude.

With the realizability conditions thus specified, $\mathcal{S}$, $\mathcal{S}'$, and $\mathcal{S}''$ are clearly dynamic systems on $T = \mathbf{R}$ in the sense of the definition given in section 2.2.

Taking any singleton set $\{\alpha\}$ as a common state set Σ for all the systems involved, we can easily write down descriptions of the form (2.1.4) for our systems. To be definite, let us take $\Sigma = \{0\}$, $0 =$ the zero of $\mathbf{R}$. A description of $\mathcal{S}$ is thus given by

3
$$(\mathscr{X}, \mathscr{W}, \Sigma, s, \mathcal{S}, h), \quad \text{with} \qquad h(0, .) = s.$$

Descriptions of $\mathcal{S}'$ and $\mathcal{S}''$ can be formed analogously.

Next we shall form three feedback interconnections of the general structure shown in Fig. 3.3. Formally, the three interconnections are

$$(\mathscr{A}, C, X), (\mathscr{A}', C, X), \quad \text{and } (\mathscr{A}'', C, X)$$

with $\mathscr{A}$, $\mathscr{A}'$, and $\mathscr{A}''$ singleton sets containing the descriptions of $\mathcal{S}$, $\mathcal{S}'$, and

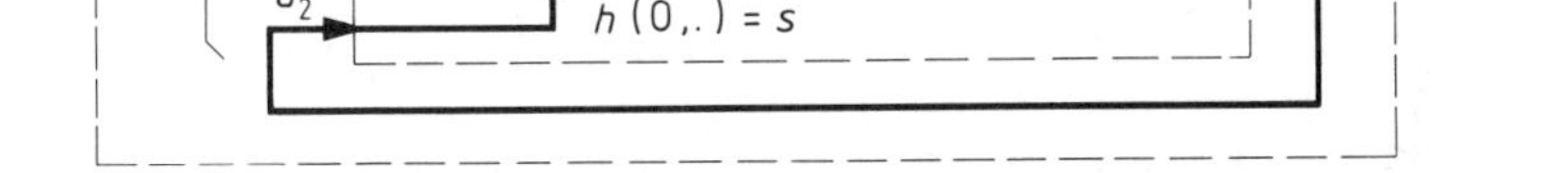

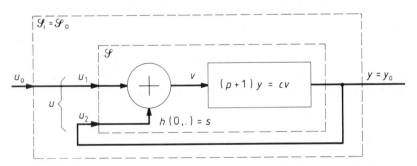

Fig. 3.3. Feedback interconnection with system $\mathcal{S}$. Corresponding interconnections with $\mathcal{S}$ replaced by $\mathcal{S}'$ and $\mathcal{S}''$ are also considered.

$\mathcal{S}''$ respectively, and with C and X given by

4
$$C = \begin{array}{c} \\ y_0 \\ u_1 \\ u_2 \end{array} \begin{array}{c} \begin{array}{cc} u_0 & y \end{array} \\ \left[\begin{array}{c:c} 0 & 1 \\ \hdashline 1 & 0 \\ 0 & 1 \end{array}\right] \end{array}, \qquad X = \Sigma = \{0\}.$$

Note that this C corresponds to $[C_3 : C_4] = I$ and $C_2 = I$ in Fig. 3.2. These matrices have therefore been deleted from Fig. 3.3.

We are now ready to form the relations $S_{i\alpha}$, $S'_{i\alpha}$, and $S''_{i\alpha}$ for $\alpha = 0$ according to (3.2.7) of all input-output pairs (u_0, y) which can be associated with our three interconnections $(\mathcal{A}, C, X)$, $(\mathcal{A}', C, X)$, and $(\mathcal{A}'', C, X)$ respectively. (3.2.7) yields in this case for $S_{i\alpha}$, $\alpha = 0$, i.e. for S_{i0}

$$S_{i0} = \{(u_0, y) | u_0, y \quad \text{are such that}$$
$$(u_0, y) \in \mathfrak{X} \quad \text{and} \quad y = s(u_0, y)\},$$

or equivalently, using the definition of s, *cf.* (1),

5
$$S_{i0} = \{(u_0, y) | u_0, y \in \mathcal{X}_F \quad \text{or}$$
$$u_0, y \in \mathcal{X}_L, \quad \text{and} \quad (p+1)y = c(u_0 + y)\}.$$

The other two relations are

6
$$S'_{i0} = \{(u_0, y) | u_0, y \in \mathcal{X}_F \quad \text{and} \quad (p+1)y = c(u_0 + y)\},$$

7
$$S''_{i0} = \{(u_0, y) | u_0, y \in \mathcal{X}_L \quad \text{and} \quad (p+1)y = c(u_0 + y)\}.$$

The explicit expressions for (5), (6), and (7) now depend on the value of the root $-(1-c)$ of the characteristic polynomial of the feedback systems.

The Fourier and Laplace transforms can again be conveniently applied for the purpose of finding the appropriate solutions. We obtain the following results.

The case $(1-c) > 0$ (i.e. a "stable" root). S_{i0}, S'_{i0}, and S''_{i0} are mappings $\mathcal{W} \to \mathcal{W}$, $\mathcal{X}_F \to \mathcal{X}_F$, and $\mathcal{X}_L \to \mathcal{X}_L$ respectively, and the output value $y(t)$ is for any input $u_0 \in \mathcal{W}$, $\mathcal{X}_F$, or $\mathcal{X}_L$, and any $t \in \mathbf{R}$ given by

8
$$y(t) = \int_{-\infty}^{t} c e^{-(1-c)(t-\tau)} u_0(\tau) d\tau.$$

S_{i0}, S'_{i0}, and S''_{i0} are clearly seen to be mappings satisfying our set of realizability conditions.

With $\mathfrak{Q}_{i0}$, $\mathfrak{Q}'_{i0}$, and $\mathfrak{Q}''_{i0}$ denoting the corresponding families of mappings according to (3.4.1), we thus have in this particular case:

9
$$\cup \mathfrak{Q}_{i0} = S_{i0} \in \mathfrak{Q}_{i0},$$
$$\cup \mathfrak{Q}'_{i0} = S'_{i0} \in \mathfrak{Q}'_{i0}, \quad \text{and}$$
$$\cup \mathfrak{Q}''_{i0} = S''_{i0} \in \mathfrak{Q}''_{i0}.$$

All the interconnections $(\mathcal{A}, C, X)$, $(\mathcal{A}', C, X)$, and $(\mathcal{A}'', C, X)$ considered thus satisfy condition (3.4.3), i.e. they are all determinate. The dynamic systems determined by these interconnections can be formed in a trivial way, *cf.* (3.4.6).

The case $(1-c)<0$ (i.e. an "unstable" root). S'_{i0} and S''_{i0} are mappings, S_{i0} is a relation but not a mapping. We have:

$$S'_{i0}:\mathscr{X}_F \to \mathscr{X}_F,$$

and for any $u_0 \in \mathscr{X}_F$ and any $t \in \mathbf{R}$ it holds that

10 $$y(t) = -\int_t^{\infty} ce^{-(1-c)(t-\tau)}u_0(\tau)d\tau$$

with y denoting $S'_{i0}(u_0)$,

$$S''_{i0}:\mathscr{X}_L \to \mathscr{X}_L,$$

and for any $u_0 \in \mathscr{X}_L$ and any $t \in \mathbf{R}$ it holds that

11 $$y(t) = \int_{-\infty}^{t} ce^{-(1-c)(t-\tau)}u_0(\tau)d\tau$$

with y denoting $S''_{i0}(u_0)$,

12 $$S_{i0} = S'_{i0} \cup S''_{i0}.$$

First consider S'_{i0}. According to (10), S'_{i0} is clearly a mapping but not a causal one. This shows that an interconnection comprising causal elements can generate noncausal solutions. Our realizability conditions are easily seen to lead to the result that

13 $$\mathscr{Q}'_{i0} = \{\phi\},$$

where $\mathscr{Q}'_{i0}$ is formed from S'_{i0} according to (3.4.1). Thus $\cup\,\mathscr{Q}'_{i0} = \phi \in \mathscr{Q}'_{i0}$ (*cf.* note (3.4.4)) and $(\mathscr{A}', C, X)$ is formally determinate. Of course, the corresponding system $\mathscr{S}'_i = \mathscr{S}'_o$ in this case consists of the empty mapping only, and this means that the interconnection does not give us any useful input-output pairs within the structure chosen. The space $\mathscr{X}_F$ is not a very suitable signal space in this unstable case.

Next consider S''_{i0}. (11) shows that S''_{i0} is a mapping satisfying our set of realizability conditions. Further it is found that (with $\mathscr{Q}''_{i0}$ formed in accordance with (3.4.1))

14 $$\cup\,\mathscr{Q}''_{i0} = S''_{i0} \in \mathscr{Q}''_{i0},$$

and $(\mathscr{A}'', C, X)$ is consequently determinate. It can thus be concluded that $\mathscr{X}_L$ is a suitable signal space both in the stable and the unstable case.

Finally, consider $S_{i0} = S'_{i0} \cup S''_{i0}$. S_{i0} is not a mapping because there are pairs $(u'_0, y') \in S'_{i0}$ and $(u''_0, y'') \in S''_{i0}$ such that $u'_0 = u''_0$ and $y' \neq y''$. S_{i0} is the union of two mappings, only one of them fulfilling our set of realizability conditions. A moment's reflection now leads to the result that ($\mathcal{Q}_{i0}$ is again formed in accordance with (3.4.1))

15 $$\cup\, \mathcal{Q}_{i0} = S''_{i0} \in \mathcal{Q}_{i0}.$$

Hence, $(\mathcal{A}, C, X)$ is determinate. Also $\mathcal{W} = \mathcal{X}_F \cup \mathcal{X}_L$ can thus be regarded as a suitable signal set in both the stable and the unstable case.

Note that $\cup\, \mathcal{Q}''_{i0} = \cup \mathcal{Q}_{i0} = S''_{i0}$.

To summarize, we have arrived at the following conclusions.

In the stable case $(1 - c) > 0$ all the interconnections $(\mathcal{A}, C, X)$, $(\mathcal{A}', C, X)$, and $(\mathcal{A}'', C, X)$ are determinate with respect to the given set of realizability conditions, and well-defined corresponding dynamic systems $\mathcal{S}_i = \mathcal{S}_o = \{S_{i0}\})$, $\mathcal{S}'_i = \mathcal{S}'_o = \{S'_{i0}\}$, and $\mathcal{S}''_i = \mathcal{S}''_o = \{S''_{i0}\}$ can consequently be associated with these interconnections.

In the unstable case $(1 - c) < 0$ all the interconnections considered are again formally determinate. However, only the empty mapping can be associated with $(\mathcal{A}', C, X)$. The other two interconnections $(\mathcal{A}, C, X)$ and $(\mathcal{A}'', C, X)$ in turn determine the same dynamic system $\mathcal{S}_i = \mathcal{S}_o = \mathcal{S}''_i = \mathcal{S}''_o = \{S''_{i0}\}$. Note that the given set of realizability conditions makes it possible to extract the dynamic part S''_{i0} from S_{i0} uniquely. The noncausal part of S_{i0} is simply discarded because this part is wholly irrelevant as far as models of real systems are concerned.

It is important to observe the role played in this context by the requirement concerning the richness of the domain of an acceptable input-output mapping which was included in the set of realizability conditions. If this requirement were deleted—leaving causality as the only realizability condition—then $(\mathcal{A}, C, X)$ and $(\mathcal{A}', C, X)$ above would be indeterminate in the unstable case $(1 - c) < 0$.

Part II

Differential Systems. The Module Structure

In this part of the book consideration will be given to systems governed by ordinary linear time-invariant differential equations on the basis of input-output relations (*cf.* section 2.1) generated by given "regular" sets of such equations. It is found that every input-output relation of this kind determines a corresponding natural dynamic (causal) system (according to the definitions given in section 2.2) on a chosen nonempty time interval. Moreover, this system proves to be linear and finite dimensional provided that the signal spaces involved are appropriately chosen. This result enables us to analyse various properties of such a system directly with the aid of the underlying input-output relation without explicitly using the system itself, i.e. the family of input-output mappings which constitutes the system.

It turns out that the appropriate algebraic structure to be used in this context is that of a *module*. The module is formed in a very natural way from a suitable signal space together with the ring of polynomials in the differentiation operator. This structure further leads to the "polynomial systems theory" applied by Rosenbrock, Wolovich and others.

This part concludes with a number of typical analysis and synthesis problems which show how the polynomial systems theory can be used in an efficient way in cases usually treated by means of state-space methods.

Realizability conditions

Throughout the rest of this book it is assumed that the set of realizability conditions (*cf.* section 2.2) comprises the causality property only.

4

Generation of differential systems

This chapter is intended to give a very brief presentation of the main method that will be used in this context to generate differential systems. A more thorough treatment will be presented in subsequent chapters.

The presentation follows the lines of chapter 4 in Blomberg, 1975, which is in turn based on an earlier paper, Blomberg *et al.*, 1969. The basic idea is that differential equations are considered as differential operator equations in which the operators are regarded as mappings between suitably chosen signal spaces. A differential equation in this way defines a relation on a Cartesian product of signal spaces, and this relation forms the starting point for the whole analysis. It turns out that the properties possessed by the operators and the corresponding relations decisively depend on the underlying signal space. The choice of signal space is therefore of crucial significance in this context. In this and subsequent chapters we shall discuss in detail a number of important properties possessed by different signal spaces as well as the consequences of these properties.

The above idea makes it possible for us to develop an efficient and precise method for treating various problems concerning systems governed by ordinary linear time-invariant differential equations—the same idea is later used in the corresponding discrete-time case. In this way it is also possible to avoid many of the ambiguities appearing in this context in the literature.

4.1 Signal spaces and differential operators

Let $\mathscr{X}$ denote a vector space over $\mathbf{C}$ of suitable complex-valued time functions (signals) defined on some open nonempty time interval $T \subset \mathbf{R}$

(natural pointwise operations are assumed). We shall require that the elements of $\mathscr{X}$ be differentiable and that $\mathscr{X}$ is invariant under the differentiation operator p, i.e. that p can be regarded as an element of the set of all linear mappings $\mathscr{X} \to \mathscr{X}$.

A function space $\mathscr{X}$ satisfying the above conditions will henceforth be called a (continuous-time) *signal space* for short.

There are many function spaces that qualify as signal spaces. One convenient and well-behaved space would be the space C^∞ of infinitely continuously differentiable complex-valued functions on T, or any subspace of C^∞ invariant under p. A more complicated space would be the space $\mathscr{D}$ of "piecewise infinitely regularly differentiable complex-valued generalized functions (distributions) on T" (*cf.* appendix A4), or any subspace of $\mathscr{D}$ invariant under p (note that C^∞ can be regarded as such a subspace).

Note that the time interval T is regarded as arbitrary but fixed, and it is therefore not explicitly indicated in the notations $\mathscr{X}$, C^∞, and $\mathscr{D}$.

If the reader feels ill at ease with generalized functions s/he may, throughout this part of the book, think of $\mathscr{X}$ as C^∞. The space $\mathscr{D}$ and various subspaces of $\mathscr{D}$ are not really needed until part III.

Accordingly, any polynomial in p with complex coefficients, i.e. any polynomial of the form

1 $$a(p) \triangleq a_0 + a_1 p + \ldots + a_n p^n$$

with $a_0, a_1, \ldots, a_n \in \mathbf{C}$ and $n \in \{0, 1, 2, \ldots\}$ represents a linear mapping $\mathscr{X} \to \mathscr{X}$ given by the assignment ($p^0 x \triangleq 1x$, $p^i x \triangleq x^{(i)}$, the ith-order derivative of x)

2 $$x \mapsto a(p)x = a_0 x + a_1 px + \ldots + a_n p^n x.$$

Note that two polynomials $a(p) \triangleq a_0 + a_1 p + \ldots + a_n p^n$ and $b(p) \triangleq a_0 + a_1 p + \ldots + a_n p^n + a_{n+1} p^{n+1} + a_{n+2} p^{n+2} + \ldots + a_m p^m$ with $a_{n+1} = a_{n+2} = \ldots = a_m = 0$ of the form (1) represent the same mapping $\mathscr{X} \to \mathscr{X}$, i.e. we have $a(p) = b(p)$.

More generally, any polynomial s by q-matrix $A(p)$ of the form ($s, q \in \{1, 2, 3, \ldots\}$)

3 $$A(p) \triangleq \begin{bmatrix} a_{11}(p) & a_{12}(p) & \ldots & a_{1q}(p) \\ a_{21}(p) & a_{22}(p) & \ldots & a_{2q}(p) \\ \vdots & \vdots & & \vdots \\ a_{s1}(p) & a_{s2}(p) & \ldots & a_{sq}(p) \end{bmatrix}$$

with the $a_{ij}(p)$ polynomials in p of the form (1) represents a linear mapping $\mathscr{X}^q \to \mathscr{X}^s$. For any $x = (x_1, x_2, \ldots, x_q) \in \mathscr{X}^q$ the corresponding $y = (y_1,$

$y_2, \ldots y_s) = A(p)x \in \mathscr{X}^s$ is evaluated by interpreting $y = A(p)x$ as a matrix equality with x and y regarded as column matrices.

Mappings of the form (1) and (3) are frequently called "differential operators". Differential operators of the form (1) and (3) will in this context be called "polynomial operators" and "polynomial matrix operators" respectively. Polynomial operators will be identified with polynomial matrix operators represented by 1 by 1-matrices.

4 **Note.** **(i)** The determinant of a square polynomial matrix operator $A(p)$ of the form (3) is a polynomial in p and it is obtained from $A(p)$ according to the usual rules for polynomial matrices (*cf.* appendix A2). The determinant of $A(p)$ is denoted by $\det A(p)$.
(ii) Let $a(p) \triangleq a_0 + a_1 p + \ldots + a_n p^n$ and $b(p) \triangleq b_0 + b_1 + \ldots + b_m p^m$, $m \geqslant n$, be two polynomial operators of the form (1) representing the same mapping $\mathscr{X} \to \mathscr{X}$, i.e. let $a(p) = b(p)$. This does not necessarily imply that $a_i = b_i$ for $i = 0, 1, \ldots, n$ and (if $m > n$) $b_{n+1} = b_{n+2} = \ldots = b_m = 0$ unless certain additional conditions are fulfilled (*cf.* chapter 5) □

5 **Remark.** **(i)** Throughout the above we have assumed complex-valued rather than real-valued quantities. As far as the behaviour of real systems is concerned, real-valued quantities would be entirely satisfactory. Complex-valued quantities have been chosen merely because this choice gives certain formal advantages (for instance: any polynomial of the form (1) with $a_n = 1$ can be expressed as a product of n monic binomials).
(ii) The concepts of a signal space could be further generalized. Let V be a normed vector space over $\mathbf{C}$, and let $\mathscr{X}$ be a vector space of V-valued functions on the open nonempty time interval $T \subset \mathbf{R}$ (pointwise operations are again assumed). $\mathscr{X}$ is then qualified to be called a "signal space of V-valued functions on T" if the elements of $\mathscr{X}$ are differentiable and if $\mathscr{X}$ is invariant under the differentiation operator p. The results obtained for complex-valued signals would in general—with some minor alterations—also hold in this generalized case. □

4.2 Matrix differential equations

Let $\mathscr{X}$ be a signal space in the sense explained in the previous section. In what follows we shall, when speaking of $\mathscr{X}$ and its members, use the terminology and notions normally used in connection with ordinary pointwise defined complex-valued functions defined on some interval of $\mathbf{R}$. If $\mathscr{X}$ is a space of generalized functions, i.e. if $\mathscr{X} = \mathscr{D}$ or if $\mathscr{X}$ is equal to some suitable subspace of $\mathscr{D}$, then some care is necessary in interpreting the various statements. Chapter 8 and appendix A4 contain further material relating to this matter.

Consider the following matrix differential equation.

1 $$A(p)y = B(p)u,$$

where $u \in \mathscr{X}^r$, $y \in \mathscr{X}^q$, and where $A(p)$ and $B(p)$ are polynomial matrix operators of sizes q by q and q by r respectively with $\det A(p) \neq 0$ (more general cases are considered in section 6.1).

Interpreting u in (1) as an input signal and y as an output signal (1) generates a relation $S \subset \mathscr{X}^r \times \mathscr{X}^q$, i.e. a set of input-output pairs in $\mathscr{X}^r \times \mathscr{X}^q$, according to

2 $$S \triangleq \{(u, y) | (u, y) \in \mathscr{X}^r \times \mathscr{X}^q \quad \text{and} \quad A(p)y = B(p)u\}.$$

A relation of this kind will be called a *differential input-output relation* (*cf.* section 6.1). We shall say that S above is *generated* by equation (1), or alternatively, by the pair $(A(p), B(p))$ of operators.

Now choose a $t_0 \in T$ and let (u_1, y_1) be an input-output pair contained in S. Suppose that we observe the "time evolution" of (u_1, y_1) from the beginning of T up to the point t_0. Let u_1^- and y_1^- denote the observed "initial segments" on $T \cap (-\infty, t_0)$ of u_1 and y_1 respectively. With (u_1^-, y_1^-) given we would next wish to consider the behaviour of those input-output pairs which belong to S and which coincide with (u_1, y_1) up to the point t_0. The set of all these pairs is denoted by $S_{(u_1^-, y_1^-)}$, and we have formally

3 $$S_{(u_1^-, y_1^-)} \triangleq \{(u, y) | (u, y) \in S \quad \text{and } u | T \cap (-\infty, t_0) = u_1^- \quad \text{and} \quad y | T \cap (-\infty, t_0) = y_1^-\}.$$

Note that the fixed pair (u_1^-, y_1^-) of initial segments also fixes the value at t_0- of every pair (u, y) contained in $S_{(u_1^-, y_1^-)}$ and of every derivative of such a pair. The above fact, together with the assumption that $\det A(p) \neq 0$, guarantees that $S_{(u_1^-, y_1^-)}$ is a mapping. This follows from the general theory for ordinary linear time-invariant differential equations as presented for instance in the book by Zadeh and Desoer, 1963, chapters 4 and 5.

Letting (u_1^-, y_1^-) range over the set of all possible pairs of initial segments we thus obtain a family of input-output mappings, i.e. a system $\mathscr{S}$ according to

4 $$\mathscr{S} \triangleq \{S_{(u_1^-, y_1^-)} | u_1^- = u | T \cap (-\infty, t_0) \quad \text{and} \quad y_1^- = y | T \cap (-\infty, t_0) \quad \text{for some } (u, y) \in S\}.$$

Evidently

5 $$S = \cup \mathscr{S},$$

and so S is indeed an input-output relation associated with $\mathcal{S}$ in the sense of the presentation in section 2.1. An appropriate description of $\mathcal{S}$ of the form $(\mathcal{U}, \mathcal{Y}, \Sigma, S, \mathcal{S}, h)$ can easily be formed with the set of all possible pairs of initial segments taken as the state set Σ.

6 **Note.** **(i)** The system $\mathcal{S}$ as given above is clearly uniquely determined by the construction employed, the differential input-output relation S, and the point $t_0 \in T$—it thus does not depend on the particular pair $(A(p), B(p))$ of operators appearing in (1) and (2). In fact, there are infinitely many pairs of operators generating the same differential input-output relation S and the same system $\mathcal{S}$. Pairs of operators generating the same differential input-output relation are said to be equivalent. This particular equivalence—called *input-output equivalence*—will be further pursued in chapter 6.

We shall call a system $\mathcal{S}$ of the kind (4) a *differential* (*input-output*) *system*.

(ii) The system $\mathcal{S}$ given above proves to be a causal system on T (*cf.* chapter 8). It is, however, generally *not* a linear system in the sense of the definition given in section 2.3. If, for instance, $u_{\bar{1}}^{-}$ is not the zero segment, then the domain of $S_{(u_{\bar{1}}^{-}, y_{\bar{1}}^{-})}$ is clearly not a vector space. Nevertheless, if $\mathcal{X}$ possesses appropriate properties, then a finite dimensional linear dynamic (causal) system on $T \cap [t_0, \infty)$ can be obtained from $\mathcal{S}$ simply by restricting the input-output pairs belonging to the members of $\mathcal{S}$ to the time interval $T \cap [t_0, \infty)$. This question will be discussed in detail in chapter 8, where the restriction is actually replaced by a corresponding projection operation.

(iii) There are evidently cases in which a differential input-output relation S of the form (2) is a mapping. This happens whenever $A(p): \mathcal{X}^q \to \mathcal{X}^q$ has an inverse $A(p)^{-1}: \mathcal{X}^q \to \mathcal{X}^q$.

(iv) The zero-input response space of $\mathcal{S}$ is according to (3), (4) given as the space of all signals of the form $S_{(0^-, y_{\bar{1}}^{-})}(0)$. This space evidently coincides with $\ker A(p)$, which is a subspace of $\mathcal{X}^q$. Referring to S according to (2), we shall also use the notation $S(0)$ for $\ker A(p)$.

(v) There are, of course, other possibilities for assigning a system to a differential relation S of the form (2). However, the possibility used here appears to be a very natural one. □

This brief presentation should convince the reader of the validity of the statement that any differential input-output relation S of the form (2) also determines a natural corresponding dynamic (causal) differential system $\mathcal{S}$ of the form (4) (with $t_0 \in T$ given). Differential systems of this kind can therefore be analysed on the basis of their corresponding differential input-output relations. The algebraic machinery needed for this purpose will be developed in the following chapter.

5

The $\mathbf{C}[p]$-module $\mathscr{X}$

The basic machinery needed for the purpose of treating differential systems on the basis of their corresponding differential input-output relations will be developed in this chapter. In subsequent chapters it will then be shown how this machinery can be used to treat a number of specific problems.

The appropriate machinery is obtained by interpreting the signal space algebraically as a *module* over the ring of polynomial operators of the form discussed in section 4.1 above. How this is done—and why—has been explained in a series of papers starting with Blomberg and Salovaara, 1968, followed by Blomberg *et al.*, 1969, Blomberg, 1975, Ylinen, 1975, Blomberg and Ylinen, 1976, 1978.

Now it is found that for the generated differential input-output relations to be representative in a certain sense, the signal space must be chosen "sufficiently rich". This in turn leads to the result that the nonzero polynomial operators are generally not invertible as mappings. As a consequence, it follows that the module mentioned above cannot be strengthened to a corresponding vector space. A special methodology must therefore be devised for the treatment of relevant problems on the basis of the relatively weak module structure obtained. This is the particular subject of chapter 7.

The module structure so constructed is, in fact, the inherent basic structure used in the "polynomial systems theory" as developed and used in particular by Rosenbrock and Wolovich (see for instance Rosenbrock, 1970, Wolovich, 1974). These authors, however, never explicitly mention this fact, neither do they give any convincing reason for this particular choice of structure. Their polynomial theory is developed formally on the basis of the Laplace transform for zero initial conditions where "cancellation is not permitted" (Rosenbrock, 1970, pp. 46 and 49). This is rather confusing.

5.1 Suitable signal spaces $\mathscr{X}$

The main condition that we imposed on the signal space $\mathscr{X}$ above was that it should be invariant under the differentiation operator p. We shall now introduce additional conditions dictated by our present aim.

The concept of a regular signal space $\mathscr{X}$

The treatment of various problems concerning structural properties of differential input-output relations and compositions of such relations will be considerably simplified if the signal space possesses suitable properties. In view of this we shall introduce the concept of a "regular" signal space.

1 **Property.** A (continuous-time) signal space $\mathscr{X}$ is said to possess property (1) if the following condition is fulfilled.
A polynomial operator $\mathscr{X} \to \mathscr{X}$ of the form

$$a_0 + a_1 p + \ldots + a_n p^n$$

as given by (4.1.1) is zero (i.e. the zero mapping) if and only if all its coefficients $a_0, a_1, \ldots, a_n$ are zero. □

2 **Property.** A (continuous-time) signal space $\mathscr{X}$ is said to possess property (2) if every nonzero polynomial operator $\mathscr{X} \to \mathscr{X}$ of the form (4.1.1) is a surjection with respect to $\mathscr{X}$. □

3 **Definition.** A (continuous-time) signal space $\mathscr{X}$ is said to be *regular* if it possesses property (1) as well as property (2). □

4 **Note.** **(i)** A (continuous-time) nontrivial signal space $\mathscr{X}$ is clearly regular if and only if every polynomial operator $\mathscr{X} \to \mathscr{X}$ of the form

$$a_0 + a_1 p + \ldots + a_n p^n, a_n \neq 0$$

as given by (4.1.1) is a surjection with respect to $\mathscr{X}$.
(ii) If $\mathscr{X}$ is a signal space possessing property (1), then p—using polynomial terminology—is an *indeterminate* over **C** (see section 5.2). Moreover, $\mathscr{X}$ is necessarily infinite dimensional.
(iii) Property (1) implies a number of important consequences. Thus let $\mathscr{X}$ be a signal space possessing property (1) and consider two polynomial operators $\mathscr{X} \to \mathscr{X}$ of the form (4.1.1) given by $a(p) \triangleq a_0 + a_1 p + \ldots + a_n p^n$, $b(p) \triangleq b_0 + b_1 p + \ldots + b_m p^m$ with $m \geqslant n$. If $a(p) = b(p)$ then $a(p) - b(p) = 0$, and consequently $a_i = b_i$ for $i = 0, 1, \ldots, n$, and (if $m > n$) $b_{n+1} = b_{n+2} = \ldots = b_m = 0$. Moreover, the *degree* of $a(p)$ is in

this case a well-defined quantity. It is denoted by $\partial(a(p))$ and it is equal to the greatest $k \in \{0, 1, \ldots, n\}$, if it exists, such that $a_k \neq 0$. The zero polynomial is formally assigned the degree $-\infty$.

(iv) Let $\mathscr{X}$ be a signal space possessing property (2) and let $a(p)$ be a nonzero polynomial operator $\mathscr{X} \rightarrow \mathscr{X}$ of the form (4.1.1). $a(p)$ is then surjective with respect to $\mathscr{X}$ and so the equation $a(p)y = x$ has at least one solution $y \in \mathscr{X}$ for every $x \in \mathscr{X}$, i.e. the domain of the differential input-output relation (*cf.* (4.2.2))

$$S \triangleq \{(x, y) \mid (x, y) \in \mathscr{X} \times \mathscr{X} \quad \text{and}\, a(p)y = x\}$$

is the whole of $\mathscr{X}$. This property simplifies to a considerable extent the treatment of problems concerning compositions of differential input-output relations (*cf.* section 7.2).

(v) The signal spaces C^∞ and $\mathscr{D}$ are regular signal spaces. This is readily seen on the basis of the general theory for ordinary linear time-invariant differential equations (*cf.* Zadeh and Desoer, 1963, chapter 4). It is namely found that the equation $a(p)y = x$, with $a(p)$ any polynomial operator of the form $a_0 + a_1p + \ldots + a_np^n$, $a_n \neq 0$, has at least one solution $y \in C^\infty$ for every $x \in C^\infty$, and at least one solution $y \in \mathscr{D}$ for every $x \in \mathscr{D}$. Additional regular signal spaces appear in section 8.2 (*cf.* also (vi) below).

(vi) Let $n \in \{0, 1, 2, \ldots\}$ be fixed and consider the space P_n of all complex-valued polynomial functions on T, i.e. for any $x \in P_n$ there are corresponding complex numbers $\alpha_0, \alpha_1, \ldots, \alpha_n$ such that

$$x(t) = \alpha_0 + \alpha_1 t + \ldots + \alpha_n t^n$$

for all $t \in T$. P_n is obviously a signal space but it is not a regular one (why not?).

If n above is permitted to range over the whole of $\{0, 1, 2, \ldots\}$, then a regular signal space $P \triangleq \cup \{P_n | n \in \{0, 1, 2, \ldots\}\}$ is obtained. □

"Sufficiently rich" signal spaces $\mathscr{X}$

Next we shall discuss another aspect relating to the choice of a suitable signal space $\mathscr{X}$.

Consider a differential input-output relation S of the form (4.2.2) and the corresponding differential system $\mathscr{S}$ as given by (4.2.4). Let S and $\mathscr{S}$ be generated by any suitable pair $(a(p), b(p))$, $a(p) \neq 0$, of polynomial operators. It is evident that the "richness" of S and $\mathscr{S}$ depends on the underlying signal space $\mathscr{X}$. Hence, if we wish S and $\mathscr{S}$ to be representative in some sense for the class of all differential relations and differential systems that can be generated by a given pair $(a(p), b(p))$, we must choose $\mathscr{X}$ rich enough. In view of the great general significance of the zero input response space (*cf.* section 2.3), we would in particular require $\mathscr{X}$ to be rich enough to "maximize" the zero-input response space $S(0)$ of $\mathscr{S}$, i.e. to contain all possible solutions y to $a(p)y = 0$ (*cf.* note (4.2.6), (iv)). We shall therefore consider a signal space $\mathscr{X}$ sufficiently rich if it possesses the following property.

5 **Property.** Let $\mathscr{X}$ be a given signal space and let $\mathscr{X}_1$ be any other signal space such that $\mathscr{X} \subset \mathscr{X}_1$. $\mathscr{X}$ is then said to possess property (5) if it holds for any nonzero polynomial operator $a(p)$ of the form (4.1.1) interpreted as a mapping $\mathscr{X} \to \mathscr{X}$ or $\mathscr{X}_1 \to \mathscr{X}_1$ respectively that

$$\ker a(p)_{\text{with respect to } \mathscr{X}} = \ker a(p)_{\text{with respect to } \mathscr{X}_1},$$

where the kernels involved are considered as sets. □

6 **Note. (i)** Let $\mathscr{X}$ be a signal space possessing property (5) and let $a(p) \triangleq a_0 + a_1p + \ldots + a_np^n$, $a_n \neq 0$, be a polynomial operator $\mathscr{X} \to \mathscr{X}$ of the form (4.1.1). It is known (see Zadeh and Desoer, 1963, chapter 4) that the set of all possible solutions y to $a(p)y = 0$, i.e. $\ker a(p)$, forms an n-dimensional subspace of $\mathscr{X}$ (the zero of $\mathscr{X}$ is regarded as a zero-dimensional subspace of $\mathscr{X}$). Moreover, every ordinary function y defined on T and satisfying $a(p)y = 0$ is necessarily infinitely continuously differentiable, and so the space C^∞ clearly qualifies as $\mathscr{X}$. From the fact that C^∞ can be regarded as a subspace of $\mathscr{D}$ it also follows that $\mathscr{D}$ qualifies as $\mathscr{X}$ (for a precise interpretation of the equation $a(p)y = 0$ with $y \in \mathscr{D}$, see Gelfand and Shilov, 1964). It is also easily seen that $\mathscr{X}$ must in this case contain all complex-valued functions on T of the form $t \mapsto ct^n e^{\alpha t}$, $t \in T$, $n \in \{0, 1, 2, \ldots\}$, α, $c \in \mathbf{C}$ and all finite sums of such functions.

Referring to note (4), (v) above it is thus seen that C^∞ and $\mathscr{D}$ are regular signal spaces possessing the additional property (5).

(ii) Property (5) implies property (1).

(iii) Let $\mathscr{X}$ be a regular signal space possessing property (5). Then any polynomial operator $\mathscr{X} \to \mathscr{X}$ of the form (4.1.1) has a degree (*cf.* note (4), (iii)). Moreover, such an operator has an inverse $\mathscr{X} \to \mathscr{X}$ if and only if it is of degree zero, i.e. if and only if it is a nonzero constant. This follows from the fact that a linear mapping $f: \mathscr{X} \to \mathscr{X}$ has an inverse $f^{-1}: \mathscr{X} \to \mathscr{X}$ if and only if it is a surjection with respect to $\mathscr{X}$ and $\ker f = \{0\}$. According to item (i) above, only nonzero constants possess the latter property. □

5.2 The ring **C**[p] of polynomial operators. The **C**[p]-module $\mathscr{X}$

Let $\mathscr{X}$ be any signal space and let $L(\mathscr{X}, \mathscr{X})$ denote the algebra of all linear mappings $\mathscr{X} \to \mathscr{X}$ with addition and scalar multiples defined pointwise, and with composition corresponding to multiplication. Let, furthermore, $\mathbf{C}[p]$ denote the set of all polynomial operators of the form (4.1.1). Then $\mathbf{C}[p]$ forms a commutative subalgebra of $L(\mathscr{X}, \mathscr{X})$, and $\mathscr{X}$ can be regarded as a (left) $\mathbf{C}[p]$-module (*cf.* appendix A3).

We can now invoke the general theory for polynomials (*cf.* appendix A2) to yield additional structure. To this end $L(\mathscr{X}, \mathscr{X})$ is regarded as a ring and the field $\mathbf{C}$ as a subring of $L(\mathscr{X}, \mathscr{X})$. We have $p \in L(\mathscr{X}, \mathscr{X})$, $ap = pa$ for every $a \in \mathbf{C}$, and $\mathbf{C}$ contains the unity ($\triangleq 1$, the identity mapping $\mathscr{X} \to \mathscr{X}$) of $L(\mathscr{X}, \mathscr{X})$. It follows that the elements of $\mathbf{C}[p]$ really are polynomials in p over the field $\mathbf{C}$, and that $\mathbf{C}[p]$ is a commutative subring of $L(\mathscr{X}, \mathscr{X})$ with unity. p is called an "indeterminate" over $\mathbf{C}$ if any polynomial in p over $\mathbf{C}$ is zero if and only if all its coefficients are zero, that is to say, if the signal space $\mathscr{X}$ possesses property (5.1.1) (*cf.* note (5.1.4), (ii)). p is thus an indeterminate over $\mathbf{C}$ in particular if $\mathscr{X}$ is regular.

1 **Note.** **(i)** The signal spaces C^∞ and $\mathscr{D}$ were found to be regular. $p \in L(\mathscr{X}, \mathscr{X})$ is consequently an indeterminate over $\mathbf{C}$ if $\mathscr{X}$ is equal to either C^∞ or $\mathscr{D}$. These spaces possess, in addition, property (5.1.5) (*cf.* note (5.1.6), (i)). These facts make C^∞ and $\mathscr{D}$ particularly suitable as signal spaces for our present purpose.
(ii) Let $\mathscr{X}$ be a signal space possessing property (5.1.1), i.e. let $p \in L(\mathscr{X}, \mathscr{X})$ be an indeterminate over $\mathbf{C}$. $\mathbf{C}[p]$ is then ring isomorphic in a natural way to any ring $\mathbf{C}[z]$ of polynomials in an indeterminate z over $\mathbf{C}$. Moreover, from the fact that $\mathbf{C}$ is a field it follows that $\mathbf{C}[p]$ is not only a commutative ring with unity but also an integral domain. It is therefore possible to construct the "field of quotients" of $\mathbf{C}[p]$, denoted by $\mathbf{C}(p)$ (*cf.* appendix A3). $\mathbf{C}(p)$ consists of equivalence classes, "quotients", on $\mathbf{C}[p] \times (\mathbf{C}[p] - \{0\})$, each class being representable by quotients of the form $b(p)/a(p)$ with $b(p) \in \mathbf{C}[p]$ and $a(p) \in \mathbf{C}[p] - \{0\}$. $\mathbf{C}[p]$ is ring isomorphic in a natural way to the subring of $\mathbf{C}(p)$ consisting of the set of all quotients representable as $b(p)/1$ for some $b(p) \in \mathbf{C}[p]$.
(iii) Let $\mathscr{X}$ be a regular signal space and consider the $\mathbf{C}[p]$-module $\mathscr{X}$. It is known (see appendix A3) that such a $\mathbf{C}[p]$-module $\mathscr{X}$ can be strengthened to a vector space $\mathscr{X}$ over the field $\mathbf{C}(p)$ of quotients of $\mathbf{C}[p]$—with the elements of $\mathbf{C}(p)$ identified in a natural way with corresponding elements of $L(\mathscr{X}, \mathscr{X})$ (for details, see section 8.3)—provided that $\mathscr{X}$ is such that every

nonzero element of $\mathbf{C}[p]$ is invertible in $L(\mathscr{X}, \mathscr{X})$. It is now evident that if the signal space $\mathscr{X}$ possesses property (5.1.5), then this strengthening cannot be performed, because in such a case only the nonzero constants satisfy the necessary invertibility condition (*cf.* note (5.1.6), (iii)). In section 8.2 we shall study regular signal spaces that permit the above vector space structure.

Thus we see that the vector space structure and the representativity property of differential relations cannot be simultaneously achieved. □

5.3 Relationship between polynomial matrix operators and matrices over C[*p*]

Let $\mathscr{X}$ be any signal space and consider a polynomial s by q-matrix $A(p)$ as given by (4.1.3). In section 4.1 $A(p)$ was interpreted as a linear mapping $\mathscr{X}^q \to \mathscr{X}^s$ and it was called a "polynomial matrix operator".

Formally, the matrix $A(p)$ can also be regarded as a mapping $\mathbf{s} \times \mathbf{q} \mapsto \mathbf{C}[p]$ with values $A(p)(i, j) = a_{ij}(p)$ i.e. as an element of the set $\mathbf{C}[p]^{s \times q}$. $A(p)$ is then customarily called a "matrix with entries from $\mathbf{C}[p]$" or a "matrix over $\mathbf{C}[p]$" (*cf.* appendix A2). Whenever this interpretation is intended, we shall also call $A(p)$ a "polynomial matrix".

If $\mathscr{X}$ is a regular signal space, then $A(p)$ can be further identified with the mapping $\mathbf{s} \times \mathbf{q} \to \mathbf{C}(p)$, $(i, j) \mapsto a_{ij}(p)/1$. In this case the polynominal matrix $A(p)$ can also be identified with the corresponding "rational matrix" $A(p)/1$.

1 **Note.** **(i)** Because $\mathbf{C}[p]$ is a (commutative) ring, sums and products of matrices over $\mathbf{C}[p]$ of appropriate sizes are again matrices over $\mathbf{C}[p]$, and they are determined according to the usual rules for matix operations. Also the determinant of a square matrix over $\mathbf{C}[p]$ is obtained in the usual way (*cf.* note (4.1.4), (i)) as an element of $\mathbf{C}[p]$.

The *rank* of a polynomial matrix operator and of the corresponding matrix $A(p)$ over $\mathbf{C}[p]$ is the greatest integer r such that $A(p)$ contains a nonzero minor of order r. It is equal to the number of linearly independent rows or columns in $A(p)$ (linear independence interpreted with respect to the ring $\mathbf{C}[p]$).

(ii) No ambiguity should arise from the fact that $A(p)$ above can be interpreted in different ways. This is so because the relevant operations of forming sums and products are compatible in the following sense.

Let $A(p)$ and $B(p)$ be two s by q-matrices over $\mathbf{C}[p]$ and let $C(p)$ be their matrix sum, i.e. $A(p) + B(p) = C(p)$. Then, if $A(p)$, $B(p)$, and

$C(p)$ are interpreted as linear mappings $\mathscr{X}^q \to \mathscr{X}^s$, it again holds that $A(p) + B(p) = C(p)$, where the sum is taken pointwise.

Correspondingly, let $A(p)$ be an s by q-, and $B(p)$ a q by r-matrix over $\mathbf{C}[p]$ and let $C(p)$ be their matrix product, i.e. $A(p)B(p) = C(p)$. Then, if $A(p)$, $B(p)$, and $C(p)$ are interpreted as linear mappings $\mathscr{X}^q \to \mathscr{X}^s$, $\mathscr{X}^r \to \mathscr{X}^q$, and $\mathscr{X}^r \to \mathscr{X}^s$ respectively, it holds that $A(p)B(p) = C(p)$, where the product denotes composition.

(iii) A polynomial matrix $A(p)$ can be identified with the corresponding rational matrix $A(p)/1$ in particular if $\mathscr{X}$ is equal to C^∞ or $\mathscr{D}$. □

5.4 Fundamental properties of polynomial matrix operators

We shall, throughout the rest of this part of the book, be interested only in regular signal spaces that possess the additional richness property (5.1.5). Therefore we shall collect below a number of properties possessed by polynomial matrix operators in this particular case.

1 **Note.** Let $\mathscr{X}$ throughout items (i), (ii), . . . , (v) below be a regular signal space possessing property (5.1.5). $\mathscr{X}$ may in particular be equal to C^∞ or $\mathscr{D}$.

(i) Let $A(p)$ be a polynomial matrix operator $\mathscr{X}^q \to \mathscr{X}^s$ of the form (4.1.3) with $q \geqslant s$. It is then easily seen that there also is the following property corresponding to property (5.1.2): For every $x \in \mathscr{X}^s$ there is at least one $y \in \mathscr{X}^q$ such that $A(p)y = x$, i.e. $A(p)$ is a surjection with respect to $\mathscr{X}^s$, if and only if $A(p)$ is of full rank s (*cf.* also note (5.1.4), (iv)).

(ii) Let $\mathscr{X}_1$ be any other regular signal space such that $\mathscr{X} \subset \mathscr{X}_1$ and let $A(p)$ be a polynomial matrix operator of the form (4.1.3) with $q = s$ interpreted as a mapping $\mathscr{X}^q \to \mathscr{X}^q$ or $\mathscr{X}_1^q \to \mathscr{X}_1^q$ respectively. Furthermore, let $A(p)$ be of full rank q, i.e. $\det A(p) \neq 0$. Then $\mathscr{X}^q$ is sufficiently rich so as to contain all possible solutions y to $A(p)y = 0$ (*cf.* section 5.1), that is, we have

$$\ker A(p)_{\text{with respect to } \mathscr{X}^q} = \ker A(p)_{\text{with respect to } \mathscr{X}_1^q},$$

where the kernels involved are considered as sets. This follows from property (5.1.5) and from the fact that the components of any element of $\ker A(p)$ also belong to $\ker(\det A(p))$ (note that $(\det A(p))I = A(p)_{\text{adj}}A(p)$, where $A(p)_{\text{adj}}$ denotes the adjoint of $A(p)$).

(iii) Still a little more can be said about $\ker A(p)$. So let $A(p)$ again be a polynomial matrix operator $\mathscr{X}^q \to \mathscr{X}^q$ of the form (4.1.3). Then $\ker A(p)$ is an infinite dimensional subspace of $\mathscr{X}^q$ if $\det A(p) = 0$, and a

$\partial(\det A(p))$-dimensional subspace of $\mathscr{X}^q$ if $\det A(p) \neq 0$ (*cf.* also note (5.1.6), (i)).

An interesting task in the latter case is the construction of an explicit basis for $\ker A(p)$. We shall return to this and related questions in chapter 7.

(iv) Let $A(p)$ be a polynomial matrix operator $\mathscr{X}^q \to \mathscr{X}^q$ of the form (4.1.3). Note that the properties mentioned in item (iii) above imply that $\ker A(p) = \{0\}$ if and only if $\ker(\det A(p)) = \{0\}$. On combining this result with item (i) above it is easily concluded that $A(p)$ has an inverse $A(p)^{-1}: \mathscr{X}^q \to \mathscr{X}^q$ if and only if $\det A(p): \mathscr{X} \to \mathscr{X}$ has an inverse $(\det A(p))^{-1}: \mathscr{X} \to \mathscr{X}$, i.e. if and only if $\det A(p)$ is a nonzero constant (*cf.* note (5.1.6), (iii)). $A(p)$ has thus an inverse if and only if the polynomial matrix representing $A(p)$ is unimodular,‡ i.e. invertible as a polynomial matrix (*cf.* MacLane and Birkhoff, 1967, chapter IX). If so then the matrix representing $A(p)^{-1}$ is just the matrix inverse of the matrix representing $A(p)$.

(v) Let $A(p)$ be an s by q, and $B(p)$ a q by r polynomial matrix operator of the form (4.1.3) such that $A(p)B(p) = 0$. Then we have the following easily derived results.

$$s \geqslant q \quad \text{and } A(p) \text{ of full rank } q \text{ imply that } B(p) = 0.$$

$$r \geqslant q \quad \text{and } B(p) \text{ of full rank } q \text{ imply that } A(p) = 0.$$ □

‡ A square matrix A over a ring R is generally called (R-) "unimodular" if $\det A$ is invertible in R.

6

Differential input-output relations. Generators

In this chapter we shall study some fundamental properties of differential input-output relations (*cf.* section 4.2). We shall, in particular, be interested in *regular* input-output relations and *regular generators* for such relations. Now, given a regular generator, this generator determines a unique corresponding regular differential input-output relation and also a unique corresponding *transfer matrix*. On the other hand, given a regular differential input-output relation, it is found that there is a whole class of regular generators generating this relation, but only one corresponding transfer matrix. Finally, given a transfer matrix, there is a whole class of regular generators, as well as a whole class of regular differential input-output relations corresponding to this transfer matrix. These relationships then lead us in a natural way to certain equivalence relations—to *input-output equivalence* on the set of all regular generators, and to *transfer equivalence* on the set of all regular generators as well as on the set of all regular differential input-output relations. *Canonical forms* for these equivalences are also considered. It turns out that the canonical forms devised for the transfer equivalence on the set of all regular differential input-output relations possess an interesting minimality property.

The material presented in this chapter is to a large extent based on results presented by Hirvonen, Blomberg, and Ylinen, 1975, and Ylinen, 1975. The transfer equivalence on the set of all regular generators and the corresponding canonical forms are also discussed in a paper by Aracil and Montes, 1976, in connection with the problem of factoring a transfer matrix in a unique way. Various forms of system equivalence introduced and discussed by Rosenbrock, 1970, Morf, 1975, and others are in a certain sense related to our input-output equivalence. The relationship between these different equivalences will be discussed in a subsequent chapter.

Choice of signal space $\mathscr{X}$

The signal space $\mathscr{X}$ is, throughout this chapter, assumed to be a regular signal space according to definition (5.1.3) possessing the richness property (5.1.5). $\mathscr{X}$ may, in particular, be equal to C^∞ or $\mathscr{D}$.

6.1 Introduction. Regular differential input-output relations and regular generators

In section 4.2 we introduced the concept of a differential input-output relation $S \subset \mathscr{X}^r \times \mathscr{X}^q$ generated by a pair $(A(p), B(p))$ of polynomial matrix operators with $A(p)$ square and $\det A(p) \neq 0$. In the sequel we shall also encounter input-output relations generated by more general pairs of polynomial matrix operators. So, consider the matrix differential equation

1 $$A(p)y = B(p)u,$$

where $u \in \mathscr{X}^r, y \in \mathscr{X}^q$, and where $A(p)$ and $B(p)$ are polynomial matrix operators (*cf.* (4.1.3)) of sizes s by q and s by r respectively. Interpreting the signal space $\mathscr{X}$ as a $\mathbf{C}[p]$-module (*cf.* section 5.2), (1) can be regarded as representing a set of module equations.

The *differential input-output relation* generated by (1), or alternatively by the pair $(A(p), B(p))$, is now defined as the set S of all input-output pairs $(u, y) \in \mathscr{X}^r \times \mathscr{X}^q$ satisfying equation (1), i.e. as the relation $S \subset \mathscr{X}^r \times \mathscr{X}^q$ given by (*cf.* (4.2.2))

2 $$S \triangleq \{(u, y) | (u, y) \in \mathscr{X}^r \times \mathscr{X}^q \text{ and } A(p)y = B(p)u\}.$$

S is of course a subspace of $\mathscr{X}^r \times \mathscr{X}^q$. Writing $A(p)y = B(p)u$ in the form of a partitioned matrix equality

3 $$[A(p) \ \vdots \ -B(p)] \begin{bmatrix} y \\ \hdashline u \end{bmatrix} = 0,$$

and interpreting $[A(p) \ \vdots \ -B(p)]$ as a partitioned polynomial matrix operator $\mathscr{X}^{q+r} \to \mathscr{X}^s$, i.e. as a linear mapping $\mathscr{X}^q \times \mathscr{X}^r \to \mathscr{X}^s$, it is seen that (2) can be written simply as

4 $$S^{-1} = \ker[A(p) \ \vdots \ -B(p)],$$

where S^{-1} denotes the converse of S, i.e. $S^{-1} \triangleq \{(y, u) | (u, y) \in S\}$. We shall also say that S is *generated* by $[A(p) \vdots -B(p)]$, and that $[A(p) \vdots -B(p)]$ is a *generator* for S. S is clearly uniquely determined by its generator, but—as was already pointed out in note (4.2.6), (i)—the same S can generally be generated by infinitely many generators.

5 **Remark.** The concept of a generator for a differential input-output relation S as introduced above is closely related to the concept of the "system matrix" as used by Rosenbrock, 1970. We shall return to this question in chapter 7. □

Regular generators for regular differential input-output relations. Order

A differential input-output relation S is said to be *regular*, if there is a generator $[A(p) \vdots -B(p)]$ for S with $A(p)$ square and $\det A(p) \neq 0$. Such a generator is likewise called *regular*, and $\partial(\det A(p))$ is called the *order* of the generator and of the corresponding regular differential input-output relation S (*cf.* section 6.2 below).

It has previously been pointed out (see note (4.2.6), (ii)) that a regular differential input-output relation possesses the remarkable property of determining a natural corresponding dynamic differential system (with respect to a fixed time point t_0). Regular input-output relations also possess other attractive properties which make them particularly suitable as mathematical models of real systems. Thus a regular differential input-output relation $S \subset \mathscr{X}^r \times \mathscr{X}^q$ possesses the property of having the whole of $\mathscr{X}^r$ as its domain (*cf.* note (5.4.1), (i)). This property is very convenient when compositions of input-output relations are studied (*cf.* chapter 7). A regular S also allows a finite dimensional *state-space representation* (*cf.* chapter 7).

We shall, in what follows, mainly be interested in regular differential input-output relations and regular generators.

6 **Note. (i)** A regular differential input-output relation $S \subset \mathscr{X}^r \times \mathscr{X}^q$ generated by the regular generator $[A(p) \vdots -B(p)]$ is clearly a linear mapping $\mathscr{X}^r \to \mathscr{X}^q$ if and only if $A(p): \mathscr{X}^q \to \mathscr{X}^q$ has an inverse $A(p)^{-1}: \mathscr{X}^q \to \mathscr{X}^q$, i.e. if and only if the matrix $A(p)$ is unimodular (*cf.* note (5.4.1), (iv)).
(ii) Let $S_1 \subset \mathscr{X}^{r_1} \times \mathscr{X}^{q_1}$ and $S_2 \subset \mathscr{X}^{r_2} \times \mathscr{X}^{q_2}$ be two differential input-output relations generated by $[A_1(p) \vdots -B_1(p)]$ and $[A_2(p) \vdots -B_2(p)]$ respectively. If $S_1 \cap S_2$ is nonempty, then $r_1 = r_2$ and $q_1 = q_2$. If, in addition, $A_1(p)$ and $A_2(p)$ are square, then both the generators are of the same size.

Let S_1 now be regular and $A_1(p)$ square with $\det A_1(p) \neq 0$. If $S_2 \subset S_1$ and $A_2(p)$ is square, then $\det A_2(p) \neq 0$. This is seen as follows. First note that $S_2 \subset S_1$ also implies that $S_2(0) \subset S_1(0)$, i.e. that $\ker A_2(p) \subset \ker A_1(p)$. Ker $A_1(p)$ is finite dimensional because $\det A_1(p) \neq 0$ (*cf.* note (5.4.1), (iii)). It is concluded that $\ker A_2(p)$ is also finite dimensional and consequently $\det A_2(p) \neq 0$.

The above properties mean that if S_1 is a regular differential input-output relation, then every other differential input-output relation S_2, which is a subset of S_1 and which is generated by some $[A_2(p) \ \vdots \ -B_2(p)]$ with $A_2(p)$ square, is also regular. Further, if $[A_1(p) \ \vdots \ -B_1(p)]$ with $A_1(p)$ square is a generator for a regular S_1, then this generator is necessarily regular.

(iii) Let S be a differential input-output relation, and let $(u, y) \in S$. Then every other $(u, y_1) \in S$ is such that $y - y_1 \in S(0)$, and conversely, every $(u, y + z)$ with $z \in S(0)$ belongs to S.

(iv) In a generator $[A(p) \ \vdots \ -B(p)]$ as given above we may even allow $B(p)$ to be equal to the empty matrix, i.e. the matrix having no rows and no columns. This case corresponds to a situation where there is no input present. □

6.2 Input-output equivalence. Complete invariants and canonical forms for input-output equivalence

Two fundamental theorems

There are a number of important relationships between differential input-output relations and their generators. The most significant of them are contained in the following fundamental theorems. The proofs of the theorems are given in section 6.5.

1 **Theorem.** *Let $S_1 \subset \mathscr{X}^r \times \mathscr{X}^q$ be the differential input-output relation generated by $[A_1(p) \ \vdots \ -B_1(p)]$, $A_1(p) \in \mathbf{C}[p]^{s_1 \times q}$, $B_1(p) \in \mathbf{C}[p]^{s_1 \times r}$, and let $S_2 \subset \mathscr{X}^r \times \mathscr{X}^q$ be the differential input-output relation generated by $[A_2(p) \ \vdots \ -B_2(p)]$, $A_2(p) \in \mathbf{C}[p]^{s_2 \times q}$, $B_2(p) \in \mathbf{C}[p]^{s_2 \times r}$. Suppose further that $[A_1(p) \ \vdots \ -B_1(p)] = L(p)[A_2(p) \ \vdots \ -B_2(p)]$ for some $L(p) \in \mathbf{C}[p]^{s_1 \times s_2}$. Then the following statements are true*:

(i) $S_2 \subset S_1$,

(ii) *if $s_1 = s_2$ and $L(p)$ is unimodular, then $S_1 = S_2$,*

(iii) *if $S_1 = S_2$, $s_1 = s_2 = q$, and $\det A_1(p) \neq 0$ (or $\det A_2(p) \neq 0$), then $\det A_2(p) \neq 0$ ($\det A_1(p) \neq 0$) and $L(p)$ is unimodular.* □

2 **Theorem.** *Let $S_1 \subset \mathscr{X}^r \times \mathscr{X}^q$ be the regular differential input-output relation generated by $[A_1(p) \ \vdots \ -B_1(p)]$, $A_1(p) \in \mathbf{C}[p]^{q \times q}$, $\det A_1(p) \neq 0$, $B_1(p) \in \mathbf{C}[p]^{q \times r}$, and let $S_2 \subset \mathscr{X}^r \times \mathscr{X}^q$ be the differential input-output rela-*

tion generated by $[A_2(p) \vdots -B_2(p)]$, $A_2(p) \in \mathbf{C}[p]^{q\times q}$, $B_2(p) \in \mathbf{C}[p]^{q\times r}$. *Then the following statement is true: If* $S_2 \subset S_1$, *then* $\det A_2(p) \neq 0$, *moreover, there exists a unique* $L(p) \in \mathbf{C}[p]^{q\times q}$, $\det L(p) \neq 0$, *such that* $[A_1(p) \vdots -B_1(p)] = L(p)[A_2(p) \vdots -B_2(p)]$, *and if, in particular,* $S_1 = S_2$, *then* $L(p)$ *is unimodular.* □

Note that statement (iii) of theorem (1) is contained in theorem (2).

Generators generating the same regular differential input-output relation

The theorems presented above lead to the following important conclusion.

Let S be the regular differential input-output relation generated by the regular generator $[A_1(p) \vdots -B_1(p)]$, i.e. with $A_1(p)$ square and $\det A_1(p) \neq 0$. Let $[A_2(p) \vdots -B_2(p)]$ be another partitioned polynomial matrix operator of the same size as $[A_1(p) \vdots -B_1(p)]$. Then $[A_2(p) \vdots -B_2(p)]$ is also a regular generator for S if and only if $[A_1(p) \vdots -B_1(p)]$ and $[A_2(p) \vdots -B_2(p)]$ are *row equivalent* (*cf.* appendix A2 for details concerning this equivalence), i.e. if and only if there exists a unimodular polynomial matrix operator $P(p)$ of proper size such that $[A_2(p) \vdots -B_2(p)] = P(p)[A_1(p) \vdots -B_1(p)]$. Row equivalent regular generators are thus of the same order = the order of the regular differential input-output relation S generated by them.

Input-output equivalence = row equivalence

The above considerations suggest the following formalization (*cf.* Fig. 6.1).

Let $\mathcal{R}$ denote the set of all regular differential input-output relations, and $\mathcal{M}$ the set of all regular generators for the elements of $\mathcal{R}$, i.e. define

3 $$\mathcal{M} \triangleq \{[A(p) \vdots -B(p)] \,|\, A(p) \in \mathbf{C}[p]^{q\times q},$$
$$\det A(p) \neq 0,\ B(p) \in \mathbf{C}[p]^{q\times r} \quad \text{for some } q, r \in \{1, 2, 3, \ldots\}\},\ddagger$$

4 $$\mathcal{R} \triangleq \{S \,|\, S^{-1} = \ker M \quad \text{for some } M \in \mathcal{M}\}.$$

Further let the assignment

5 $$M \mapsto S,\ S^{-1} = \ker M,$$

define the mapping (a surjection with respect to $\mathcal{R}$)

6 $$f: \mathcal{M} \to \mathcal{R}.$$

The generators M_1, $M_2 \in \mathcal{M}$ are now said to be *input-output equivalent* if they generate the same input-output relation $S \in \mathcal{R}$, i.e. if $f(M_1) = f(M_2)$. The result quoted above then implies that input-output equivalence in the

‡ The case $r = 0$ would require special treatment.

present case means the same thing as row equivalence. These terms are therefore used as synonyms in the sequel as far as regular generators are concerned.

Complete invariant for input-output equivalence on $\mathcal{M}$

Input-output equivalence is of course an equivalence relation on $\mathcal{M}$, it is in fact the natural equivalence determined by the mapping f above. According to the terminology of MacLane and Birkhoff, 1967, chapter VIII, f can thus be called a *complete invariant* for input-output equivalence.

Canonical forms for input-output equivalence on $\mathcal{M}$

From the fact that input-output equivalence in this case coincides with row equivalence, it also follows that the set of CUT-forms ("canonical upper triangular"-forms) of the elements of $\mathcal{M}$ constitute a set of *canonical forms*, denoted by $\mathcal{M}^*$, for input-output equivalence (see appendix A2 for a detailed discussion of the CUT-form of polynomial matrices). This means that every equivalence class on $\mathcal{M}$ for input-output equivalence contains a unique element of $\mathcal{M}^*$. Let, for $M \in \mathcal{M}$, the corresponding CUT-form

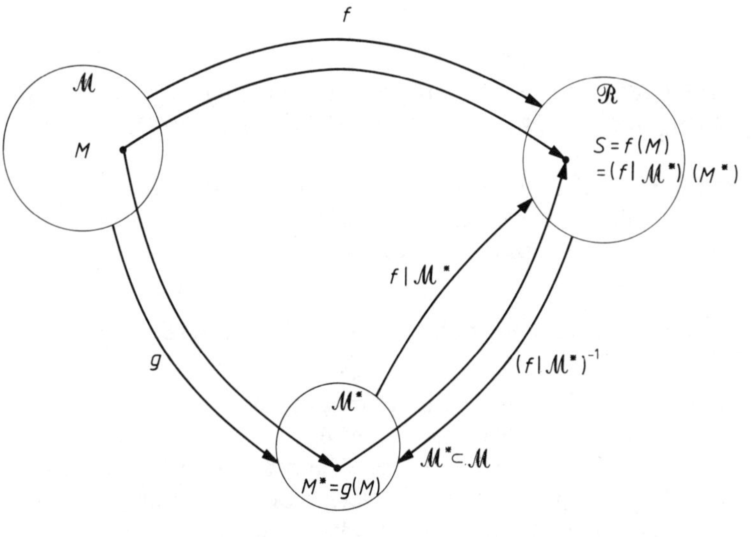

Fig. 6.1. Commutative diagram.
$\mathcal{R}$ is the set of all regular differential input-output relations,
$\mathcal{M}$ is the set of all regular generators for the elements of $\mathcal{R}$,
$\mathcal{M}^*$ is the set of all CUT-forms contained in $\mathcal{M}$.

be denoted by M^*, and let

7 $$g: \mathcal{M} \to \mathcal{M}^*$$

be the mapping (a surjection with respect to $\mathcal{M}^*$) defined by the assignment

8 $$M \mapsto M^*.$$

g is then also qualified to be called a complete invariant for input-output equivalence on $\mathcal{M}$.

Finally, we note that the restriction of f to $\mathcal{M}^*$, i.e. $f|\mathcal{M}^*$, is the mapping assigning a unique element of $\mathcal{R}$ to every element of $\mathcal{M}^*$. $f|\mathcal{M}^*$ is clearly surjective with respect to $\mathcal{R}$ and injective—it thus has an inverse $(f|\mathcal{M}^*)^{-1}: \mathcal{R} \to \mathcal{M}^*$.

The situation is illustrated by the commutative diagram shown in Fig. 6.1.

9 **Note.** Above we defined input-output equivalence only for regular generators. Of course, also nonregular generators can be said to be input-output equivalent if they generate the same differential input-output relation. Quite generally then it holds that row equivalent generators are always also input-output equivalent (*cf.* theorem (1), (ii)). The converse is, however, not generally true. □

6.3 The transfer matrix. Proper and strictly proper transfer matrices, generators, and differential input-output relations

The transfer function and transfer matrix are well-known concepts in the operational calculus based on the use of the Laplace transform. We shall here introduce generalized forms of these concepts.

The transfer matrix determined by a regular differential input-output relation

Consider a regular differential input-output relation $S \subset \mathcal{X}^r \times \mathcal{X}^q$, and let $[A(p) \vdots -B(p)]$, $A(p) \in \mathbf{C}[p]^{q\times q}$, $\det A(p) \neq 0$, $B(p) \in \mathbf{C}[p]^{q\times r}$, be a regular generator for S. Next let $A(p)$ and $B(p)$ be identified with the corresponding rational matrices over $\mathbf{C}(p)$, i.e. over the field of quotients of $\mathbf{C}[p]$ (*cf.* note (5.2.1), (ii)). We thus have $A(p) \in \mathbf{C}(p)^{q\times q}$, $\det A(p) \neq 0$, and $B(p) \in \mathbf{C}(p)^{q\times r}$. But then $A(p)$ is invertible in $\mathbf{C}(p)^{q\times q}$, and the rational matrix

1 $$\mathcal{G}(p) = A(p)^{-1} B(p)$$

is well-defined as an element of $\mathbf{C}(p)^{q\times r}$. We shall call $\mathcal{G}(p)$ the *transfer matrix* determined by $[A(p) \vdots -B(p)]$. It turns out (*cf.* section 6.4) that $\mathcal{G}(p)$ is, in fact, also uniquely determined by S. $\mathcal{G}(p)$ is therefore alternatively said to be the transfer matrix determined by S. In the single input–single output case the term "transfer ratio" is occasionally used instead of the term "transfer matrix".

It should be emphasized that no mapping in which signal spaces are involved is assigned to the transfer matrix at this stage—here the transfer matrix is just a matrix with entries from the field $\mathbf{C}(p)$. Later on (*cf.* part III of the book) the transfer matrix will be interpreted, under certain circumstances, as a mapping between Cartesian products of suitable signal spaces.

Proper and strictly proper transfer matrices, generators and differential input-output relations

According to appendix A2, a transfer matrix $\mathcal{G}(p)$ as given by (1) above can be written in a unique way as

2 $$\mathcal{G}(p) = \mathcal{G}^0(p) + K(p),$$

where $K(p)$ is a polynomial matrix and where every entry of $\mathcal{G}^0(p)$ is such that its numerator is of lower degree than its denominator (the zero is supposed to fulfil this condition).

$\mathcal{G}(p)$ is now said to be *proper* if $K(p)$ is a constant matrix and *strictly proper* if $K(p)$ is the zero matrix. Generators and differential input-output relations are likewise said to be proper or strictly proper depending on the properties of the transfer matrix determined by the generators and input-output relations in question. $\mathcal{G}^0(p)$ in (2) is called the "strictly proper part" of $\mathcal{G}(p)$.

Conditions for a regular generator $[A(p) \vdots -B(p)]$ to be proper are discussed in appendix A2.

It is a well-known fact that a realistic description of the behaviour of real systems at high frequencies implies the use of models based on proper or strictly proper input-output relations.

6.4 Transfer equivalence. Complete invariants and canonical forms for transfer equivalence. Controllability

Transfer equivalence on the set $\mathcal{M}$ of all regular generators

Consider the set $\mathcal{M}$ of all regular generators as given by (6.2.3). In section 6.2 we defined the input-output equivalence = row equivalence on this set.

Now there is also another important equivalence on $\mathcal{M}$, called "transfer equivalence". This transfer equivalence and related concepts are discussed in detail in appendix A2. We shall recapitulate here briefly the relevant results.

Let $M_1 \triangleq [A_1(p) \,\vdots\, -B_1(p)]$ and $M_2 \triangleq [A_2(p) \,\vdots\, -B_2(p)]$ be two elements of $\mathcal{M}$. These elements are said to be *transfer equivalent* if they determine the same transfer matrix $\mathcal{G}(p)$ (*cf.* (6.3.1)), i.e. if

1 $$A_1(p)^{-1}B_1(p) = A_2(p)^{-1}B_2(p).$$

Clearly, if M_1 and M_2 are input-output equivalent (=row equivalent) then they are also transfer equivalent (the converse does not generally hold).

Complete invariant for the transfer equivalence on $\mathcal{M}$

Let $\mathcal{T}$ denote the set of all transfer matrices determined by the elements of $\mathcal{M}$ (*cf.* Fig. 6.2), i.e.

2 $$\mathcal{T} \triangleq \{\mathcal{G}(p) | \mathcal{G}(p) = A(p)^{-1}B(p) \quad \text{for some } [A(p) \,\vdots\, -B(p)] \in \mathcal{M}\},$$

and let the assignment

3 $$[A(p) \,\vdots\, -B(p)] \mapsto A(p)^{-1}B(p)$$

define the mapping (a surjection with respect to $\mathcal{T}$)

4 $$f_t : \mathcal{M} \to \mathcal{T}.$$

The transfer equivalence on $\mathcal{M}$ is then the natural equivalence determined by f_t, and f_t is thus qualified to be called a complete invariant for this equivalence.

Canonical forms for transfer equivalence on $\mathcal{M}$

It is further shown in appendix A2 that there is a set $\mathcal{M}_t^* \subset \mathcal{M}^* \subset \mathcal{M}$ of CUT-forms which qualifies as a set of canonical forms for the transfer equivalence on $\mathcal{M}$, i.e. there exists a mapping (a surjection with respect to $\mathcal{M}_t^*$)

5 $$g_t : \mathcal{M} \to \mathcal{M}_t^*$$

which is a complete invariant for the transfer equivalence on $\mathcal{M}$. The image $g_t([A(p) \,\vdots\, -B(p)]) \in \mathcal{M}_t^*$ of $[A(p) \,\vdots\, -B(p)] \in \mathcal{M}$, denoted by $[A(p) \,\vdots\, -B(p)]_t^*$, is obtained as follows. Write

$$[A(p) \,\vdots\, -B(p)] = L(p)\,[A_1(p) \,\vdots\, -B_1(p)],$$

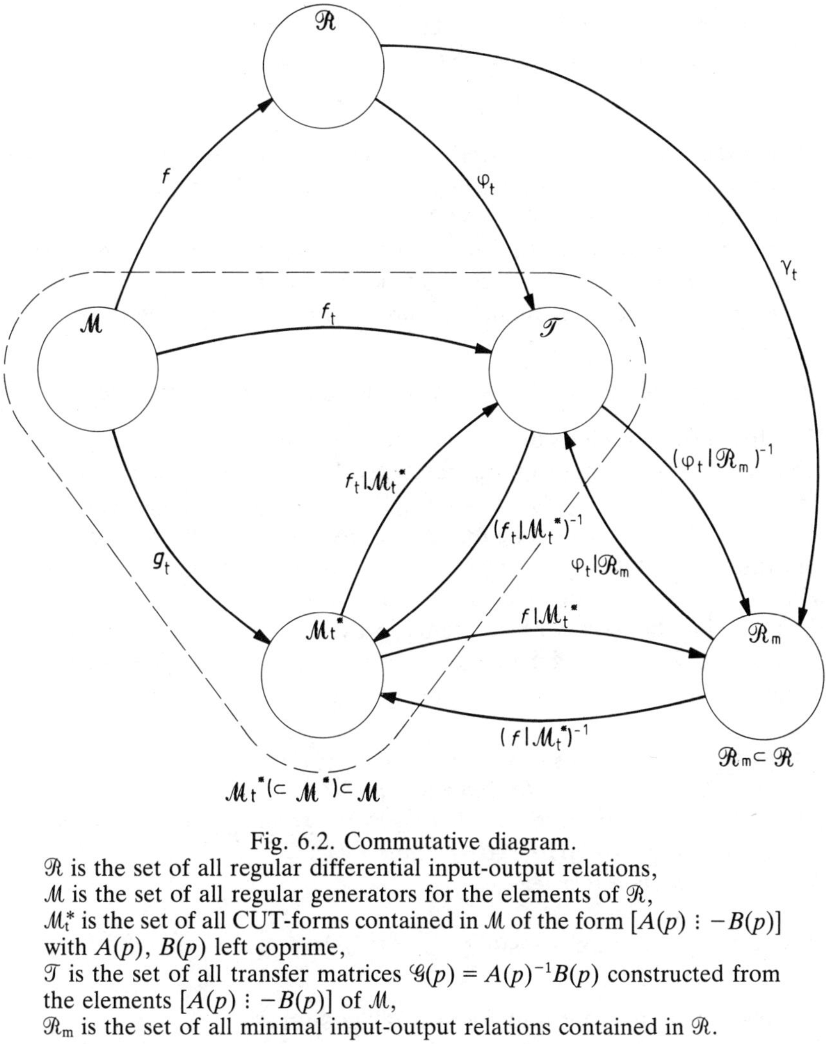

Fig. 6.2. Commutative diagram.
$\mathcal{R}$ is the set of all regular differential input-output relations,
$\mathcal{M}$ is the set of all regular generators for the elements of $\mathcal{R}$,
$\mathcal{M}_t^*$ is the set of all CUT-forms contained in $\mathcal{M}$ of the form $[A(p) \vdots -B(p)]$ with $A(p)$, $B(p)$ left coprime,
$\mathcal{T}$ is the set of all transfer matrices $\mathcal{G}(p) = A(p)^{-1}B(p)$ constructed from the elements $[A(p) \vdots -B(p)]$ of $\mathcal{M}$,
$\mathcal{R}_m$ is the set of all minimal input-output relations contained in $\mathcal{R}$.

where $L(p)$ is a GCLD (greatest common left divisor) of $A(p)$ and $B(p)$ (a method for finding an $L(p)$ is given in appendix A2). Then it holds that

6 $$g_t([A(p) \vdots -B(p)]) \triangleq [A(p) \vdots -B(p)]_t^* = [A_1(p) \vdots -B_1(p)]^*,$$

where $[\cdot]^*$ denotes the CUT-form of $[\cdot]$ (*cf.* section 6.2).

The restriction $f_t|\mathcal{M}_t^*$ of f_t to $\mathcal{M}_t^*$ is again the mapping assigning a unique element of $\mathcal{T}$ to every element of $\mathcal{M}_t^*$. $f_t|\mathcal{M}_t^*$ has clearly an inverse

$(f_t|\mathcal{M}_t^*)^{-1}:\mathcal{T}\to\mathcal{M}_t^*$. Given $\mathcal{G}(p)\in\mathcal{T}$, the corresponding element $M\in\mathcal{M}_t^*$ can be found by factoring $\mathcal{G}(p)$ as

7
$$\mathcal{G}(p)=A(p)^{-1}B(p),$$

where $A(p)$, $B(p)$ are left coprime polynomial matrices (applicable factorization procedures are described in appendix A2), and then taking the CUT-form $[A(p)\ \vdots\ -B(p)]^*$ of $[A(p)\ \vdots\ -B(p)]$ as M.

The situation is illustrated by Fig. 6.2, where the broken line encircles the quantities and relationships explained so far in this section. Note that $\mathcal{M}$, $\mathcal{R}$, $\mathcal{M}^*$, and f denote quantities already introduced in section 6.2.

Transfer equivalence on the set $\mathcal{R}$ of all regular input-output relations

We shall now also introduce a transfer equivalence on the set $\mathcal{R}$ of all regular differential input-output relations. Let S_1, $S_2\in\mathcal{R}$ be generated by $M_1\triangleq[A_1(p)\ \vdots\ -B_1(p)]$ and $M_2\triangleq[A_2(p)\ \vdots\ -B_2(p)]\in\mathcal{M}$ respectively. Then S_1 and S_2 are said to be *transfer equivalent* if M_1 and M_2 are transfer equivalent, i.e. if $A_1(p)^{-1}B_1(p)=A_2(p)^{-1}B_2(p)$. This new transfer equivalence is indeed an equivalence relation on $\mathcal{R}$. This follows readily from the following theorem. The theorem is proved in section 6.5, and the proof of theorem (6.2.2) given in the same section is actually based on this result.

8 **Theorem.** *Let S_1 and $S_2\subset\mathcal{X}^r\times\mathcal{X}^q$ be the regular differential input-output relations generated by the regular generators $[A_1(p)\ \vdots\ -B_1(p)]$ and $[A_2(p)\ \vdots\ -B_2(p)]$ respectively. Let the corresponding transfer matrices be denoted by $\mathcal{G}_1(p)\triangleq A_1(p)^{-1}B_1(p)$ and $\mathcal{G}_2(p)\triangleq A_2(p)^{-1}B_2(p)$. Then the following statement is true:*

If $S_2\subset S_1$, or $S_1\subset S_2$, then $\mathcal{G}_1(p)=\mathcal{G}_2(p)$. □

Complete invariants for transfer equivalence on $\mathcal{R}$

Using the notations of the above theorem it follows that the assignments $S_1\mapsto\mathcal{G}_1(p)$ and $S_2\mapsto\mathcal{G}_2(p)$ define a mapping (a surjection with respect to $\mathcal{T}$)

9
$$\varphi_t:\mathcal{R}\to\mathcal{T},$$

and that, by construction,

10
$$f_t=\varphi_t\circ f.$$

It can thus be seen that the transfer equivalence on $\mathcal{R}$ according to the definition given above is just the natural equivalence determined by φ_t. φ_t is accordingly a complete invariant for the transfer equivalence on $\mathcal{R}$.

Another complete invariant for this equivalence is the mapping (*cf*. Fig. 6.2)

11 $$(f_t|\mathcal{M}_t^*)^{-1} \circ \varphi_t,$$

which assigns a unique element of $\mathcal{M}_t^*$ to every element of $\mathcal{R}$.

Canonical forms for the transfer equivalence on $\mathcal{R}$

It remains to find a suitable set of canonical forms for the transfer equivalence on $\mathcal{R}$, i.e. a mapping $\mathcal{R} \to \mathcal{R}$ such that the transfer equivalence on $\mathcal{R}$ is the natural equivalence determined by this mapping. To begin with, let $\mathcal{R}_m \subset \mathcal{R}$ denote the range of $f|\mathcal{M}_t^*$, i.e. the set of all regular differential input-output relations that can be generated by regular generators $[A(p) \vdots -B(p)]$ of CUT-form with $A(p)$, $B(p)$ left coprime. Clearly (*cf*. (10))

12 $$f_t|\mathcal{M}_t^* = (\varphi_t|\mathcal{R}_m) \circ (f|\mathcal{M}_t^*),$$

and the mapping (a surjection with respect to $\mathcal{R}_m$)

13 $$\gamma_t \triangleq (\varphi_t|\mathcal{R}_m)^{-1} \circ \varphi_t : \mathcal{R} \to \mathcal{R}_m \subset \mathcal{R}$$

fulfils the requirement mentioned above. This means that γ_t is a complete invariant for the transfer equivalence on $\mathcal{R}$ and that $\mathcal{R}_m$ constitutes a set of canonical forms for the transfer equivalence on $\mathcal{R}$. Fig. 6.2 is now completely explained.

Minimal differential input-output relations

The elements of $\mathcal{R}_m$ possess a minimality property as stated in the following theorem.

14 **Theorem.** *Let $\mathcal{E}_t \subset \mathcal{R}$ denote an equivalence class for the transfer equivalence on $\mathcal{R}$, and let $S_m \in \mathcal{R}_m$ denote the canonical form determined by $\mathcal{E}_t$, i.e. $S_m \in \mathcal{E}_t$ and $S_m = \gamma_t\,(S)$ for an arbitrary $S \in \mathcal{E}_t$. Finally let $\mathcal{E}_t$ be partially ordered by set inclusion. Then S_m is the first element of $\mathcal{E}_t$. Moreover, $\mathcal{R}_m$ constitutes the set of all first elements of the equivalence classes for the transfer equivalence on $\mathcal{R}$.* □

15 **Proof.** For S_m to be the first element of $\mathcal{E}_t$ we shall have $S_m \in \mathcal{E}_t$ and $S_m \subset S$ for every $S \in \mathcal{E}_t$. Now the first condition is fulfilled by construction, and we have to consider only the second one.

So, let $\mathscr{G}(p) \in \mathscr{T}$ be the transfer matrix determined by $\mathscr{E}_t$, i.e. $\mathscr{G}(p) = \varphi_t(S)$ for an arbitrary $S \in \mathscr{E}_t$, and let $M_m \quad [A_m(p) \vdots -B_m(p)] \in \mathscr{M}_t^*$ be the canonical form determined by $\mathscr{G}(p)$, i.e. $M_m \quad [A_m(p) \vdots -B_m(p)] = (f_t|\mathscr{M}_t^*)^{-1}(\mathscr{G}(p))$, with $A_m(p)$, $B_m(p)$ left coprime. Then let $S \in \mathscr{E}_t$ be arbitrary, and let $M \triangleq [A(p) \vdots -B(p)] \in \mathscr{M}$ be a regular generator for S. M and M_m are then transfer equivalent by construction, and according to appendix A2 it holds that

16 $$[A(p) \vdots -B(p)] = L(p)\,[A_m(p) \vdots -B_m(p)]$$

for some square polynomial matrix operator $L(p)$. From theorem (6.2.1), (i) it then follows that $S_m \subset S$.

The last part of the theorem follows by construction. □

The element $S_m \in \mathscr{R}_m$ determined by a given $S \in \mathscr{R}$ or a given $\mathscr{G}(p) \in \mathscr{T}$ according to $S_m = \gamma_t(S)$, or $S_m = (\varphi_t|\mathscr{R}_m)^{-1}(\mathscr{G}(p))$ respectively, is also called the *minimal input-output relation* determined by S or $\mathscr{G}(p)$ respectively.

The concept of a controllable differential input-output relation

We shall use the following terminology. Let S be a regular differential input-output relation generated by the regular generator $[A(p) \vdots -B(p)]$, and let S_m be the minimal input-output relation determined by S. We shall then say that S is *controllable* if $S = S_m$, or equivalently according to the above results, if $A(p)$ and $B(p)$ are left coprime.

The controllability concept introduced above turns out to be closely related to the ordinary controllability concept used in the state-space theory. It will be shown in chapter 7 that a regular differential input-output relation is controllable in the above sense if and only if any observable state-space representation of S is controllable in the ordinary state-space sense.

17 **Note.** Let S be a regular differential input-output relation, and let S_m be the minimal input-output relation determined by S. According to theorem (14) above it then holds that $S_m \subset S$ and $S_m(0) \subset S(0)$. In addition it can easily be shown that $S = (\{0\} \times S(0)) + S_m$. We shall comment on this result in chapter 7 below. □

6.5 Proofs of theorems (6.2.1), (6.2.2), and (6.4.8)

This section is devoted to the proofs of theorems (6.2.1), (6.2.2), and (6.4.8) presented above. We shall need the following simple fact.

1 **Note.** Let U, V, W be vector spaces over $\mathbf{C}$, and let $f: U \to V$ and $g: V \to W$ be linear mappings. Then it holds that $\ker g \circ f = \{x | x \in U$ and $f(x) \in \ker g\} \supset \ker f$, and so $\ker f = \ker g \circ f$ if and only if $\ker g \cap \mathsf{R}f = \{0\}$. If, in particular, $\mathsf{R}f = V$, then $\ker f = \ker g \circ f$ if and only if $\ker g = \{0\}$, i.e. if and only if g is an injection. □

2 **Proof of theorem (6.2.1).** Now let f, g, and $g \circ f$ in the above note be identified with $[A_2(p) \vdots -B_2(p)] : \mathscr{X}^q \times \mathscr{X}^r \to \mathscr{X}^{s_2}$, $L(p): \mathscr{X}^{s_2} \to \mathscr{X}^{s_1}$, and $L(p)[A_2(p) \vdots -B_2(p)] = [A_1(p) \vdots -B_1(p)]: \mathscr{X}^q \times \mathscr{X}^r \to \mathscr{X}^{s_1}$ respectively of theorem (6.2.1).

Item (i) of the theorem then follows directly from $\ker f \subset \ker g \circ f$, i.e. $S_2 \subset S_1$.

$s_1 = s_2$ and $L(p)$ unimodular correspond to $\ker g = \{0\}$ implying $\ker f = \ker g \circ f$, i.e. $S_2 = S_1$. This is item (ii) of the theorem.

Suppose next that $S_1 = S_2$, $s_1 = s_2 = q$ and $\det A_1(p) \neq 0$ (or $\det A_2(p) \neq 0$). This implies that $S_1 = S_2$ is regular and that both the generators $[A_1(p) \vdots -B_1(p)]$ and $[A_2(p) \vdots -B_2(p)]$ must be regular (*cf.* note (6.1.6), (ii)). Consequently $\det A_2(p) \neq 0$ ($\det A_1(p) \neq 0$) and thus also $\det L(p) \neq 0$. $[A_1(p) \vdots -B_1(p)]$ is accordingly of full rank q, and $\mathsf{R}\,[A_1(p) \vdots -B_1(p)]$ is the whole of $\mathscr{X}^q$ (*cf.* note (5.4.1), (i)) corresponding to $\mathsf{R}f = V$ in the above note. $S_1 = S_2$, i.e. $\ker f = \ker g \circ f$, then implies that $\ker g = \{0\}$, i.e. that $L(p)$ is an injection. Because $\det L(p) \neq 0$, $\mathsf{R}\,L(p)$ is the whole of $\mathscr{X}^q$, and $L(p)$ thus has an inverse $\mathscr{X}^q \to \mathscr{X}^q$. It follows that $L(p)$ is unimodular as asserted in item (iii) of the theorem. □

Theorem (6.2.2) seems to be only a slightly strengthened form of theorem (6.2.1), (iii). Therefore it is somewhat surprising to notice that the proof of theorem (6.2.2) is quite involved. Our proof of theorem (6.2.2) is based on theorem (6.4.8). We shall therefore start with this proof.

3 **Proof of theorem (6.4.8).** We only need to consider the case $S_2 \subset S_1$—the case $S_1 \subset S_2$ is then only a matter of notation.

Note to begin with that the rational matrices $\mathscr{G}_1(p)$ and $\mathscr{G}_2(p)$ mentioned in the theorem are equal if and only if $\mathscr{G}_1(s) = \mathscr{G}_2(s)$ for almost all $s \in \mathbf{C}$, where $\mathscr{G}_1(s)$ and $\mathscr{G}_2(s)$ are obtained from $\mathscr{G}_1(p)$ and $\mathscr{G}_2(p)$ respectively simply by replacing p everywhere by the complex variable s.

Next we recall that our signal space $\mathscr{X}$ contains all complex-valued functions on T of the form $t \mapsto u(t) = ce^{st}$, $t \in T$, $c, s \in \mathbf{C}$ (*cf.* note (5.1.6), (i)). Consider then an input-output pair $(u, y) \in \mathscr{X}^r \times \mathscr{X}^q$ of the form

4 $$u = c_1 e^{s(\cdot)}, y = c_2 e^{s(\cdot)},$$

$c_1 \in \mathbf{C}^r, c_2 \in \mathbf{C}^q$, $s \in C \subset \mathbf{C}$, where C is the set of all $s \in \mathbf{C}$ such that $\det A_1(s) \neq 0$ and $\det A_2(s) \neq 0$.

It follows that

5 $$\begin{aligned} A_1(p)c_2 e^{s(\cdot)} &= A_1(s)c_2 e^{s(\cdot)}, \\ A_2(p)c_2 e^{s(\cdot)} &= A_2(s)c_2 e^{s(\cdot)}, \\ B_1(p)c_1 e^{s(\cdot)} &= B_1(s)c_1 e^{s(\cdot)}, \\ B_2(p)c_1 e^{s(\cdot)} &= B_2(s)c_1 e^{s(\cdot)}, \end{aligned}$$

where $A_1(s)$ etc. are again obtained from $A_1(p)$ etc. by replacing p everywhere by the complex variable s. Now choose c_1 and c_2 so that

6 $$c_2 = A_2(s)^{-1}B_2(s)c_1 = \mathscr{G}_2(s)c_1.$$

It is then easily seen that the pair (u, y) so obtained satisfies $A_2(p)y = B_2(p)u$, i.e. we have $(u, y) \in S_2$. This same (u, y) is now also an element of S_1 if and only if

7 $$\mathscr{G}_2(s)c_1 = \mathscr{G}_1(s)c_1.$$

The assumption $S_2 \subset S_1$ thus implies that (7) must hold for all $s \in C$ and all $c_1 \in \mathbf{C}^r$ implying that $\mathscr{G}_2(s) = \mathscr{G}_1(s)$ for all $s \in C$, i.e. for almost all $s \in \mathbf{C}$. Consequently $\mathscr{G}_1(p) = \mathscr{G}_2(p)$. □

8 **Proof of theorem (6.2.2).** Note first that $S_2 \subset S_1$ and $\det A_1(p) \neq 0$ also implies that $\det A_2(p) \neq 0$ (*cf.* note (6.1.6), (ii)). Further, if there exists an $L(p)$ such that $[A_1(p) \vdots -B_1(p)] = L(p)[A_2(p) \vdots -B_2(p)]$, then this $L(p)$ is unique and $\det L(p) \neq 0$. This follows from the equality $A_1(p) = L(p)A_2(p)$ and note (5.4.1), (v).

Hence, let S_1 and S_2 be regular differential input-output relations generated by the regular generators $[A_1(p) \vdots -B_1(p)]$ and $[A_2(p) \vdots -B_2(p)]$ respectively, and let $S_2 \subset S_1$. According to theorem (6.4.8) it then follows that

9 $$\mathscr{G}_1(p) = A_1(p)^{-1}B_1(p) = \mathscr{G}_2(p) = A_2(p)^{-1}B_2(p),$$

i.e. $[A_1(p) \vdots -B_1(p)]$ and $[A_2(p) \vdots -B_2(p)]$ are transfer equivalent. They thus have a common canonical form (CUT-form; *cf.* appendix A2)

10 $$[A(p) \vdots -B(p)]$$

with $A(p), B(p)$ left coprime and $A(p)^{-1}B(p) = \mathscr{G}_1(p) = \mathscr{G}_2(p)$. Moreover, there exist square polynomial matrices $M_1(p), M_2(p)$, $\det M_1(p) \neq 0$, $\det M_2(p) \neq 0$, such that

11 $$[A_1(p) \;\vdots\; -B_1(p)] = M_1(p)[A(p) \;\vdots\; -B(p)]$$

and

12 $$[A_2(p) \;\vdots\; -B_2(p)] = M_2(p)[A(p) \;\vdots\; -B(p)].$$

Next let $M(p)$, $\det M(p) \neq 0$, denote a GCRD of $M_1(p)$ and $M_2(p)$ so that

13 $$M_1(p) = N_1(p)M(p)$$

and

14 $$M_2(p) = N_2(p)M(p)$$

with $N_1(p)$, $N_2(p)$ right coprime and $\det N_1(p) \neq 0$, $\det N_2(p) \neq 0$. Now $S_2 \subset S_1$ is equivalent to $S_2 = S_1 \cap S_2$, and $S_2 = S_1$ is equivalent to $S_2 = S_1 \cap S_2$, and $S_1 = S_1 \cap S_2$. $S_1 \cap S_2$ is by definition the set of all pairs $(u, y) \in \mathscr{X}^r \times \mathscr{X}^q$ satisfying

$$A_1(p)y = B_1(p)u$$

and

$$A_2(p)y = B_2(p)u,$$

i.e. a generator for $S_1 \cap S_2$ is given by

15 $$\left[\begin{array}{c:c} A_1(p) & -B_1(p) \\ \hdashline A_2(p) & -B_2(p) \end{array}\right] = \left[\begin{array}{c} N_1(p) \\ \hdashline N_2(p) \end{array}\right] [M(p)\,A(p) \;\vdots\; -M(p)\,B(p)].$$

According to theorem (6.2.1), (ii), other generators for $S_1 \cap S_2$ can be constructed from the generator (15) by premultiplication by a unimodular matrix. It is also known (*cf.* appendix A2) that

$$\left[\begin{array}{c} N_1(p) \\ \hdashline N_2(p) \end{array}\right],$$

with $N_1(p)$ and $N_2(p)$ square and right coprime, can be brought to the Smith-form (which here coincides with the CUT-form)

$$\left[\begin{array}{c} I \\ \hdashline 0 \end{array}\right]$$

by premultiplication by a suitable unimodular matrix. Combining these results it is seen that a generator for $S_1 \cap S_2$ of the form

$$\begin{bmatrix} I \\ \hdashline 0 \end{bmatrix} [M(p)\,A(p) \,\vdots\, -M(p)\,B(p)],$$

or equivalently

16 $$[M(p)\,A(p) \,\vdots\, -M(p)\,B(p)],$$

is obtained from (15). Note that (16) is a regular generator. $S_1 \cap S_2$ is thus a regular differential input-output relation.

Summarizing then, S_1 is generated by

17 $$[A_1(p) \,\vdots\, -B_1(p)] = N_1(p)[M(p)A(p) \,\vdots\, -M(p)\,B(p)],$$

S_2 by

18 $$[A_2(p) \,\vdots\, -B_2(p)] = N_2(p)[M(p)\,A(p) \,\vdots\, -M(p)\,B(p)],$$

and $S_1 \cap S_2$ by

19 $$[M(p)A(p) \,\vdots\, -M(p)B(p)].$$

Note that all these generators are regular.

Theorem (6.2.1), (iii) is now applicable and leads to the following conclusions.

$S_2 \subset S_1$, or equivalently $S_2 = S_1 \cap S_2$, implies that $N_2(p)$ is unimodular and so

20 $$[A_1(p) \,\vdots\, -B_1(p)] = N_1(p)N_2(p)^{-1}[A_2(p) \,\vdots\, -B_2(p)]$$

with $N_1(p)N_2(p)^{-1}$ a polynomial matrix corresponding to the $L(p)$ mentioned in the theorem.

$S_2 = S_1$, or equivalently $S_2 = S_1 \cap S_2$ and $S_1 = S_1 \cap S_2$, implies that both $N_1(p)$ and $N_2(p)$ are unimodular. This corresponds to a unimodular $N_1(p)N_2(p)^{-1}$ in (20) and to a unimodular $L(p)$. □

6.6 Comments on canonical forms. Canonical row proper forms

We have in the preceding paragraphs considered canonical forms for input-output equivalence and transfer equivalence based on the CUT-form of polynomial matrices. This CUT-form proves very convenient in connection with the treatment of various analysis and synthesis problems as discussed in the next chapter.

However, there are certainly also other possibilities for choosing suitable canonical forms for input-output and transfer equivalence—recall that a set $C \subset X$ of canonical forms for an equivalence relation E on a set X is generally a collection of representatives of the equivalence classes on X determined by E so that C contains one and only one representative of each class (MacLane and Birkhoff, 1967, chapter VIII). In the preceding paragraphs the set of CUT-forms could of course be replaced everywhere by any other suitable set of canonical forms for input-output equivalence.

One important possibility in choosing the canonical forms for input-output and transfer equivalence would be to base them on the CRP-form ("canonical row proper" form) of polynomial matrices (*cf.* appendix A2). This form was discussed independently by Forney, 1975, and Guidorzi, 1975, (*cf.* also Beghelli and Guidorzi, 1976) and it is based on the concept of a "row proper" polynomial matrix as introduced by Wolovich, 1974. This form offers a very convenient representation of generators determining proper and strictly proper transfer matrices (*cf.* appendix A2). We shall discuss the application of this particular canonical form in chapter 8.

7

Analysis and synthesis problems

We are now ready to demonstrate the power of the machinery which we have constructed in the previous paragraphs, and which provides the fundamental framework for the main topic of this book, i.e. for the "polynomial systems theory". We shall do this by discussing a number of typical problems particularly well suited for treatment by means of this theory.

The problems considered are of a kind usually treated by means of state-space methods. Our presentation shows the straightforward way in which the polynomial systems theory works—the algorithms and results arrived at are in many respects more transparent and flexible than the corresponding algorithms and results derived with the aid of state-space methods. Moreover, the polynomial systems theory very naturally leads to interesting and useful generalizations of many concepts originally introduced within the state-space theory. Special mention may here be made of the concepts of observability and stabilizability. Algorithms for the synthesis of observers and feedback controllers are also devised in this connection.

In dealing with the problems on hand two main tools are employed. The first tool utilizes the existence of a set of equivalent generators for a given regular differential input-output relation. In applying this tool we simply replace a given regular generator by an input-output equivalent generator of a particularly suitable form. We have already studied this equivalence in detail in section 6.2. The results obtained now prove useful. The second tool is based on the *decomposition* principle, i.e. on the possibility of replacing a given regular differential input-output relation by a suitable *composition* of a family of other regular differential input-output relations. We shall for this purpose study compositions and decompositions of regular differential input-output relations in some detail. Our considerations lead to an equivalence, again called *input-output equivalence*, on certain classes of compositions. The relationship between this input-output equivalence and various kinds of "system equivalence" as defined first by Rosenbrock, 1970, is also considered in this context.

One crucial question arises: What is the exact relationship between a composition of a family of regular differential input-output relations and a corresponding interconnection of a family of linear dynamic (causal) systems? Recall that a natural linear dynamic system can be assigned to any regular differential input-output relation provided that certain conditions are fulfilled (*cf.* section 4.2 and the more detailed discussion to follow in chapter 8). This question proves far from trivial and in order to get the final answer we need a rather delicate machinery. Chapters 8 and 9 are devoted to this subject.

The material presented in this chapter is to a large extent new and previously either unpublished or else presented only in laboratory reports. The results are based on research carried out by the authors and their co-workers over the last few years. This applies in particular to the generalized observer outlined in section 7.11 and the feedback controller discussed in section 7.12 as well as to the algorithms suggested for the synthesis of these devices. It should be emphasized that the well-known state observer (estimator) and the state feedback controller merely appear as special cases in this rather general setting.

The signal space $\mathscr{X}$

We shall here make the same assumption concerning the signal space as in the previous chapter, i.e. the signal space $\mathscr{X}$ is, throughout this chapter, assumed to be a regular signal space according to definition (5.1.3) and possessing the richness property (5.1.5). $\mathscr{X}$ may, in particular, be equal to C^∞ or $\mathscr{D}$.

7.1 An elimination procedure

We shall frequently come up against problems of the following kind. Suppose that there is a given regular differential input-output relation $S_1 \subset \mathscr{X}^r \times (\mathscr{X}^{q_1} \times \mathscr{X}^{q_2})$ generated by the equation

1 $$\underbrace{\left[\begin{array}{c:c} A_1(p) & A_2(p) \\ \hdashline A_3(p) & A_4(p) \end{array}\right]}_{A(p)} \left[\begin{array}{c} y_1 \\ \hdashline y_2 \end{array}\right] = \left[\begin{array}{c} B_1(p) \\ \hdashline B_2(p) \end{array}\right] u$$

with $\det A(p) \neq 0$, and $A_1(p), A_4(p)$ square, i.e. by the regular generator

2
$$\begin{array}{c} \\ \end{array}\begin{bmatrix} \overset{y_1}{A_1(p)} & \overset{y_2}{A_2(p)} & \overset{u}{-B_1(p)} \\ A_3(p) & A_4(p) & -B_2(p) \end{bmatrix}.$$

Now suppose further that we are particularly interested in the input-output relation $S_2 \subset \mathscr{X}^r \times \mathscr{X}^{q_2}$ given as all the pairs (u, y_2) that can be formed from the pairs $(u, (y_1, y_2))$ contained in S_1. Formally then

3
$$S_2 \triangleq \{(u, y_2) \,|\, u, y_2 \quad \text{are such that } (u, (y_1, y_2)) \in S_1 \text{ for some } y_1\}.$$

We shall say that the elements $(u, y_2) \in S_2$ are obtained through "elimination" of the variable y_1 from the elements $(u, (y_1, y_2))$ of S_1.

In order to obtain more information about S_2 we proceed in the following way.

We replace the generator (2) for S_1 by an input-output equivalent generator (*cf.* section 6.2) of the form

4
$$\begin{bmatrix} \overset{y_1}{\tilde{A}_1(p)} & \overset{y_2}{\tilde{A}_2(p)} & \overset{u}{-\tilde{B}_1(p)} \\ 0 & \tilde{A}_4(p) & -\tilde{B}_2(p) \end{bmatrix}$$

with $\det \tilde{A}_1(p) \neq 0$, $\det \tilde{A}_4(p) \neq 0$. A generator such as this can always be found for S_1, for instance by transforming (2) to a row equivalent upper triangular form (*cf.* appendix A2).

Note in particular that $\tilde{A}_1(p)$ appearing in (4) is a GCRD of $A_1(p)$ and $A_3(p)$ in (1) and (2) (for details, see appendix A2).

Comparing (4) and (3) it is now clear that we must have $S_2^{-1} \subset \ker[\tilde{A}_4(p) \; \vdots \; -\tilde{B}_2(p)]$. On the other hand it follows from $\det \tilde{A}_1(p) \neq 0$ that for every $(y_2, u) \in \ker[\tilde{A}_4(p) \; \vdots \; -\tilde{B}_2(p)]$ a corresponding y_1 can always be found such that $(u, (y_1, y_2)) \in S_1$. In fact, with (u, y_2) given, it follows from (4) that any y_1 satisfying

5
$$\tilde{A}_1(p)y_1 = -\tilde{A}_2(p)y_2 + \tilde{B}_1(p)u$$

will do. Consequently

6
$$S_2^{-1} = \ker[\tilde{A}_4(p) \; \vdots \; -\tilde{B}_2(p)].$$

Hence it is found that S_2 is a regular differential input-output relation generated by the regular generator $[\tilde{A}_4(p) \; \vdots \; -\tilde{B}_2(p)]$.

An interesting special case arises if $\tilde{A}_1(p)$ in (5) is unimodular, i.e. if $A_1(p)$ and $A_3(p)$ in (1) and (2) are right coprime (see again appendix A2). If so then the generator (4) for S_1 can be chosen so that $\tilde{A}_1(p) = I$.

This means that with $(u, y_2) \in S_2$ given, there is a *unique* corresponding y_1 such that $(u, (y_1, y_2)) \in S_1$, and this y_1 is obtained directly from (5) (with $\tilde{A}_1(p) = I$) as $y_1 = -\tilde{A}_2(p)y_2 + \tilde{B}_1(p)u$.

The above elimination procedure proves very useful in the sequel.

A number of related results are collected in the following

7 **Note.** **(i)** The above elimination procedure can be generalized to cases where the original relation S_1 is not regular, i.e. where (2) is not a regular generator. The difference is only that (4) is replaced by a generator with $\tilde{A}_1(p)$ and $\tilde{A}_4(p)$ possibly nonsquare (*cf.* appendix A2). S_2 defined by (3) is still a differential input-output relation and $[\tilde{A}_4(p) \vdots -\tilde{B}_2(p)]$ is a generator for S_2. The regularity of S_2 depends on this generator.

(ii) Consider the regular differential input-output relation S_1 as generated by (1) or (2). Suppose that there are polynomial matrices $\tilde{A}_2(p)$ and $\tilde{B}_1(p)$ of appropriate sizes so that every $(u, (y_1, y_2)) \in S_1$ satisfies an equation of the form (5) with $\tilde{A}_1(p) = I$. Then there exists a regular generator for S_1 of the form (4) with $\tilde{A}_1(p) = I$, implying that $A_1(p)$ and $A_3(p)$ in (1) and (2) must be right coprime. This can be seen as follows.

Let S_1' be the (nonregular) differential input-output relation generated by $[I \vdots \tilde{A}_2(p) \,\vdots\vdots\, -\tilde{B}_1(p)]$. According to our assumption $S_1 \subset S_1'$, which is equivalent to $S_1 \cap S_1' = S_1$. This means that S_1 is also generated by

8
$$\begin{bmatrix} I & \tilde{A}_2(p) & -\tilde{B}_1(p) \\ A_1(p) & A_2(p) & -B_1(p) \\ A_3(p) & A_4(p) & -B_2(p) \end{bmatrix}.$$

It is immediately seen that (8) can be transformed by means of elementary row operations to a generator of the form (*cf.* also (7.3.3), . . . , (7.3.8))

9
$$\begin{bmatrix} I & \tilde{A}_2(p) & -\tilde{B}_1(p) \\ 0 & \tilde{A}_4(p) & -\tilde{B}_2(p) \\ 0 & 0 & -\tilde{B}_3(p) \end{bmatrix}.$$

Because S_1 was assumed regular, its domain is the whole of $\mathscr{X}^r$, implying that $\tilde{B}_3(p)$ in (9) must be zero. This gives the desired result.

(iii) Let $S \subset \mathscr{X}^r \times \mathscr{X}^q$ be a regular differential input-output relation generated by the equation

10
$$A(p)y = B(p)u,$$

where $A(p)$, without loss of generality, can be assumed to be of row proper form (*cf.* appendix A2). Moreover, the rows of $A(p)$ can be assumed to be ordered in such a way that the first q_1 rows ($q_1 \geqslant 0$) of $A(p)$ are nonzero constant rows, i.e. their row degrees are equal to 0 (there can be no zero rows), whereas the remaining $q_2 = q - q_1$ rows have row degrees $\geqslant 1$. Note that the number q_1 of constant rows is uniquely determined by S. Thus with y partitioned correspondingly as $y = (y_1, y_2)$, (10) can be written in partitioned form as

11 $$\underbrace{\begin{bmatrix} A_1 & A_2 \\ A_3(p) & A_4(p) \end{bmatrix}}_{A(p)} \begin{bmatrix} y_1 \\ y_2 \end{bmatrix} = \underbrace{\begin{bmatrix} B_1(p) \\ B_2(p) \end{bmatrix}}_{B(p)} u,$$

where $[A_1 \vdots A_2]$ is a q_1 by $(q_1 + q_2)$ constant matrix of full rank q_1. In this case (10) and (11) are said to contain q_1 equations purely algebraic in $y = (y_1, y_2)$. If $B_1(p)$ also happens to be a constant matrix, then (10) and (11) contain q_1 equations purely algebraic in y and u.

Finally, if the components of y are ordered in a suitable way, we can always obtain an equation (11) where A_1 is equal to the q_1 by q_1 identity matrix, which in turn implies that $A_3(p)$ can be made equal to the zero matrix. Thus with suitably ordered output components S can always be generated by an equation (11) of the particular form

12 $$\underbrace{\begin{bmatrix} I & A_2 \\ 0 & A_4(p) \end{bmatrix}}_{A(p)} \begin{bmatrix} y_1 \\ y_2 \end{bmatrix} = \underbrace{\begin{bmatrix} B_1(p) \\ B_2(p) \end{bmatrix}}_{B(p)} u,$$

where $A(p)$ is row proper and where $A_4(p)$ does not contain any constant rows. □

7.2 Compositions and decompositions of regular differential input-output relations. Observability

The concept of a composition of a family of input-output relations is closely related to the concept of an interconnection of a family of systems as presented in chapter 3. We shall therefore closely follow the lines of that chapter here. Further only compositions of *regular* differential input-output relations (*cf.* section 6.1) are considered.

Formally, we define a *composition of a family of regular differential input-output relations* as a pair

1 $$(\mathcal{B}, C)$$

which satisfies certain conditions as given below.

A family of regular generators

We have a given (indexed) family $\{S_1, S_2, \ldots, S_N\} \subset \mathcal{R}$ (*cf.* section 6.2) of regular differential input-output relations involved in the composition. This family is represented in (1) by a family $\mathcal{B} \subset \mathcal{M}$ (*cf.* section 6.2) of corresponding regular generators for $S_1, S_2, \ldots, S_N$. Thus

2 $$\mathcal{B} = \{M_1, M_2, \ldots, M_N\}$$

with

3 $$S_j^{-1} = \ker M_j \quad \text{for } j = 1, 2, \ldots, N.$$

For every j we have $S_j \subset \mathcal{X}^{r_j} \times \mathcal{X}^{q_j}$ with $\mathsf{D}S_j = \mathcal{X}^{r_j}$ (*cf.* section 6.1). S_j is said to have r_j distinct input components and q_j distinct output components. Graphically, S_j is represented by a block diagram of the form shown in Fig. 7.1 (*cf.* Fig. 3.1), where M_j is a regular generator for S_j, and where (u_j, y_j) symbolizes some input-output pair contained in S_j.

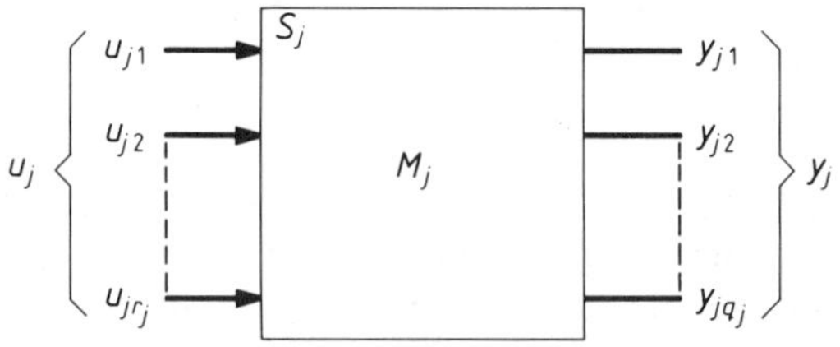

Fig. 7.1. Block diagram representing the regular differential input-output relation S_j. $u_j = (u_{j1}, u_{j2}, \ldots, u_{jr_j})$, $y_j = (y_{j1}, y_{j2}, \ldots, y_{jq_j})$.

An interconnection matrix

C is an *interconnection matrix* which describes a set of interconnection constraints. The role and structure of this interconnection matrix coincide (with obvious minor alterations) with the role and structure of the interconnection matrix discussed in section 3.1. The reader is therefore asked to consult that paragraph.

The interconnection constraints can accordingly again be written in a compact matrix form in the following way (*cf.* (3.1.6)):

4 $$\begin{bmatrix} y_0 \\ \hdashline u \end{bmatrix} = \begin{matrix} \\ y_0 \\ u \end{matrix} \underbrace{\overset{\begin{matrix} u_0 & \;\; y \end{matrix}}{\left[\begin{array}{c:c} 0 & C_2 \\ \hdashline C_3 & C_4 \end{array}\right]}}_{C} \begin{bmatrix} u_0 \\ \hdashline y \end{bmatrix},$$

where $u = (u_1, u_2, \ldots, u_N)$, $y = (y_1, y_2, \ldots, y_N)$, and where $(u_0, y_0) \in \mathscr{X}^{r_0} \times \mathscr{X}^{q_0}$ represents an input-output pair contained in the "overall" input-output relation S_o which we are going to associate with the composition $(\mathscr{B}, C)$ under consideration. Note that the product appearing in (4) can here be interpreted as a true matrix product and not just as a formal product as in (3.1.6).

5 **Note.** Referring to note (3.1.8), it is again evident that the given family $\{S_1, S_2, \ldots, S_N\}$ of regular differential input-output relations involved in the composition and represented by the family $\mathscr{B}$ of regular generators (*cf.* (2) above) could equally well be replaced by a singleton family comprising a suitably chosen single regular differential input-output relation S. An obvious choice would be to take S as generated by the regular generator $[A(p) \vdots -B(p)]$ given by

6 $$\left[\begin{array}{cccc|cccc} A_1(p) & 0 & \cdots & 0 & -B_1(p) & 0 & \cdots & 0 \\ 0 & A_2(p) & \cdots & 0 & 0 & -B_2(p) & \cdots & 0 \\ \vdots & \vdots & \ddots & \vdots & \vdots & \vdots & \ddots & \vdots \\ 0 & 0 & \cdots & A_N(p) & 0 & 0 & \cdots & -B_N(p) \end{array}\right],$$

where $[A_j(p) \vdots -B_j(p)]$ for $j = 1, 2, \ldots, N$ denotes the regular generator M_j for S_j. Other choices are also possible (*cf.* note (24), (ii) below).

We shall in the sequel feel free to replace the given family $\mathscr{B}$ of regular generators by a suitable corresponding singleton family $\{[A(p) \vdots -B(p)]\}$ whenever this seems convenient. The definitions and results given for a composition $(\mathscr{B}, C)$ with $\mathscr{B}$ a singleton family generally carry over in an obvious way to more general compositions of the form (1). □

Note that remark (3.1.10), (i) is in a slightly modified form applicable in this context, too.

Input-output relations determined by a composition of a family of regular differential input-output relations

Now consider a composition of a family of regular differential input-output relations given by

7 $$(\mathscr{B}, C),$$

where $\mathscr{B}$ is a singleton family of regular generators according to

8 $$\mathscr{B} = \{[A(p) \vdots -B(p)]\}$$

with $A(p) \in \mathbf{C}[p]^{q \times q}$, $\det A(p) \neq 0$, and $B(p) \in \mathbf{C}[p]^{q \times r}$, and with C as given by (4).

Let $S \subset \mathscr{X}^r \times \mathscr{X}^q$ further denote the regular differential input-output relation generated by the single member of $\mathscr{B}$, i.e.

9 $$S^{-1} = \ker\,[A(p) \,\vdots\, -B(p)].$$

Suppose that u_0, y_0, u, and y are such that $(u_0, y_0) \in \mathscr{X}^{r_0} \times \mathscr{X}^{q_0}$, $(u, y) \in S$, and u_0, y_0, u, and y satisfy the interconnection constraints (4), i.e. (*cf.* (3.2.6))

10 $$u = [C_3 \,\vdots\, C_4]\begin{bmatrix} u_0 \\ \hdashline y \end{bmatrix} = C_3 u_0 + C_4 y,$$

$$y_0 = C_2 y, \quad \text{and}$$

$$A(p)y = B(p)u.$$

Note here that the algebraic structures involved allow us to write u in (10) as a true sum.

(10) gives the fundamental relations associated with the composition under consideration. They are schematically represented by the block diagram shown in Fig. 7.2 (*cf.* Fig. 3.2).

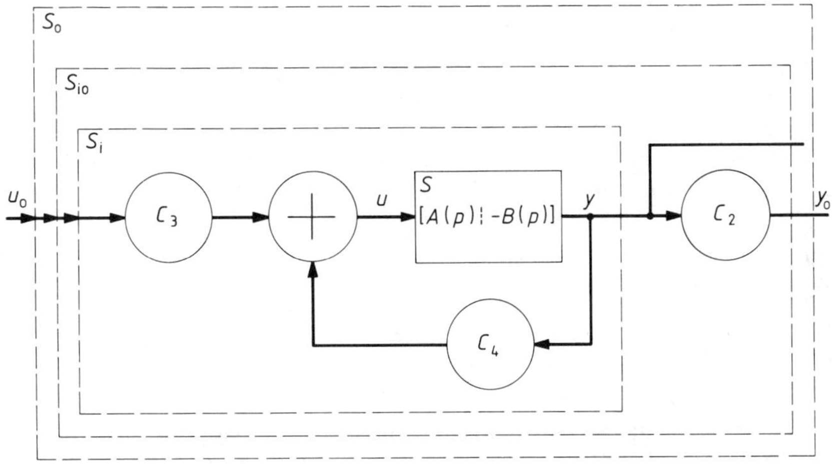

Fig. 7.2. Block diagram representing the composition (7.2.7) and the corresponding relations (7.2.10).

The internal input-output relation S_i

For reasons already explained in section 3.2, we are in the first place interested in the internal behaviour of our composition. So let us start with the "internal" input-output relation S_i determined by the composition (7) in accordance with (10). By definition (*cf.* Fig. 7.2) we have

11 $$S_i \triangleq \{(u_0, y) | (u_0, y) \in \mathscr{X}^{r_0} \times \mathscr{X}^q \text{ and } A(p)y = B(p)(C_3 u_0 + C_4 y)\}.$$

If nothing else is explicitly stated we shall in what follows assume that the dimensions of the various quantities appearing in (11) are compatible. S_i is then never empty—S_i is then, in fact, a differential input-output relation generated by the equation

12 $$(A(p) - B(p)C_4)y = B(p)C_3 u_0,$$

or equivalently, by the generator

13 $$[A(p) - B(p)C_4 \vdots -B(p)C_3] \triangleq [A_i(p) \vdots -B_i(p)].$$

Clearly this result is also directly applicable to more general compositions $(\mathscr{B}, C)$ of the form (1).

The concept of a regular composition

It follows that S_i is regular if and only if (13) is regular, i.e. if and only if $\det A_i(p) \neq 0$. A composition $(\mathscr{B}, C)$ determining a regular internal input-output relation S_i is said to be *regular*. In the sequel we are mainly interested only in regular compositions.

There is a close relationship between the regularity of a composition $(\mathscr{B}, C)$ and the determinateness of a corresponding interconnection of a family of linear dynamic (causal) systems, *cf.* chapter 9.

Equivalent compositions

Let $(\mathscr{B}_1, C)$ and $(\mathscr{B}_2, C)$ be two regular compositions with $\mathscr{B}_1$ represented by $\{[A_1(p) \vdots -B_1(p)]\}$ and $\mathscr{B}_2$ by $\{[A_2(p) \vdots -B_2(p)]\}$. We shall say that $(\mathscr{B}_1, C)$ and $(\mathscr{B}_2, C)$ are *equivalent* if $[A_1(p) \vdots -B_1(p)]$ and $[A_2(p) \vdots -B_2(p)]$ are input-output equivalent in the sense of the definition given in section 6.2, i.e. if $[A_1(p) \vdots -B_1(p)]$ and $[A_2(p) \vdots -B_2(p)]$ generate the same regular differential input-output relation.

14 **Note.** **(i)** Consider two regular compositions $(\mathscr{B}_1, C)$ and $(\mathscr{B}_2, C)$ as given by (7) with $\mathscr{B}_1 = \{[A_1(p) \vdots -B_1(p)]\}$ and $\mathscr{B}_2 = \{[A_2(p) \vdots -B_2(p)]\}$, and let the following relations be introduced.

$$S_1^{-1} = \ker[A_1(p) \vdots -B_1(p)],$$

$$S_2^{-1} = \ker[A_2(p) \vdots -B_2(p)],$$
$$S_{i1}^{-1} = \ker[A_1(p) - B_1(p)C_4 \vdots -B_1(p)C_3],$$
$$S_{i2}^{-1} = \ker[A_2(p) - B_2(p)C_4 \vdots -B_2(p)C_3],$$

where S_{i1} and S_{i2} denote the internal input-output relations determined by $(\mathcal{B}_1, C)$ and $(\mathcal{B}_2, C)$ respectively.

Now if $(\mathcal{B}_1, C)$ and $(\mathcal{B}_2, C)$ are equivalent, i.e. if $S_1 = S_2$, then it is easily confirmed that also $S_{i1} = S_{i2}$. The converse, however, does not necessarily hold.

(ii) A pair $(\mathcal{B}, C)$ as given by (1) should strictly speaking be regarded merely as a description of the composition of input-output relations determined by it and not as the composition itself. Correspondingly, equivalent pairs $(\mathcal{B}_1, C)$ and $(\mathcal{B}_2, C)$ should be regarded as descriptions of one and the same composition. For the sake of simplicity, we have here identified a description with the composition determined by it. Accordingly, we shall also frequently identify mutually equivalent descriptions with each other. □

The overall input-output relation S_o

With the internal input-output relation S_i determined by the composition (7) given by (11), . . . , (13), it is now easy to form the corresponding "overall" input-output relation S_o. By definition (*cf.* Fig. 7.2)

15 $$S_o \triangleq \{(u_0, y_0) \mid (u_0, y_0) = (u_0, C_2 y) \quad \text{for some } (u_0, y) \in S_i\}.$$

It can quite generally be shown that S_o is always a differential input-output relation, although not necessarily a regular one; S_o turns out to be regular if S_i is regular (*cf.* below).

Hence we would like to find a generator for S_o. This requires some effort. We shall proceed in the following way.

The internal-overall input-output relation S_{io}

To begin with we shall form the "internal-overall" input-output relation S_{io} determined by the composition (7) as follows (*cf.* Fig. 7.2):

16 $$S_{io} \triangleq \{(u_0, (y, y_0)) \mid (u_0, (y, y_0)) = (u_0, (y, C_2 y)) \quad \text{for some } (u_0, y) \in S_i\}.$$

It is now easy to see that S_{io} is a differential input-output relation generated by the equation

17 $$\begin{bmatrix} A(p) - B(p)C_4 & 0 \\ -C_2 & I \end{bmatrix} \begin{bmatrix} y \\ y_0 \end{bmatrix} = \begin{bmatrix} B(p)C_3 \\ 0 \end{bmatrix} u_0,$$

or equivalently, by the generator (*cf.* (13))

18
$$\begin{array}{c} \\ \end{array}\begin{bmatrix} \overset{y}{A(p)-B(p)C_4} & \overset{y_0}{0} & \overset{u_0}{-B(p)C_3} \\ -C_2 & I & 0 \end{bmatrix}$$

$$= \begin{bmatrix} \overset{y}{A_i(p)} & \overset{y_0}{0} & \overset{u_0}{-B_i(p)} \\ -C_2 & I & 0 \end{bmatrix} \triangleq [\overset{(y,\,y_0)}{A_{io}(p)} \;\vdots\; \overset{u_0}{-B_{io}(p)}].$$

Note that both S_i and S_o can be obtained from S_{io} in a trivial way; in particular we have for S_o:

19
$$S_o = \{(u_0, y_0)\,|\,u_0, y_0 \quad \text{are such that } (u_0, (y, y_0)) \in S_{io} \quad \text{for some } y\}.$$

Referring to section 7.1 it is now seen that the elements (u_0, y_0) of S_o are obtained through "elimination" of the variable y from the elements $(u_0, (y, y_0))$ of S_{io}.

From (18) it follows that S_{io} is regular if and only if S_i is regular, i.e. if and only if our composition $(\mathscr{B}, C)$ is regular.

Let us now suppose that $(\mathscr{B}, C)$ is regular. Application of the elimination procedure presented in section 7.1 then leads to the conclusion that S_o is a regular differential input-output relation and a regular generator for S_o can be obtained on transforming (18) to a row equivalent upper triangular form (*cf.* (7.1.4)). This is done in the following way. Let $P(p)$ denote a unimodular polynomial matrix of appropriate size corresponding to a suitable sequence of elementary row operations so that premultiplication of (18) by $P(p)$ yields:

20
$$\underbrace{\begin{bmatrix} P_1(p) & P_2(p) \\ P_3(p) & A_0(p) \end{bmatrix}}_{P(p)} \begin{bmatrix} \overset{y}{A_i(p)} & \overset{y_0}{0} & \overset{u_0}{-B_i(p)} \\ -C_2 & I & 0 \end{bmatrix}$$

21
$$\triangleq \begin{bmatrix} \overset{y}{\tilde{A}_{io1}(p)} & \overset{y_0}{\tilde{A}_{io2}(p)} & \overset{u_0}{-\tilde{B}_{io1}(p)} \\ 0 & A_0(p) & -B_0(p) \end{bmatrix}$$

with $\tilde{A}_{io1}(p)$ a GCRD of $A_i(p)$ and C_2 (*cf.* section 7.1), $\det \tilde{A}_{io1}(p) \neq 0$,

$\det A_0(p) \neq 0$, and

22
$$\begin{aligned} P_1(p)A_i(p) - P_2(p)C_2 &= \tilde{A}_{io1}(p), \\ P_3(p)A_i(p) - A_0(p)C_2 &= 0, \\ P_3(p)B_i(p) &= B_0(p), \\ \det A_i(p) \propto \det \tilde{A}_{io1}(p) \det A_0(p) &\neq 0. \end{aligned}$$

(21) is thus a regular generator for S_{io} of upper triangular form, and a regular generator for the overall input-output relation S_o determined by $(\mathscr{B}, C)$ is obtained from (21) as $[A_0(p) \vdots -B_0(p)]$.

The order of a composition

We shall, in the regular case, define the *order* of the composition $(\mathscr{B}, C)$ to be equal to the order of S_i, i.e. equal to $\partial(\det A_i(p)) = \partial(\det (A(p) - B(p)C_4))$.

The concept of an observable composition

As in section 7.1, it is concluded that if $\tilde{A}_{io1}(p)$ happens to be unimodular, i.e. if $A_i(p)$ and C_2 in (18) are right coprime, then (21) can be chosen so that $\tilde{A}_{io1}(p) = I$. Further, with $(u_0, y_0) \in S_o$ given, there is then and only then a *unique* corresponding y such that $(u_0, (y, y_0)) \in S_{io}$, and this y is given by (*cf.* (7.1.5))

23
$$y = -\tilde{A}_{io2}(p)y_0 + \tilde{B}_{io1}(p)\,u_0.$$

A regular composition $(\mathscr{B}, C)$ possessing this remarkable property is said to be *observable.* This concept turns out to be a generalization of the well-known observability concept used in connection with state-space representations. We shall return to this question in a moment.

From the above considerations it follows that the order of S_o $(=\partial(\det A_0(p)))$ is generally at most equal to the order of $(\mathscr{B}, C)$ $(=\partial(\det(A(p) - B(p)C_4)))$, and that the equality is attained if and only if $(\mathscr{B}, C)$ is observable.

Input-output equivalent compositions

We shall need yet another equivalence relation on the set of all regular compositions $(\mathscr{B}, C)$. So let $(\mathscr{B}_1, C_1)$ and $(\mathscr{B}_2, C_2)$ be two regular compositions of the form discussed above, and let S_{o1} and S_{o2} denote the corresponding overall input-output relations. We shall then say that ($\mathcal{B}_1$, C_1) and ($\mathcal{B}_2$, C_2) are *input-output equivalent* if $S_{o1} = S_{o2}$. Clearly, if two compositions are equivalent, then they are also input-output equivalent. The converse, however, does not generally hold.

Decompositions of a regular differential input-output relation

Consider an arbitrary regular differential input-output relation S_o, and let ($\mathcal{B}$, C) be a regular composition as presented above. We shall say that ($\mathcal{B}$, C) is a *decomposition* of S_o into a regular composition of a family of regular differential input-output relations if the overall input-output relation determined by ($\mathcal{B}$, C) coincides with S_o. If this ($\mathcal{B}$, C) is, in addition, observable, then it is said to be an *observable* decomposition of S_o. If ($\mathcal{B}_1$, C_1) and ($\mathcal{B}_2$, C_2) are two decompositions of S_o then they are, by definition, input-output equivalent compositions.

It turns out that many analysis and synthesis problems concerning systems governed by linear time-invariant differential equations amount to finding a suitable decomposition of a given regular differential input-output relation. The following paragraphs illustrate the point.

24 **Note.** (i) There are also interesting cases of *partially observable* compositions. To illustrate this, consider a regular composition ($\mathcal{B}$, C) of the form (7) with the corresponding internal-overall input-output relation S_{io} generated by a regular generator of the form (21). Suppose that this generator is given with $\tilde{A}_{io1}(p)$ in CUT-form (*cf.* section 6.2 and appendix A2), and let it be renamed as

25
$$\begin{array}{c} \begin{matrix} \quad y \quad & \quad y_0 \quad & \quad u_0 \quad \end{matrix} \\ \left[\begin{array}{c:c:c} A_1(p) & A_2(p) & -B_1(p) \\ \hdashline 0 & A_0(p) & -B_0(p) \end{array}\right] \end{array} = j \left[\begin{array}{c:c:c} y & y_0 & u_0 \\ \hdashline A_1^j(p) & A_2^j(p) & -B_1^j(p) \\ \hdashline \\ \hdashline 0 & A_0(p) & -B_0(p) \end{array}\right],$$

where $A_1^j(p)$, $A_2^j(p)$, and $-B_1^j(p)$ denote the jth row of $A_1(p)$, $A_2(p)$, and $-B_1(p)$ respectively, and where $A_1^j(p)$ is of the form

26
$$\begin{array}{rl} & \begin{matrix} 1 \ldots j-1 & j & \quad j+1 & \quad \ldots \quad q \end{matrix} \\ A_1^j(p) = & [0 \ldots \quad 0 \; a_{jj}(p) \quad a_{j(j+1)}(p) \ldots a_{jq}(p)]. \end{array}$$

Given any $(u_0, y_0) \in S_o$ with $S_o^{-1} = \ker[A_0(p) \vdots -B_0(p)]$ there is then according to the above results always a $y = (y_1, y_2, \ldots, y_j, \ldots, y_q)$ such that $(u_0, (y, y_0)) \in S_{io}$. Now it is found that the jth component y_j of this

y is *unique* if and only if $a_{jj}(p)$ is a (nonzero) constant (implying that all the entries of $A_1(p)$ below and above $a_{jj}(p)$ are zeros), and $a_{j(j+1)}(p) = a_{j(j+2)}(p) = \ldots = a_{jq}(p) = 0$. The appropriate y_j can then be solved in an obvious way from the equality

27 $$A_1^j(p)y + A_2^j(p)y_0 - B_1^j(p)u_0 = 0.$$

If $A_2^j(p)$ and $B_1^j(p)$ happen to be constant matrices, then y_j as given by (27) is a linear combination of the components of y_0 and u_0.

A regular composition $(\mathcal{B}, C)$ determining a unique y_j as explained above is said to be *observable with respect to the jth component of the internal output.*

Note that any regular composition is observable with respect to those components of the internal output y that form the overall output y_0.

(ii) The above considerations can formally be somewhat simplified by a suitable (re-)ordering of the output components. So suppose that $(\mathcal{B}, C)$ is a regular composition of the form (7) with $\mathcal{B} = \{[A(p) \vdots -B(p)]\}$ and with the corresponding internal input-output relation S_i generated by the equation (*cf.* (12), (13))

28 $$A_i(p)y = B_i(p)u_0.$$

Without any significant loss of generality we may here assume that the components of y above have been ordered in such a way that y can be written as $y = (y_1, y_0)$, where y_0 represents those components of y which appear as outputs of the overall input-output relation S_o determined by $(\mathcal{B}, C)$. Next let the generator $[A_i(p) \vdots -B_i(p)]$ for S_i (*cf.* (13)) be replaced by an input-output equivalent generator of the form

29 $$\begin{array}{c} \\ \\ \end{array}\begin{bmatrix} \overset{y_1}{\tilde{A}_{i1}(p)} & \overset{y_0}{\tilde{A}_{i2}(p)} & \overset{u_0}{-\tilde{B}_{i1}(p)} \\ 0 & A_0(p) & -B_0(p) \end{bmatrix}$$

with $\det \tilde{A}_{i1}(p) \neq 0$ and $\det A_0(p) \neq 0$. It should now be evident that all the considerations presented above and based on the generator (21) could be repeated here with (21) replaced by (29). In particular it is found that $(\mathcal{B}, C)$ is observable if and only if $\tilde{A}_{i1}(p)$ in (29) is unimodular. Moreover, with $\tilde{A}_{i1}(p)$ of CUT-form, $(\mathcal{B}, C)$ is observable with respect to the jth component of y_1 if and only if the only nonzero entry in the jth row of $\tilde{A}_{i1}(p)$ is the diagonal entry and this entry is a constant. The components of y_1 can thus also be partitioned into two groups, one group consisting of all those components with respect to which the composition is observable, the other group consisting of the remaining components. Occasionally, the

elements of the first group are called "observable components" of y_1 and the elements of the second group "unobservable components" of y_1.
(iii) Consider again a regular composition ($\mathcal{B}$, C) of the form depicted in Fig. 7.2, and suppose that y contains components which are not connected anywhere, that is to say the matrix

$$\begin{bmatrix} C_2 \\ \hdashline C_4 \end{bmatrix}$$

contains one or more zero columns. It seems clear that such a composition in many cases cannot be observable. It is left as an exercise for the reader to work out the details.
(iv) Transfer matrices associated with various regular compositions are most directly determined by applying the multidimensional version of the classical "block diagram algebra" (*cf.* e.g. Truxal, 1955, chapter 2) to the corresponding interconnections. It should thus be evident that the transfer matrix $\mathcal{G}_o(p)$ determined by the overall input-output relation S_o associated with a regular composition ($\mathcal{B}$, C) of the form shown in Fig. 7.2 should read

30 $$\mathcal{G}_o(p) = C_2(I - \mathcal{G}(p)C_4)^{-1}\mathcal{G}(p)C_3$$

with $\mathcal{G}(p) \triangleq A(p)^{-1}B(p)$.

It must, of course, be possible to derive this expression also on the basis of definitions and results valid in this context. So, by definition (*cf.* section 6.3) we have

31 $$\mathcal{G}_o(p) = A_0(p)^{-1}B_0(p)$$

with $A_0(p)$ and $B_0(p)$ as given by (21). From (22) it can be seen that $B_0(p) = P_3(p)B_i(p)$, and $A_0(p)^{-1}P_3(p) = C_2A_i(p)^{-1}$. Using these expressions (31) can be written

32 $$\mathcal{G}_o(p) = C_2A_i(p)^{-1}B_i(p)$$

with $A_i(p)$ and $B_i(p)$ given by (13). Substitution of these expressions in (32) then leads to the result (30).

Regular compositions determining the same transfer matrix are said to be *transfer equivalent.*
(v) Let us once again consider a composition ($\mathcal{B}$, C) as shown in Fig. 7.2 using the above notations. Suppose that S is proper (*cf.* section 6.3), and that the transfer matrix $\mathcal{G}(p)$ determined by S thus can be written as $\mathcal{G}(p) = \mathcal{G}^0(p) + K$ (*cf.* (6.3.2)) with K a constant matrix and with $\mathcal{G}^0(p)$

representing the strictly proper part of $\mathscr{G}(p)$. Suppose furthermore that $\det(I - KC_4) \neq 0$ (note that this happens in particular if $K = 0$, i.e. if S is strictly proper). Then the following statements can be shown to be true.

$\det(A(p) - B(p)C_4) \neq 0$ (i.e. $(\mathscr{B}, C)$ is regular). The order of $(\mathscr{B}, C)$ $(=\partial(\det(A(p) - B(p)C_4)))$ is equal to the order of S $(=\partial(\det A(p)))$. S_i as well as S_o are proper. If S happens to be strictly proper, then so are S_i and S_o.

The above assertion can easily be proved by using row proper forms of the generators involved (*cf.* appendix A2 and the proof of theorem (9.3.8)).

We shall further discuss problems concerning transfer matrices and related matters later on in chapters 8 and 9 of this book. □

We shall conclude this paragraph with a theorem concerning the relationships between the controllability (in the sense of section 6.4) of various input-output relations associated with a given composition.

33 **Theorem.** *Let $(\mathscr{B}, C)$ be a regular composition as given by* (7), *and let the associated quantities S, S_i, S_o, etc. be as in* (8), . . . , (13), *and* (15). *Then the following statements are true.*
(i) *If S_i is controllable, then so are S and S_o.*
(ii) *Suppose that $(\mathscr{B}, C)$ is observable. Then S_o is controllable if and only if S_i is controllable.* □

34 **Proof.** To begin with it is noted that the observability of a regular composition and the controllability of a regular differential input-output relation are invariant with respect to the reordering of the components of the input and output signals. Without loss of generality we may therefore here assume that the output components involved in our composition $(\mathscr{B}, C)$ mentioned in the theorem are ordered as explained in note (24), (ii).

Consider first item (i). Suppose that S is not controllable. According to section 6.4, $A(p)$ and $B(p)$ are then not left coprime, and we may write

35
$$A(p) = L(p)\,A_1(p),$$
$$B(p) = L(p)\,B_1(p),$$

where $L(p)$ is a GCLD of $A(p)$ and $B(p)$. From (13) it then follows that $L(p)$ is also a common left divisor of $A_i(p)$ and $B_i(p)$, i.e. $A_i(p)$ and $B_i(p)$ are not left coprime, and S_i is not controllable. This shows that controllability of S_i implies controllability of S. Note that the converse does not necessarily hold.

Suppose next that a regular generator for S_i is given in an upper triangular form according to (29), and let this generator be renamed as follows:

36
$$\begin{array}{c} \\ \\ \end{array}\begin{bmatrix} \overset{y_1}{A_{i1}(p)} & \overset{y_0}{A_{i2}(p)} & \overset{u_0}{-B_{i1}(p)} \\ 0 & A_0(p) & -B_0(p) \end{bmatrix} \triangleq [A_i(p) \;\vdots\; -B_i(p)].$$

Suppose that S_o is not controllable. $A_0(p)$ and $B_0(p)$ are then not left coprime, and we may write

37
$$A_0(p) = L_0(p)\,A_{01}(p),$$
$$B_0(p) = L_0(p)\,B_{01}(p),$$

where $L_0(p)$ is a GCLD of $A_0(p)$ and $B_0(p)$. It follows from (36) that

$$\begin{bmatrix} I & 0 \\ 0 & L_0(p) \end{bmatrix}$$

is a common left divisor of $A_i(p)$ and $B_i(p)$, i.e. $A_i(p)$ and $B_i(p)$ are not left coprime, and S_i is not controllable. This shows that controllability of S_i implies controllability of S_o. The converse is again not generally true.

Consider then item (ii). We have evidently to show only that controllability of S_o implies controllability of S_i in the case of an observable composition $(\mathcal{B}, C)$. So suppose that $(\mathcal{B}, C)$ is observable, i.e. that $A_{i1}(p)$ in (36) is unimodular, and let the generator (36) for S_i be chosen so that $A_{i1}(p) = I$. Suppose furthermore that S_i is not controllable, i.e. that $A_i(p)$ and $B_i(p)$ are not left coprime. According to the results presented in appendix A2, a GCLD of $A_i(p)$ and $B_i(p)$ can be constructed from (36) by transforming this matrix to a lower triangular form, for instance to a canonical lower triangular form (*cf.* appendix A2), with the aid of elementary column operations. Evidently, the resulting matrix will have the following structure:

38
$$\begin{bmatrix} I & 0 & 0 \\ 0 & L_0(p) & 0 \end{bmatrix}.$$

Consequently,

$$\begin{bmatrix} I & 0 \\ 0 & L_0(p) \end{bmatrix}$$

is a GCLD of $A_i(p)$ and $B_i(p)$, and $L_0(p)$ a GCLD of $A_0(p)$ and $B_0(p)$. Thus S_o is not controllable, and controllability of S_o implies controllability of S_i. This completes the proof of the theorem. □

7.3 A parallel composition

Let $(\mathscr{B}, C)$ be a simple *parallel composition* of the form depicted in Fig. 7.3. $S_1 \subset \mathscr{X}^r \times \mathscr{X}^q$ and $S_2 \subset \mathscr{X}^r \times \mathscr{X}^q$ are here regular differential input-output relations generated by the regular generators

1 $$[A_1(p) \;\vdots\; -B_1(p)] \quad \text{and} \; [A_2(p) \;\vdots\; -B_2(p)]$$

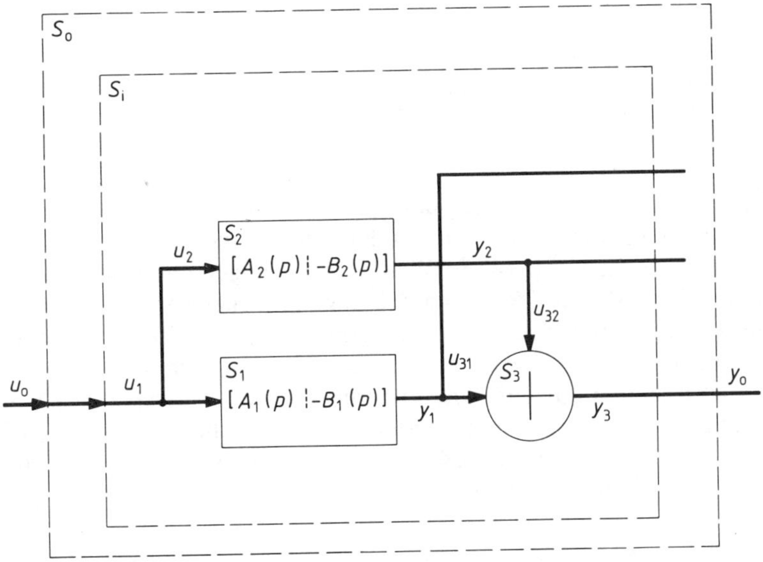

Fig. 7.3. A simple parallel composition $(\mathscr{B}, C)$ comprising two regular differential input-output relations S_1 and S_2 and an adder.

respectively. There is also an adder representing a regular (degenerated) differential input-output relation $S_3 \subset (\mathscr{X}^q \times \mathscr{X}^q) \times \mathscr{X}^q$ generated by the regular generator

2 $$\begin{matrix} y_3 & u_{31} & u_{32} \\ [I \;\vdots\vdots & -I & \vdots\; -I]. \end{matrix}$$

The composition $(\mathscr{B}, C)$ under consideration is a very simple one, and the equation generating the associated internal input-output relation S_i can be written down directly by inspection as follows (it is left as an exercise for the reader to work out a more formal derivation along the lines presented in the previous paragraph):

3
$$\begin{bmatrix} -I & -I & I \\ A_1(p) & 0 & 0 \\ 0 & A_2(p) & 0 \end{bmatrix} \begin{bmatrix} y_1 \\ y_2 \\ y_0 \end{bmatrix} = \begin{bmatrix} 0 \\ B_1(p) \\ B_2(p) \end{bmatrix} u_0.$$

A generator for S_i is thus

4
$$\begin{array}{c} \begin{matrix} y_1 & \quad y_2 & \quad y_0 & \quad u_0 \end{matrix} \\ \left[\begin{array}{ccc:c} -I & -I & I & 0 \\ A_1(p) & 0 & 0 & -B_1(p) \\ 0 & A_2(p) & 0 & -B_2(p) \end{array}\right] \end{array} \triangleq [A_i(p) \;\vdots\; -B_i(p)].$$

We have $\det A_i(p) \propto \det A_1(p) \det A_2(p) \neq 0$. Hence S_i is always regular and so is the composition $(\mathcal{B}, C)$ under consideration.

The regular generator (4) for S_i is now of the general form mentioned in note (7.2.24), (ii). To obtain a regular generator for the overall input-output relation $S_o \subset \mathcal{X}^r \times \mathcal{X}^q$ determined by $(\mathcal{B}, C)$, we thus have to transform (4) to a row equivalent upper triangular form.

The first step is obvious. The uppermost group of rows in (4) is premultiplied by $A_1(p)$ (note the compatibility of the dimensions of the matrices involved), and the result is added to the second group of rows. The new regular generator for S_i thus obtained is

5
$$\begin{array}{c} \begin{matrix} y_1 & \quad y_2 & \quad y_0 & \quad u_0 \end{matrix} \\ \left[\begin{array}{ccc:c} -I & -I & I & 0 \\ 0 & -A_1(p) & A_1(p) & -B_1(p) \\ 0 & A_2(p) & 0 & -B_2(p) \end{array}\right] \end{array}.$$

The final form now depends on whether $A_1(p)$ and $A_2(p)$ are right coprime or not. So let $M(p)$ be a GCRD of $A_1(p)$ and $A_2(p)$ and write

6
$$A_1(p) = A_{11}(p)M(p),$$
$$A_2(p) = A_{21}(p)M(p).$$

There is then a unimodular polynomial matrix $P(p)$ corresponding to a suitable sequence of elementary row operations so that premultiplication of (5) by $P(p)$ yields:

7 $$\underbrace{\begin{bmatrix} I & 0 & 0 \\ 0 & P_1(p) & P_2(p) \\ 0 & P_3(p) & P_4(p) \end{bmatrix}}_{P(p)} \begin{bmatrix} \overset{y_1}{-I} & \overset{y_2}{-I} & \overset{y_0}{I} & \overset{u_0}{0} \\ 0 & -A_1(p) & A_1(p) & -B_1(p) \\ 0 & A_2(p) & 0 & -B_2(p) \end{bmatrix}$$

8 $$= \begin{bmatrix} \overset{y_1}{-I} & \overset{y_2}{-I} & \overset{y_0}{I} & \overset{u_0}{0} \\ 0 & M(p) & P_1(p)A_1(p) & -P_1(p)B_1(p) - P_2(p)B_2(p) \\ 0 & 0 & P_3(p)A_1(p) & -P_3(p)B_1(p) - P_4(p)B_2(p) \end{bmatrix}$$

implying

9 $$\begin{aligned} -P_1(p)A_1(p) + P_2(p)A_2(p) &= M(p), \\ -P_3(p)A_1(p) + P_4(p)A_2(p) &= 0, \end{aligned}$$

and

10 $$\begin{aligned} &\det M(p) \det P_3(p) \det A_1(p) \\ &\propto \det A_1(p) \det A_2(p) = \det A_1(p) \det A_{21}(p) \det M(p), \end{aligned}$$

that is

11 $$\det P_3(p) \propto \det A_{21}(p).$$

(8) is thus a regular generator for S_i of upper triangular form, and a regular generator $[A_0(p) \vdots -B_0(p)]$ for the overall input-output relation S_o determined by $(\mathcal{B}, C)$ is accordingly

12 $$[A_0(p) \vdots -B_0(p)] = [P_3(p)A_1(p) \vdots -P_3(p)B_1(p) - P_4(p)B_2(p)].$$

13 **Note. (i)** The parallel composition $(\mathcal{B}, C)$ considered above is observable in the sense of section 7.2 if and only if $M(p)$ in (6) is unimodular, i.e. if and only if $A_1(p)$ and $A_2(p)$ in (1) are right coprime.

The overall input-output relation S_o determined by $(\mathcal{B}, C)$ is controllable in the sense of section 6.4 if and only if $A_0(p)$ and $B_0(p)$ as given by (12) are left coprime. If in particular, $B_2(p) = 0$, then S_o is controllable if and only if S_1 is controllable and $A_{21}(p)$ in (6) is unimodular. Note that a unimodular $A_{21}(p)$ is equivalent to a unimodular $P_3(p)$ (*cf.* (11)) as well as to $A_2(p)$ being a right divisor of $A_1(p)$ (*cf.* (6)).

(ii) The transfer matrix $\mathcal{G}_o(p)$ determined by S_o above should clearly be (*cf.* Fig. 7.3)

14 $$\mathcal{G}_o(p) = \mathcal{G}_1(p) + \mathcal{G}_2(p)$$

with $\mathcal{G}_1(p) \triangleq A_1(p)^{-1}B_1(p)$ and $\mathcal{G}_2(p) \triangleq A_2(p)^{-1}B_2(p)$. Formally, this result can be obtained as follows. By definition and using (12) we have

15 $$\begin{aligned}\mathcal{G}_o(p) &= A_0(p)^{-1}B_0(p)\\ &= A_1(p)^{-1}P_3(p)^{-1}(P_3(p)B_1(p) + P_4(p)B_2(p)).\end{aligned}$$

The final result (14) then follows on noting that, according to (9), $P_3(p)^{-1}P_4(p) = A_1(p)A_2(p)^{-1}$. □

In the next section we shall apply the results obtained above to two special cases.

7.4 Parallel decompositions of regular differential input-output relations

Let $S_o \subset \mathcal{X}^r \times \mathcal{X}^q$ be a regular differential input-output relation generated by the regular generator

1 $$[A_0(p) \;\vdots\; -B_0(p)].$$

Let $L_0(p)$ further be a GCLD of $A_0(p)$ and $B_0(p)$ and write

2 $$\begin{aligned}A_0(p) &= L_0(p)A_{01}(p),\\ B_0(p) &= L_0(p)B_{01}(p).\end{aligned}$$

Finally, let $\mathcal{G}_o(p)$ be the transfer matrix (*cf.* section 6.3) determined by S_o, i.e.

3 $$\mathcal{G}_o(p) = A_0(p)^{-1}B_0(p) = A_{01}(p)^{-1}B_{01}(p).$$

$\mathcal{G}_o(p)$ can, according to section 6.3, be written in a unique way as

4 $$\mathcal{G}_o(p) = \mathcal{G}_o^0(p) + K_o(p),$$

where $\mathcal{G}_o^0(p)$ is a strictly proper transfer matrix representing the strictly proper part of $\mathcal{G}_o(p)$, and $K_o(p)$ is a polynomial matrix (*cf.* (6.3.2)).

Referring to previous results, S_o is controllable if and only if $A_0(p)$ and $B_0(p)$ in (1) are left coprime, i.e. if and only if $L_0(p)$ in (2) is unimodular (*cf*. section 6.4). Furthermore, S_o is strictly proper by definition (*cf*. section 6.3) if $K_o(p)$ in (4) is the zero matrix.

S_o is not strictly proper

Suppose that S_o is not strictly proper and that $K_o(p)$ in (4) is thus nonzero. Premultiplication of (4) by $A_0(p)$ now yields

5 $$A_0(p)\mathcal{G}_o(p)$$

6 $$= B_0(p) = B_0^0(p) + A_0(p)K_o(p),$$

where $B_0^0(p)$ denotes the polynomial matrix

7 $$B_0^0(p) \triangleq A_0(p)\mathcal{G}_o^0(p).$$

Note that (6) represents a matrix counterpart to the usual division algorithm for ordinary polynomials.

Next consider a parallel composition $(\mathcal{B}, C)$ of the form discussed in the previous paragraph with $S_1 \subset \mathcal{X}^r \times \mathcal{X}^q$ generated by the regular generator

8 $$[A_0(p) \,\vdots\, -B_0^0(p)],$$

and $S_2 \subset \mathcal{X}^r \times \mathcal{X}^q$ by the regular generator

9 $$[I \,\vdots\, -K_o(p)].$$

S_2 generated by (9) is, of course, always controllable in the sense of section 6.4. S_1 generated by (8) is clearly controllable if and only if S_o generated by (1) is controllable.

The composition $(\mathcal{B}, C)$ thus obtained is shown in Fig. 7.4. It turns out that this $(\mathcal{B}, C)$ is actually a decomposition of S_o. Note that the transfer matrix determined by S_1 is just $\mathcal{G}_o^0(p)$, i.e. the strictly proper part of $\mathcal{G}_o(p)$. S_2 is in turn a mapping, in fact, $S_2 = K_o(p)$.

Substitution of (8) and (9) for $[A_1(p) \,\vdots\, -B_1(p)]$ and $[A_2(p) \,\vdots\, -B_2(p)]$ respectively in (7.3.5) gives a regular generator for S_i in Fig. 7.4 of the form

10 $$\begin{array}{c} \begin{array}{cccc} y_1 & y_2 & y_0 & u_0 \end{array} \\ \left[\begin{array}{c:c:c::c} -I & -I & I & 0 \\ \hdashline 0 & -A_0(p) & A_0(p) & -B_0^0(p) \\ \hdashline 0 & I & 0 & -K_o(p) \end{array}\right]. \end{array}$$

$A_0(p)$ and I are right coprime, and the final upper triangular form of (10)

corresponding to (7.3.8) is easily found to be

11

$$\begin{array}{c} \begin{array}{cccc} y_1 & y_2 & y_0 & u_0 \end{array} \\ \left[\begin{array}{c:c:c::c} -I & -I & I & 0 \\ \hdashline 0 & I & 0 & -K_o(p) \\ \hdashline 0 & 0 & A_0(p) & -B_0^0(p) - A_0(p)K_o(p) \end{array}\right] \end{array}$$

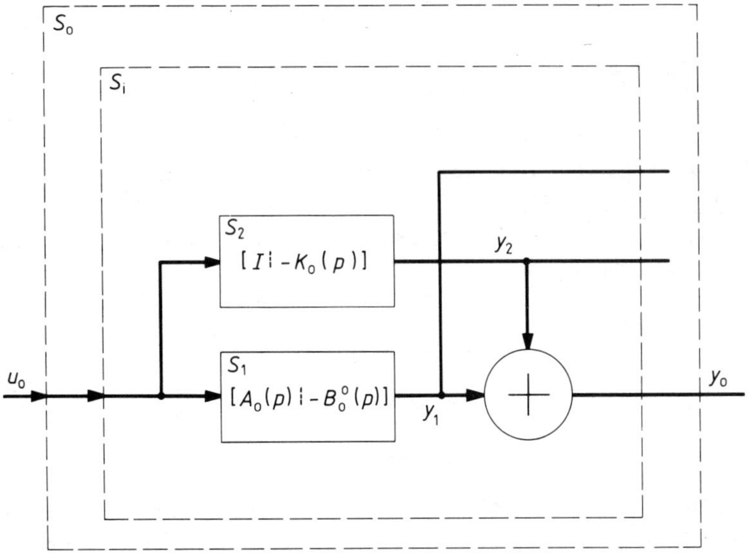

Fig. 7.4. Parallel decomposition ($\mathcal{B}$, C) of a regular differential input-output relation S_o that is not strictly proper.

with $B_0^0(p) + A^0(p)K_o(p)$ equal to $B_0(p)$ according to (6).

(11) is thus regular and gives the regular generator

$$[A_0(p) \vdots -B_0^0(p) - A_0(p)K_o(p)] = [A_0(p) \vdots -B_0(p)]$$

for the overall input-output relation determined by ($\mathcal{B}$, C) as given in Fig. 7.4. This relation is consequently equal to the given S_o (*cf.* (1)), and ($\mathcal{B}$, C) is a decomposition of S_o in the sense of section 7.2. Moreover, this ($\mathcal{B}$, C) is always observable in the sense of section 7.2, which fact can be seen from (11) (*cf.* also note (7.3.13), (i)). We shall say that ($\mathcal{B}$, C) above is a parallel decomposition of S_o into a strictly proper part and a polynomial matrix operator.

S_o is not controllable

Next suppose that S_o is not controllable and that $L_0(p)$ in (2) is thus not unimodular. Then consider a parallel composition $(\mathcal{B}, C)$ of the form discussed in the previous paragraph with $S_1 \subset \mathcal{X}^r \times \mathcal{X}^q$ generated by the regular generator (*cf.* (2))

12 $$[A_{01}(p) \; \vdots \; -B_{01}(p)],$$

and $S_2 \subset \mathcal{X}^r \times \mathcal{X}^q$ by the regular generator

13 $$[A_2(p) \; \vdots \; 0],$$

where $A_2(p)$ must fulfil suitable conditions. We shall discuss these conditions in a moment.

The composition $(\mathcal{B}, C)$ thus obtained is depicted in Fig. 7.5. We shall

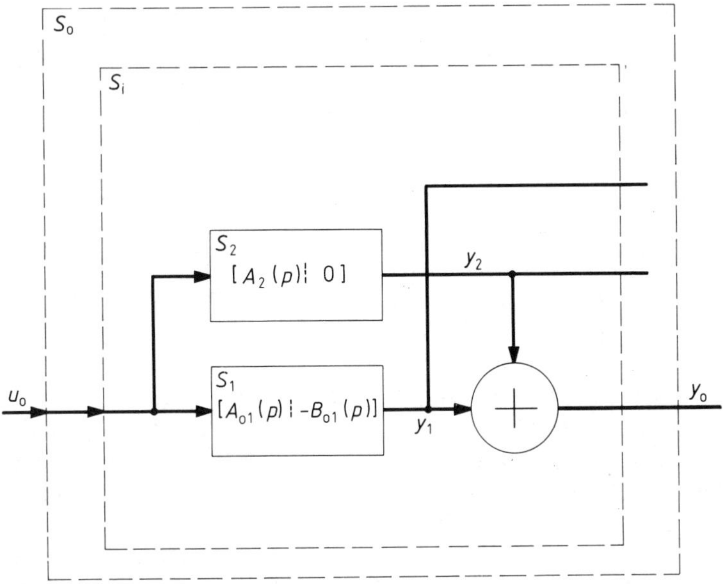

Fig. 7.5. Parallel decomposition $(\mathcal{B}, C)$ of a regular differential input-output relation S_o that is not controllable.

show that this $(\mathcal{B}, C)$—with a suitably chosen $A_2(p)$—is actually a decomposition of S_o. Note that S_1 can be regarded as the controllable part of S_o. S_1 is, in fact, the minimal input-output relation determined by S_o (*cf.* section 6.4). Further, $A_2(p)$ cannot be unimodular in the present case (see note (18), (ii) below), and S_2 is thus not controllable.

Substitution of (12) and (13) for $[A_1(p) \vdots -B_1(p)]$ and $[A_2(p) \vdots -B_2(p)]$ respectively in (7.3.5) gives a regular generator for S_i in Fig. 7.5 of the form

14
$$\begin{array}{c} \begin{matrix} y_1 & y_2 & y_0 & u_0 \end{matrix} \\ \left[\begin{array}{c:c:c::c} -I & -I & I & 0 \\ \hdashline 0 & -A_{01}(p) & A_{01}(p) & -B_{01}(p) \\ \hdashline 0 & A_2(p) & 0 & 0 \end{array}\right]. \end{array}$$

The main condition imposed on $A_2(p)$ is now as follows. $A_2(p)$ must be such that there exists a unimodular polynomial matrix $P(p)$ of appropriate size as indicated in (7.3.7) with $P_3(p) = L_0(p)$ (*cf.* (2)) such that pre-multiplication of (14) by $P(p)$ gives the result (*cf.* (7.3.7), (7.3.8))

15
$$\underbrace{\left[\begin{array}{c:c:c} I & 0 & 0 \\ \hdashline 0 & P_1(p) & P_2(p) \\ \hdashline 0 & L_0(p) & P_4(p) \end{array}\right]}_{P(p)} \begin{array}{c} \begin{matrix} y_1 & y_2 & y_0 & u_0 \end{matrix} \\ \left[\begin{array}{c:c:c::c} -I & -I & I & 0 \\ \hdashline 0 & -A_{01}(p) & A_{01}(p) & -B_{01}(p) \\ \hdashline 0 & A_2(p) & 0 & 0 \end{array}\right] \end{array}$$

16
$$= \begin{array}{c} \begin{matrix} y_1 & y_2 & y_0 & u_0 \end{matrix} \\ \left[\begin{array}{c:c:c::c} -I & -I & I & 0 \\ \hdashline 0 & M(p) & P_1(p)A_{01}(p) & -P_1(p)B_{01}(p) \\ \hdashline 0 & 0 & L_0(p)A_{01}(p) & -L_0(p)B_{01}(p) \end{array}\right], \end{array}$$

where $M(p)$ is a GCRD of $A_{01}(p)$ and $A_2(p)$ (*cf.* (7.3.6)).

It follows that (16) is a regular generator for S_i and that $[L_0(p)A_{01}(p) \vdots -L_0(p)B_{01}(p)] = [A_0(p) \vdots -B_0(p)]$ is a regular generator for the overall input-output relation determined by $(\mathcal{B}, C)$. This input-output relation is consequently equal to S_o, and $(\mathcal{B}, C)$ is a decomposition of S_o in the sense of section 7.2. $(\mathcal{B}, C)$ is observable in the sense of section 7.2 if and only if $M(p)$ is unimodular. We shall say that $(\mathcal{B}, C)$ above is a parallel decomposition of S_o into a controllable part, and a part that is not controllable.

One evident choice of $A_2(p)$ above would be to take

17
$$A_2(p) = A_0(p) = L_0(p)A_{01}(p)$$

with $P_1(p) = -I$, $P_2(p) = 0$, $P_4(p) = I$, and $M(p) = A_{01}(p)$ in (15), (16). The resulting decomposition $(\mathcal{B}, C)$ is then observable if and only if $A_{01}(p)$ is unimodular.

18 **Note. (i)** Suppose that S_o is a given regular differential input-output relation that is neither controllable, nor strictly proper. One possible decomposition of S_o can then be found in the following way. First we construct a parallel decomposition of S_o into a strictly proper part and a polynomial matrix operator as shown in Fig. 7.4. Then we construct a parallel decomposition of the strictly proper part into a controllable part, and a part that is not controllable according to Fig. 7.5.

The above constructions can, of course, also be carried out in the reversed order. It is easily seen that the order in which the constructions are performed is immaterial with regard to the result.

(ii) Consider the case of an S_o that is not controllable as discussed above, and let the quantities involved be denoted as in (12), . . . , (17); see also Fig. 7.5.

It was found that a possible choice of $A_2(p)$ would be (*cf.* (17)) $A_2(p) = A_0(p) = L_0(p)A_{01}(p)$, giving $\partial(\det A_2(p)) = \partial(\det A_0(p))$. The question arises: Is there possibly a feasible $A_2(p)$ fulfilling the necessary conditions such that $\partial(\det A_2(p)) < \partial(\det A_0(p))$?

Now for every feasible $A_2(p)$ we have the equality (follows from (7.3.10) with $P_3(p)$ replaced by $L_0(p)$)

19 $$\det M(p) \det L_0(p) \propto \det A_2(p),$$

that is

20 $$\partial(\det M(p)) + \partial(\det L_0(p)) = \partial(\det A_2(p)).$$

Hence $\partial(\det A_2(p)) \geqslant \partial(\det L_0(p))$ holds, and equality is attained if and only if our decomposition $(\mathcal{B}, C)$ is observable.

We thus arrive at the following conclusion:

There exists a "minimal" feasible $A_2(p)$—here denoted by $A_{2m}(p)$—such that $\partial(\det A_{2m}(p)) \leqslant \partial(\det A_2(p))$ for every feasible $A_2(p)$, and satisfying

21 $$\partial(\det L_0(p)) \leqslant \partial(\det A_{2m}(p)) \leqslant \partial(\det A_0(p)).$$

Further, $\partial(\det L_0(p)) = \partial(\det A_{2m}(p))$ if and only if the decomposition $(\mathcal{B}, C)$ depicted in Fig. 7.5, with $A_{2m}(p)$ inserted for $A_2(p)$, is observable.

(iii) Let us continue the discussion of the matter of item (ii) above, our aim being to find a method for constructing a minimal $A_2(p)$.

Consider the condition imposed on $A_2(p)$ in Fig. 7.5. As stated above, this condition is equivalent to the possibility of passing from (14) to (16). This condition can now be restated in a number of different ways.

First it can be read directly from Fig. 7.5 that our decomposition $(\mathcal{B}, C)$ of S_o is required to act as follows. For every $(u_0, y_1) \in S_1$ and every

$(u_0, y_2) \in S_2$ the pair $(u_0, y_1 + y_2) = (u_0, y_1) + (0, y_2)$ is formed, and this pair is required to be an element (u_0, y_0) contained in the given S_o, and conversely, for every $(u_0, y_0) \in S_o$ there must exist $(u_0, y_1) \in S_1$ and $(u_0, y_2) \in S_2$ such that $(u_0, y_0) = (u_0, y_1) + (0, y_2)$. We shall, in other words, have

22 $$S_o = S_1 + (\{0\} \times S_2(0))$$

or equivalently

23 $$S_o(0) = S_1(0) + S_2(0)$$

with $S_o(0) = \ker A_0(p)$, $S_1(0) = \ker A_{01}(p)$, and $S_2(0) = \ker A_2(p)$, all of them subspaces of the vector space $\mathscr{X}^q$ over $\mathbf{C}$. The dimensions of these spaces are $\partial(\det A_0(p))$, $\partial(\det A_{01}(p))$, and $\partial(\det A_2(p))$ respectively (*cf.* note (5.4.1), (iii)).

Note that (22) is clearly satisfied with $S_2(0) = S_o(0)$, corresponding to the statement contained in note (6.4.17).

A minimal $A_2(p)$ should thus be such that $S_2(0) = \ker A_2(p)$ satisfies (23), and $\partial(\det A_2(p))$ is minimal. It follows that if $\partial(\det A_2(p))$ attains the lower bound $\partial(\det L_0(p))$, then the sums in (22) and (23) are direct sums.

Another way to restate the condition imposed on $A_2(p)$ can be read from (7.3.9). Using the quantities relevant in this context it is found that passing from (14) to (16) is equivalent to the construction of a polynomial matrix $P_4(p)$ such that

24 $$L_0(p)A_{01}(p) = P_4(p)A_2(p)$$

with $L_0(p)$ and $P_4(p)$ left coprime. Note here that according to the general theory for polynomial matrices (*cf.* appendix A2) it is always possible to complete $P(p)$ in (15) with suitable polynomial matrices $P_1(p)$ and $P_2(p)$ as soon as the left coprime polynomial matrices $L_0(p)$ and $P_4(p)$ are specified.

Again the "factorization" condition (24) is clearly satisfied with $P_4(p) = I$ and $A_2(p) = A_0(p) = L_0(p)A_{01}(p)$.

In order to find a minimal $A_2(p)$ we have thus to construct a $P_4(p)$ such that (24) is fulfilled and $\partial(\det P_4(p))$—the dimension of $\ker P_4(p)$—is maximal.

The problem of finding a minimal $A_2(p)$ can accordingly be formulated either on the basis of (23), or on the basis of (24). A solution to one of these problems then also solves the other problem.‡

‡ The factorization problem contained in (24) turns out to be closely related to a corresponding problem concerning so-called skew coprime polynomial matrices, see Wolovich, 1977.

Now a problem formulation on the basis of (23) seems to be the most convenient, and it leads to a rather straightforward construction. In order to avoid an excessive notational burden, we shall in the next paragraph present this construction with the aid of an illustrative example. □

7.5 Illustrative example

In what follows we shall, without detailed explanations, use the notations introduced in the previous section.

Preliminary observations

So let $S_o \subset \mathscr{X}^r \times \mathscr{X}^q$ be a given regular differential input-output relation that is not controllable, and let a regular generator for S_o be given by (7.4.1) and (7.4.2) with $L_0(p)$ not unimodular. Furthermore, let $(\mathscr{B}, C)$ be a parallel decomposition of S_o into a controllable part S_1 and a part S_2 that is not controllable according to Fig. 7.5. The condition imposed on S_2 is given by (7.4.23) or, equivalently, by

1 $$\ker A_0(p) = \ker A_{01}(p) + \ker A_2(p).$$

Given $A_{01}(p)$ and $A_0(p) = L_0(p)A_{01}(p)$ our problem is to construct a minimal $A_2(p)$, i.e. an $A_2(p)$ satisfying (1) and being such that $\partial(\det A_2(p))$—the dimension of $\ker A_2(p)$—is minimal. For the sake of convenience we may also require that $A_2(p)$ be of CUT-form.

Now in view of the results presented in section 6.2 it is clear that $\ker A_0(p) = \ker P_0(p)A_0(p)$, $\ker A_{01}(p) = \ker P_1(p)A_{01}(p)$, and $\ker A_2(p) = \ker P_2(p)A_2(p)$ for any unimodular $P_0(p)$, $P_1(p)$, and $P_2(p)$ of appropriate sizes. Also, (1) holds if and only if

2 $$\ker A_0(p)Q_1(p) = \ker A_{01}(p)\,Q_1(p) + \ker A_2(p)Q_1(p)$$

holds for any unimodular $Q_1(p)$ of appropriate size, and the dimensions of the subspaces of $\mathscr{X}^q$ appearing in (1) are equal to the dimensions of the corresponding subspaces of $\mathscr{X}^q$ appearing in (2). It follows that $A_2(p)$ as determined by (1) is minimal if and only if $A_2(p)Q_1(p)$ as determined by (2) is minimal.

Next consider the equality

3 $$\ker \tilde{A}_0(p) = \ker \tilde{A}_{01}(p) + \ker \tilde{A}_2(p)$$

with $\tilde{A}_0(p) = P_0(p)A_0(p)Q_1(p)$, $\tilde{A}_{01}(p) = P_1(p)A_{01}(p)Q_1(p)$, and

$\tilde{A}_2(p) = P_2(p)A_2(p)Q_1(p)$ for arbitrary unimodular $P_0(p), P_1(p), P_2(p)$, and $Q_1(p)$ of appropriate sizes. The facts referred to above then imply that $A_2(p)$ as determined by (1) is minimal if and only if $\tilde{A}_2(p)$ as determined by (3) is minimal.

A scheme for solving the given problem

It is thus possible to treat the given problem according to the scheme outlined below. Further details of the scheme are given later on in connection with a numerical example.

Step 1. Choose unimodular $P_0(p)$ and $Q_1(p)$ so that

4 $$\tilde{A}_0(p) = P_0(p)A_0(p)Q_1(p)$$

is of diagonal form—for instance of Smith-form (*cf.* appendix A2). Construct a basis $\mathcal{A}_0$ for the vector space $\ker \tilde{A}_0(p)$. This turns out to be a simple matter owing to the diagonal form of $\tilde{A}_0(p)$. Note in passing that a basis for $\ker A_0(p)$ can be obtained from $\mathcal{A}_0$ as $Q_1(p)\mathcal{A}_0$, where the elements of $Q_1(p)\mathcal{A}_0$ are the images under $Q_1(p)$ of the corresponding elements of $\mathcal{A}_0$.

Step 2. Determine $A_{01}(p)Q_1(p)$ and choose a unimodular $P_1(p)$ so that

5 $$\tilde{A}_{01}(p) = P_1(p)A_{01}(p)Q_1(p)$$

is of convenient form—for instance of CUT-form. Construct a basis $\mathcal{A}_{01}$ for the subspace $\ker \tilde{A}_{01}(p)$ of $\ker \tilde{A}_0(p)$.

Step 3. Construct a new basis $\mathcal{A}_0'$ for $\ker \tilde{A}_0(p)$ according to

6 $$\mathcal{A}_0' = \mathcal{A}_{01} \cup \mathcal{V},$$

where $\mathcal{V} \subset \mathcal{A}_0$.

Step 4. Construct a polynomial matrix operator $\tilde{A}_2(p)$ such that $\mathrm{Sp}\,\mathcal{V} \subset \ker \tilde{A}_2(p)$ and $\ker \tilde{A}_2(p)$ is of minimal dimension. It turns out that the $\tilde{A}_2(p)$ so constructed automatically satisfies the condition $\ker \tilde{A}_2(p) \subset \ker \tilde{A}_0(p)$. In fact, this $\ker \tilde{A}_2(p)$ will have a basis of the form $\mathcal{V} \cup \mathcal{W} \subset \mathcal{A}_0$.

Step 5. Determine $\tilde{A}_2(p)Q_1(p)^{-1}$ and choose a unimodular $P_2(p)$ so that

7 $$P_2(p)^{-1}\tilde{A}_2(p)Q_1(p)^{-1}$$

is of CUT-form. Choosing (7) as $A_2(p)$ then solves our problem.

Next we shall apply the above scheme to a numerical example.

Numerical example

Step 1. Let the Smith-form of the given $A_0(p)$ be

8 $$\tilde{A}_0(p) = P_0(p)A_0(p)Q_1(p) = \begin{bmatrix} (p+2)(p+3) & 0 \\ 0 & (p+2)^2(p+3) \end{bmatrix}.$$

Clearly, a basis for ker $(p+2)(p+3)$ is given by $\{e^{-2(.)}, e^{-3(.)}\}$, and a basis for ker $(p+2)^2(p+3)$ by $\{(.)e^{-2(.)}, e^{-2(.)}, e^{-3(.)}\}$. A basis $\mathscr{A}_0$ for ker $\tilde{A}_0(p)$ is thus obtained as

9 $$\mathscr{A}_0 = \left\{ \begin{bmatrix} e^{-2(.)} \\ 0 \end{bmatrix}, \begin{bmatrix} e^{-3(.)} \\ 0 \end{bmatrix}, \begin{bmatrix} 0 \\ (.)e^{-2(.)} \end{bmatrix}, \begin{bmatrix} 0 \\ e^{-2(.)} \end{bmatrix}, \begin{bmatrix} 0 \\ e^{-3(.)} \end{bmatrix} \right\}.$$

Step 2. Let the given $A_{01}(p)$ be such that the CUT-form of $A_{01}(p)Q_1(p)$ is

10 $$\tilde{A}_{01}(p) = P_1(p)A_{01}(p)Q_1(p) = \begin{bmatrix} p+2 & p+2 \\ 0 & (p+2)(p+3) \end{bmatrix}.$$

Note that $A_0(p) = L_0(p)A_{01}(p)$ implies that $\tilde{A}_0(p) = \tilde{L}_0(p)\tilde{A}_{01}(p)$ with det $L_0(p) \propto$ det $\tilde{L}_0(p)$. In this case it is seen that

11 $$\tilde{L}_0(p) = \begin{bmatrix} p+3 & -1 \\ 0 & p+2 \end{bmatrix},$$

implying that $\partial(\det \tilde{L}_0(p)) = \partial(\det L_0(p)) = 2$.

A basis for ker $\tilde{A}_{01}(p)$ can now be constructed in several ways. One possibility would be to use the Smith-form of $\tilde{A}_{01}(p)$ (*cf.* step 1). Another possibility is outlined in Zadeh and Desoer, 1963, section 5.7. It would also be possible to apply results obtained below in section 8.5. Here we shall employ a more direct method.

It is known that the components of the elements of ker $\tilde{A}_{01}(p)$ are also elements of ker $(\det \tilde{A}_{01}(p)) =$ ker $(p+2)^2(p+3)$ (*cf.* note (5.4.1), (ii)). It is thus clear that a basis $\mathscr{A}_{01}$ for ker $\tilde{A}_{01}(p)$ can be formed as a set of suitable linear combinations of the elements of the set (this set could be replaced by a smaller set constructed on the basis of theorem (5.7.8) in Zadeh and Desoer, 1963)

12 $$\left\{ \begin{bmatrix} (.)e^{-2(.)} \\ 0 \end{bmatrix}, \begin{bmatrix} e^{-2(.)} \\ 0 \end{bmatrix}, \begin{bmatrix} e^{-3(.)} \\ 0 \end{bmatrix}, \begin{bmatrix} 0 \\ (.)e^{-2(.)} \end{bmatrix}, \begin{bmatrix} 0 \\ e^{-2(.)} \end{bmatrix}, \begin{bmatrix} 0 \\ e^{-3(.)} \end{bmatrix} \right\}.$$

$\tilde{A}_{01}(p) \times$ the set (12), i.e. the set of images under $\tilde{A}_{01}(p)$ of the elements of (12) (the order of the images is here the same as the order of the corresponding elements of (12)), is found to be

13 $$\left\{\begin{bmatrix}(-2(.)+3)e^{-2(.)}\\0\end{bmatrix},\begin{bmatrix}0\\0\end{bmatrix},\begin{bmatrix}-e^{-3(.)}\\0\end{bmatrix},\begin{bmatrix}(-2(.)+3)e^{-2(.)}\\e^{-2(.)}\end{bmatrix},\begin{bmatrix}0\\0\end{bmatrix},\begin{bmatrix}-e^{-3(.)}\\0\end{bmatrix}\right\}.$$

From (13) it is readily concluded that a basis $\mathscr{A}_{01}$ for $\ker \tilde{A}_{01}(p)$ is obtained as the set {the second element of (12), the fifth element of (12), the third element of (12)−the sixth element of (12)}, i.e.

14 $$\mathscr{A}_{01}=\left\{\begin{bmatrix}e^{-2(.)}\\0\end{bmatrix},\begin{bmatrix}0\\e^{-2(.)}\end{bmatrix},\begin{bmatrix}e^{-3(.)}\\-e^{-3(.)}\end{bmatrix}\right\}.$$

Step 3. A new basis $\mathscr{A}_0'=\mathscr{A}_{01}\cup\mathscr{V}$ for $\ker \tilde{A}_0(p)$ is now obtained from (9) and (14) with

15 $$\mathscr{V}=\left\{\begin{bmatrix}e^{-3(.)}\\0\end{bmatrix},\begin{bmatrix}0\\(.)e^{-2(.)}\end{bmatrix}\right\}.$$

Step 4. First note the crucial fact that if our $\tilde{A}_2(p)$ is such that

$$\begin{bmatrix}0\\(.)^n e^{-\alpha(.)}\end{bmatrix}\in\ker \tilde{A}_2(p)$$

for some $\alpha\in\mathbf{C}$ and $n\in\{1, 2, 3, \ldots\}$, then necessarily also

$$\begin{bmatrix}0\\(.)^i e^{-\alpha(.)}\end{bmatrix}\in\ker \tilde{A}_2(p)$$

for $i=0, 1, 2, \ldots, n-1$. Accordingly, a minimal $\tilde{A}_2(p)$ such that $\mathrm{Sp}\,\mathscr{V}\subset\ker \tilde{A}_2(p)$ can be written down by inspection:

16 $$\tilde{A}_2(p)=\begin{bmatrix}p+3 & 0\\0 & (p+2)^2\end{bmatrix}.$$

A basis for $\ker \tilde{A}_2(p)$ is thus given by $\mathscr{V}\cup\mathscr{W}$ with

17 $$\mathscr{W}=\left\{\begin{bmatrix}0\\e^{-2(.)}\end{bmatrix}\right\}\subset\mathscr{A}_0.$$

Note at this stage that $\tilde{A}_2(p)$ can always be chosen diagonal. This follows from the fact that $\mathscr{V}\cup\mathscr{W}$ consists of elements of $\mathscr{A}_0$.

Step 5. The final step as explained above then gives the final solution to our problem, i.e. a minimal $A_2(p)$ and a corresponding parallel decomposition $(\mathcal{B}, C)$ of the given S_o. In this particular example we have $\partial(\det A_2(p)) = \partial(\det \tilde{A}_2(p)) = 3$, and $\partial(\det L_0(p)) = 2$. The decomposition $(\mathcal{B}, C)$ obtained is consequently not observable (*cf.* note (7.4.18), (ii)).

The above example should make it possible for the reader to solve any particular problem of this kind.

7.6 A series composition

Let $(\mathcal{B}, C)$ be a simple *series composition* of the form depicted in Fig. 7.6. $S_1 \subset \mathcal{X}^{r_1} \times \mathcal{X}^{q_1}$ and $S_2 \subset \mathcal{X}^{r_2} \times \mathcal{X}^{q_2}$ with $r_2 = q_1$ are here regular differential

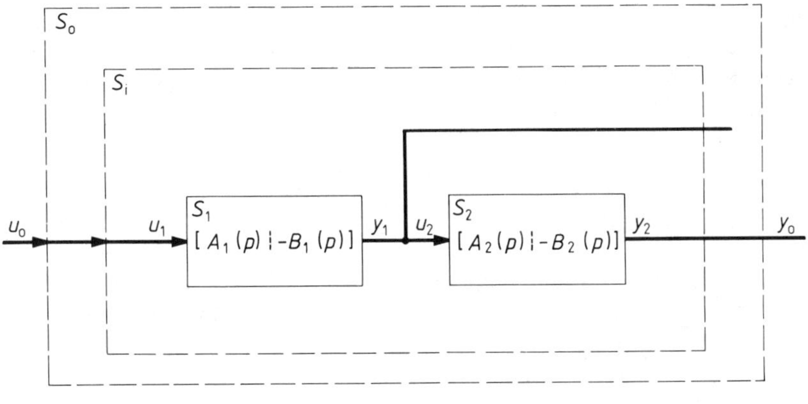

Fig. 7.6. A simple series composition $(\mathcal{B}, C)$ comprising two regular differential input-output relations S_1 and S_2.

input-output relations generated by the regular generators

1 $$[A_1(p) \;\vdots\; -B_1(p)] \quad \text{and}$$

2 $$[A_2(p) \;\vdots\; -B_2(p)]$$

respectively.

The composition $(\mathcal{B}, C)$ under consideration is again a very simple one, and the equation generating the associated internal input-output relation S_i can be written down directly by inspection (*cf.* the parallel composition considered in section 7.3):

3 $$\left[\begin{array}{c:c} A_1(p) & 0 \\ \hdashline -B_2(p) & A_2(p) \end{array}\right]\left[\begin{array}{c} y_1 \\ \hdashline y_0 \end{array}\right] = \left[\begin{array}{c} B_1(p) \\ \hdashline 0 \end{array}\right] u_0.$$

A generator for S_i is thus

4 $$\begin{array}{c} \begin{array}{ccc} \quad y_1 \quad & \quad y_0 \quad & \quad u_0 \quad \end{array} \\ \left[\begin{array}{c:c:c} A_1(p) & 0 & -B_1(p) \\ \hdashline -B_2(p) & A_2(p) & 0 \end{array}\right] \end{array} \triangleq [A_i(p) \,\vdots\, -B_i(p)].$$

It follows that $\det A_i(p) = \det A_1(p) \det A_2(p) \neq 0$. Hence S_i is always regular and so is the composition $(\mathcal{B}, C)$ under consideration.

The regular generator (4) for S_i is of the general form mentioned in note (7.2.24), (ii). To obtain a regular generator for the overall input-output relation $S_o \subset \mathcal{X}^{r_1} \times \mathcal{X}^{q_2}$ determined by $(\mathcal{B}, C)$, we thus have to transform (4) to a row equivalent upper triangular form.

The final form again depends on whether $A_1(p)$ and $B_2(p)$ are right coprime or not. So let $M(p)$ be a GCRD of $A_1(p)$ and $B_2(p)$ and write

5 $$\begin{aligned} A_1(p) &= A_{11}(p)M(p), \\ B_2(p) &= B_{21}(p)M(p). \end{aligned}$$

There is then a unimodular polynomial matrix $P(p)$ corresponding to a suitable sequence of elementary row operations so that premultiplication of (4) by $P(p)$ yields:

6 $$\underbrace{\left[\begin{array}{c:c} P_1(p) & P_2(p) \\ \hdashline P_3(p) & P_4(p) \end{array}\right]}_{P(p)} \begin{array}{c} \begin{array}{ccc} \quad y_1 \quad & \quad y_0 \quad & \quad u_0 \quad \end{array} \\ \left[\begin{array}{c:c:c} A_1(p) & 0 & -B_1(p) \\ \hdashline -B_2(p) & A_2(p) & 0 \end{array}\right] \end{array}$$

7 $$= \begin{array}{c} \begin{array}{ccc} \quad y_1 \quad & \qquad y_0 \qquad & \qquad u_0 \qquad \end{array} \\ \left[\begin{array}{c:c:c} M(p) & P_2(p)A_2(p) & -P_1(p)B_1(p) \\ \hdashline 0 & P_4(p)A_2(p) & -P_3(p)B_1(p) \end{array}\right] \end{array}$$

implying

8 $$\begin{aligned} P_1(p)A_1(p) - P_2(p)B_2(p) &= M(p), \\ P_3(p)A_1(p) - P_4(p)B_2(p) &= 0, \end{aligned}$$

and

9
$$\det M(p) \det P_4(p) \det A_2(p)$$
$$\propto \det A_1(p) \det A_2(p) = \det A_{11}(p) \det M(p) \det A_2(p),$$

that is

10
$$\det P_4(p) \propto \det A_{11}(p).$$

(7) is thus a regular generator for S_i of upper triangular form, and a regular generator $[A_0(p) \vdots -B_0(p)]$ for the overall input-output relation S_o determined by $(\mathcal{B}, C)$ is accordingly

11
$$[A_0(p) \vdots -B_0(p)] = [P_4(p)A_2(p) \vdots -P_3(p)B_1(p)].$$

12 **Note. (i)** The overall input-output relation S_o generated by (11) is, of course, equal to the ordinary composition $S_2 \circ S_1$ of S_1 and S_2.
(ii) The transfer matrix $\mathcal{G}_o(p)$ determined by S_o above is, according to (11), given by

13
$$\mathcal{G}_o(p) = A_0(p)^{-1}B_0(p) = A_2(p)^{-1}P_4(p)^{-1}P_3(p)B_1(p),$$

where $P_4(p)$ and $P_3(p)$ are left coprime (*cf.* appendix A2). From (8) and (5) it follows that

14
$$P_4(p)^{-1}P_3(p) = B_2(p)A_1(p)^{-1} = B_{21}(p)A_{11}(p)^{-1}$$

with $B_{21}(p)$ and $A_{11}(p)$ right coprime. On substituting (14) in (13) we get

15
$$\mathcal{G}_o(p) = \mathcal{G}_2(p)\mathcal{G}_1(p)$$

with $\mathcal{G}_1(p) \triangleq A_1(p)^{-1}B_1(p)$ and $\mathcal{G}_2(p) \triangleq A_2(p)^{-1}B_2(p)$. The result (15) is, of course, precisely as expected on the basis of Fig. 7.6.

The above results suggest the following procedure for finding the regular generator (11) for S_o.

Step 1. A greatest common right divisor $M(p)$ is "cancelled" from $A_1(p)$ and $B_2(p)$ yielding $A_{11}(p)$ and $B_{21}(p)$. Nontrivial cancellations occur at this stage ($M(p)$ not unimodular) whenever our composition $(\mathcal{B}, C)$ is not observable (*cf.* section 7.2).

Step 2. $B_{21}(p)A_{11}(p)^{-1}$ is factored as $P_4(p)^{-1}P_3(p)$ with $P_4(p)$ and $P_3(p)$ left coprime (*cf.* (A2.86)).

Step 3. The regular generator (11) for S_o is formed.

Note that $A_2(p)$ and $B_1(p)$ appear in (11) in their original form without any cancellations. It may therefore happen that additional nontrivial cancellations can be made in the expression (13) for $\mathcal{G}_o(p)$. This happens, of course, whenever S_o is not controllable (*cf.* section 6.4).

(iii) The series composition $(\mathcal{B}, C)$ considered above is observable in the sense of section 7.2 if and only if $M(p)$ in (5) is unimodular, i.e. if and only if $A_1(p)$ and $B_2(p)$ in (1), (2) are right coprime. Suppose that our composition $(\mathcal{B}, C)$ is not observable. Then the following observations can be made.

Referring to note (7.2.24), (ii), it is first noted that the components of y_1 in Fig. 7.6 can be assumed to be ordered in such a way that y_1 can be expressed as $y_1 = (z_1, z_2)$, where z_1 comprises the unobservable components of y_1 and z_2 the observable components of y_1. In this case the $M(p)$-block in (7) can be given the structure

16 $$M(p) = \overset{\overbrace{\begin{matrix} z_1 & \quad z_2 \end{matrix}}^{y_1}}{\left[\begin{array}{c:c} M_1(p) & 0 \\ \hdashline 0 & I \end{array}\right]}$$

with $M_1(p)$ not unimodular.

Consider the series composition $(\mathcal{B}_1, C)$ depicted in Fig. 7.7.

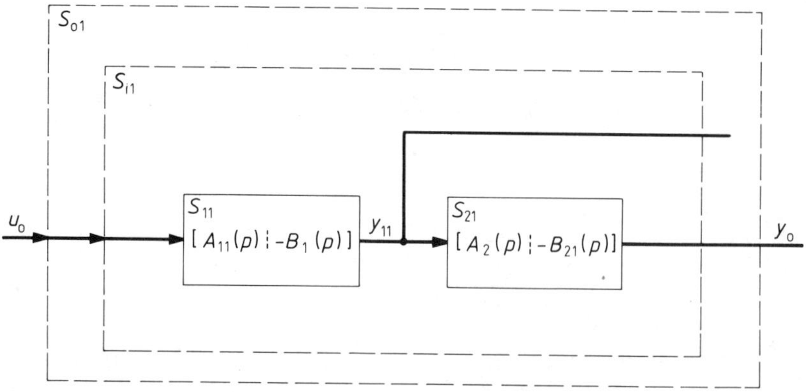

Fig. 7.7. A series composition $(\mathcal{B}_1, C)$ forming an observable series decomposition of the S_o determined by the series composition $(\mathcal{B}, C)$ shown in Fig. 7.6.

This composition is obtained from $(\mathcal{B}, C)$ according to Fig. 7.6 simply by replacing S_1 and S_2 by the corresponding regular differential input-output relations S_{11} and S_{21} generated by the regular generators

17 $$[A_{11}(p) \vdots -B_1(p)] \quad \text{and}$$

18 $$[A_2(p) \vdots -B_{21}(p)]$$

respectively, where $A_{11}(p), B_1(p), A_2(p)$, and $B_{21}(p)$ are given by (1), (2), and (5). Of course, $(\mathscr{B}_1, C)$ and $(\mathscr{B}, C)$ coincide if $A_1(p)$ and $B_2(p)$ (*cf.* (5)) are right coprime.

A regular generator for the internal input-output relation S_{i1} determined by $(\mathscr{B}_1, C)$ is clearly given by (*cf.* (4))

19 $$\begin{array}{c} \begin{matrix} y_{11} & \quad y_0 & \quad u_0 \end{matrix} \\ \left[\begin{array}{c:c:c} A_{11}(p) & 0 & -B_1(p) \\ \hdashline -B_{21}(p) & A_2(p) & 0 \end{array}\right] \end{array} \triangleq [A_{i1}(p) \vdots -B_{i1}(p)].$$

$A_{11}(p)$ and $B_{21}(p)$ are right coprime by definition. It follows that $(\mathscr{B}_1, C)$ is an observable regular composition. Comparing (19) with (4), and noting the results (6) and (7), it is further seen that the overall input-output relation S_{o1} determined by $(\mathscr{B}_1, C)$ coincides with the overall input-output relation S_o determined by the composition $(\mathscr{B}, C)$ shown in Fig. 7.6. $(\mathscr{B}_1, C)$ is consequently by definition an observable series decomposition of S_o.

Next consider the composition $(\mathscr{B}_2, C_2)$ depicted in Fig. 7.8, where S_{11} and S_{21} are as in Fig. 7.7 and where $M(p)$ is given by (5).

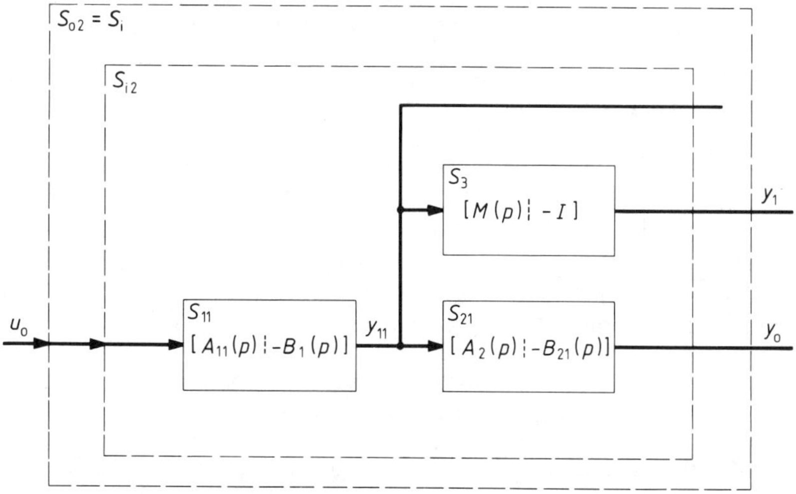

Fig. 7.8. A series-parallel composition $(\mathscr{B}_2, C_2)$ forming an observable decomposition of the S_i determined by the series composition $(\mathscr{B}, C)$ shown in Fig. 7.6.

A generator for the internal input-output relation S_{i2} determined by $(\mathscr{B}_2, C_2)$ is readily found to be given by

20
$$\begin{array}{c} \\ \\ \end{array}\begin{matrix} y_{11} & y_1 & y_0 & u_0 \end{matrix}$$
$$\left[\begin{array}{c:c:c::c} -I & M(p) & 0 & 0 \\ \hdashline 0 & A_{11}(p)M(p) & 0 & -B_1(p) \\ \hdashline 0 & -B_{21}(p)M(p) & A_2(p) & 0 \end{array}\right] \triangleq [A_{i2}(p) \vdots -B_{i2}(p)].$$

It follows that $(\mathscr{B}_2, C_2)$ is an observable regular composition. Note that the overall input-output relation S_{o2} determined by $(\mathscr{B}_2, C_2)$ consists of pairs $(u_0, (y_1, y_0)) \in \mathscr{X}^{r_1} \times (\mathscr{X}^{q_1} \times \mathscr{X}^{q_2})$. Now comparing (20) with (4) it is concluded that S_{o2} coincides with the internal input-output relation S_i determined by our series composition $(\mathscr{B}, C)$. So, $(\mathscr{B}_2, C_2)$ is, by definition, an observable decomposition of S_i. The decomposition could be given further structural properties by using structure (16) for $M(p)$ in (20).

(iv) Concerning the controllability in the sense of section 6.4 of various input-output relations associated with the series composition $(\mathscr{B}, C)$ considered above and shown in Fig. 7.6 we have, of course, the general results given in theorem (7.2.33). In particular, it follows from theorem (7.2.33), (i) that (for the sake of brevity, we shall here use the notations introduced and explained above without any additional attributes):

21 If S_i generated by (4) is controllablc, thcn so arc S_o, S_1, and S_2 generated by (11), (1), and (2) respectively.

Note that in applying theorem (7.2.33) in the above case, S can be interpreted as generated by

$$\left[\begin{array}{c:c::c:c} A_1(p) & 0 & -B_1(p) & 0 \\ \hdashline 0 & A_2(p) & 0 & -B_2(p) \end{array}\right].$$

Noting the fact that the composition $(\mathscr{B}_1, C)$ shown in Fig. 7.7 is an observable decomposition of S_o, and applying theorem (7.2.33), (ii), it is concluded that the following statements are equivalent:

22 S_o generated by (11) is controllable,
S_{i1} generated by (19) is controllable.

Consider then the special case with $A_2(p) = I$ in (1). This case plays an important role in the discussions concerning state-space representations and Rosenbrock's decompositions to follow.

Now, on putting $A_2(p) = I$ in the various generators appearing above it is easily concluded that

$$\left[\begin{array}{c:c} L_1(p) & 0 \\ \hdashline 0 & I \end{array}\right],$$

where $L_1(p)$ is a GCLD of $A_1(p)$ and $B_1(p)$ in (1), is a GCLD of $A_i(p)$ and $B_i(p)$ as given by (4). It follows that the following statements are equivalent for $A_2(p) = I$:

23 S_i generated by (4) is controllable,
S_1 generated by (1) is controllable.

Analogously it is concluded for $A_2(p) = I$ that S_{i1} generated by (19) is controllable if and only if S_{11} generated by (17) is controllable. Combining this result with (22) above we arrive at the following equivalent statements in the case $A_2(p) = I$:

24 S_o generated by (11) is controllable,
S_{i1} generated by (19) is controllable,
S_{11} generated by (17) is controllable. □

In the next few paragraphs we shall apply the results obtained above to a number of special cases.

7.7 Series and series-parallel decompositions of regular differential input-output relations

Let $S_o \subset \mathscr{X}^r \times \mathscr{X}^q$ be a regular differential input-output relation generated by the regular generator

1 $$[A_0(p) \; \vdots \; -B_0(p)],$$

and let $\mathscr{G}_o(p)$ be the transfer matrix (*cf.* section 6.3) determined by S_o, i.e.

2 $$\mathscr{G}_o(p) = A_0(p)^{-1}B_0(p).$$

The general class of series decompositions of S_o

Consider then the set of all (regular) series compositions $(\mathscr{B}, C)$ of the form depicted in Fig. 7.6 determining the given S_o, i.e. the set of all series decompositions of S_o. This set also contains trivial decompositions, for instance decompositions, corresponding to $[A_1(p) \; \vdots \; -B_1(p)]$ and $[A_2(p) \; \vdots \; -B_2(p)]$ in Fig. 7.6 equal to (1) and $[I \; \vdots \; -I]$ respectively. This general class of series decompositions of S_o is, in fact, too general to be interesting. We shall therefore comment here only on more special series decompositions.

Series decomposition of a controllable S_o

Now let us suppose that S_o is controllable in the sense of section 6.4. Accordingly, $A_0(p)$ and $B_0(p)$ in (1) are left coprime. Then consider a series composition $(\mathcal{B}, C)$ of the kind shown in Fig. 7.6, with $B_1(p) = I, A_2(p) = I$, and with $A_1(p)$, $\det A_1(p) \neq 0$, and $B_2(p)$ right coprime and chosen so that

3 $$B_0(p)A_1(p) - A_0(p)B_2(p) = 0,$$

i.e. so that

4 $$\mathcal{G}_0(p) = A_0(p)^{-1}B_0(p) = B_2(p)A_1(p)^{-1}.$$

Such a pair $A_1(p)$, $B_2(p)$ can easily be found, see (A1.85), . . . , (A1.92), and (A2.86). It is then also easy to show that the composition $(\mathcal{B}, C)$ so constructed is an observable series decomposition of the given controllable S_o (an easy exercise for the reader!). This particular series decomposition has been used extensively by Wolovich for solving various estimation and feedback problems (see e.g. Wolovich, 1974, sections 7.3 and 7.4). The related factorization problem as expressed by (4) has been discussed by many authors (see e.g. Rosenbrock and Hayton, 1974; Rosenbrock, 1970, chapter 3, section 4; Wolovich, 1974; Forney, 1975).

Series-parallel decomposition of an S_o that is not controllable

If S_o above is not controllable, then we can first construct a parallel decomposition of S_o into a controllable part and a part that is not controllable, as described in section 7.4. A series decomposition of the controllable part can then be performed as explained above. The result is a series-parallel decomposition, possibly not observable, of the given, not controllable S_o.

This leads us in a very natural way to considerations concerning the general class of series-parallel compositions and decompositions.

The state-space representation

"Modern" control theory relies heavily on a special kind of series-parallel compositions and decompositions, called "state-space representations". The literature on this subject is immense—let us here refer only to the book by Padulo and Arbib, 1974. Because a general presentation of the theory of the state-space representation is outside the scope of this book, we shall in this context only discuss a number of special aspects of this theory of particular relevance with respect to the polynomial systems theory.

The Rosenbrock representation

Before doing this we shall, however, study a form of series-parallel compositions and decompositions which are a little more general than the state-space representation mentioned above. We shall call a composition or decomposition of this kind a "Rosenbrock representation". This representation was originally introduced by Rosenbrock (see e.g. Rosenbrock, 1970), and it has since then been studied extensively in the literature. We shall devote the next few paragraphs to this subject.

Terminology

In this context the term "input-output equivalence" will be given a meaning which differs somewhat from the meaning assigned to this term in section 6.2 (*cf.* section 7.10). To avoid ambiguity, we shall in sections 7.8, . . . , 7.10 below therefore use throughout the term "row equivalence" for the "input-output equivalence" = "row equivalence" as introduced in section 6.2.

7.8 The Rosenbrock representation

In this paragraph we shall study some aspects of the Rosenbrock representation mentioned in the previous paragraph. For this purpose we shall start with a series-parallel composition ($\mathscr{B}$, C) of the form shown in Fig. 7.9.

$S_1 \subset \mathscr{X}^l \times \mathscr{X}^r$, $S_2 \subset \mathscr{X}^r \times \mathscr{X}^m$, $S_3 \subset \mathscr{X}^l \times \mathscr{X}^m$, and $S_4 \subset (\mathscr{X}^m \times \mathscr{X}^m) \times \mathscr{X}^m$ are here regular differential input-output relations generated by regular generators as follows:

1 $$S_1^{-1} = \ker[T(p) \,\vdots\, -U(p)],$$

2 $$S_2^{-1} = \ker[I \,\vdots\, -V(p)],$$

3 $$S_3^{-1} = \ker[I \,\vdots\, -W(p)],$$

4 $$S_4^{-1} = \ker[I \,\vdots\vdots\, -I \,\vdots\, -I] \text{ } (cf. \text{ } (7.3.2)).$$

S_2, S_3, and S_4 are thus mappings with $S_2 = V(p)$ and $S_3 = W(p)$.

Note that, for easy reference and comparison, we have here adopted the standard notations used in Rosenbrock, 1970.

A regular generator for the internal input-output relation S_i determined by $(\mathcal{B}, C)$ can again be written down by inspection (*cf.* the steps from

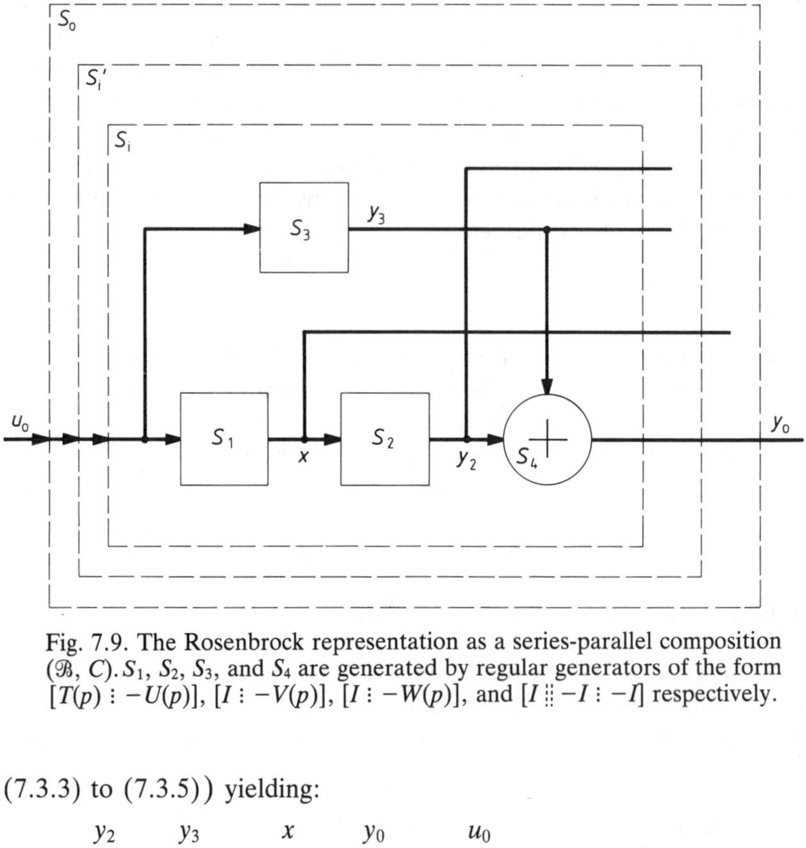

Fig. 7.9. The Rosenbrock representation as a series-parallel composition $(\mathcal{B}, C)$. S_1, S_2, S_3, and S_4 are generated by regular generators of the form $[T(p) \vdots -U(p)]$, $[I \vdots -V(p)]$, $[I \vdots -W(p)]$, and $[I \,\vdots\vdots\, -I \vdots -I]$ respectively.

(7.3.3) to (7.3.5)) yielding:

5

$$\begin{array}{c} \begin{array}{ccccc} y_2 & y_3 & x & y_0 & u_0 \end{array} \\ \left[\begin{array}{c:c:c:c:c} -I & -I & 0 & I & 0 \\ \hdashline 0 & I & 0 & 0 & -W(p) \\ \hdashline 0 & 0 & T(p) & 0 & -U(p) \\ \hdashline 0 & 0 & -V(p) & I & -W(p) \end{array}\right] \end{array} \triangleq [A_i(p) \vdots -B_i(p)].$$

We have $\det A_i(p) \propto \det T(p) \neq 0$. Hence S_i is always regular, and so is the composition $(\mathcal{B}, C)$ under consideration.

The reduced internal input-output relation S_i'

From (5) it is seen that $(\mathcal{B}, C)$ is always observable with respect to the components of (y_2, y_3). It is also concluded that the variables y_2, y_3 have no influence on controllability and stability aspects in this context. The original internal input-output relation S_i can therefore here be replaced by a "reduced" internal input-output relation S_i' obtained from S_i through elimination of the variables y_2, y_3 from the elements $(u_0, (y_2, y_3, x, y_0))$ of S_i (*cf.* section 7.1). Thus we shall here consider S_i' rather than S_i.

A regular generator for S_i' can be read directly from (5):

6 $$\begin{array}{c} \begin{array}{ccc} x & y_0 & u_0 \end{array} \\ \left[\begin{array}{c:c:c} T(p) & 0 & -U(p) \\ \hdashline -V(p) & I & -W(p) \end{array}\right] \end{array} \triangleq [A_i'(p) \; \vdots \; -B_i'(p)].$$

The regular generator (6) for S_i' is of the general form mentioned in note (7.2.24), (ii). To obtain a regular generator for the overall input-output relation $S_o \subset \mathcal{X}^l \times \mathcal{X}^m$ determined by $(\mathcal{B}, C)$, we thus have to transform (6) to a row equivalent upper triangular form. So let $M(p)$ be a GCRD of $T(p)$ and $V(p)$ and write

7 $$\begin{aligned} T(p) &= T_1(p)M(p), \\ V(p) &= V_1(p)M(p). \end{aligned}$$

There is then a unimodular polynomial matrix $P(p)$ of appropriate size and corresponding to a suitable sequence of elementary row operations so that premultiplication of (6) by $P(p)$ yields:

8 $$\underbrace{\left[\begin{array}{c:c} P_1(p) & A_2(p) \\ \hdashline P_3(p) & A_0(p) \end{array}\right]}_{P(p)} \begin{array}{c} \begin{array}{ccc} x & y_0 & u_0 \end{array} \\ \left[\begin{array}{c:c:c} T(p) & 0 & -U(p) \\ \hdashline -V(p) & I & -W(p) \end{array}\right] \end{array}$$

9 $$= \begin{array}{c} \begin{array}{ccc} x & y_0 & u_0 \end{array} \\ \left[\begin{array}{c:c:c} M(p) & A_2(p) & -P_1(p)U(p) - A_2(p)W(p) \\ \hdashline 0 & A_0(p) & -P_3(p)U(p) - A_0(p)W(p) \end{array}\right] \end{array}$$

implying that $A_2(p)$ and $A_0(p)$ are right coprime, and that

10 $$\begin{aligned} P_1(p)T(p) - A_2(p)V(p) &= M(p), \\ P_3(p)T(p) - A_0(p)V(p) &= 0, \end{aligned}$$

and

11 $$\det M(p) \det A_0(p) \propto \det T(p).$$

(9) is thus a regular generator for S_i' of upper triangular form, and a regular generator for the overall input-output relation $S_o \subset \mathscr{X}^l \times \mathscr{X}^m$ determined by $(\mathscr{B}, C)$ is accordingly obtained as

12 $$[A_0(p) \vdots -P_3(p)U(p) - A_0(p)W(p)] \triangleq [A_0(p) \vdots -B_0(p)].$$

The Rosenbrock representation. The system matrix and the modified system matrix

A series-parallel composition $(\mathscr{B}, C)$ having the above general structure will henceforth be called a "Rosenbrock representation" (R-representation for short), because this particular composition structure is the basic starting point for Rosenbrock's studies (*cf.* Rosenbrock, 1970, chapter 2). Consider then the matrix

13 $$\left[\begin{array}{c|c} T(p) & U(p) \\ \hline -V(p) & W(p) \end{array}\right]$$

formed from the generator (6) for S_i' in an obvious way (the essential thing being that the columns associated with y_0 are deleted). If p in (13) is everywhere replaced by the complex variable s, then we obtain a "system matrix" as defined by Rosenbrock for the R-representation $(\mathscr{B}, C)$ under consideration. The Rosenbrock system matrix (R-system matrix) so obtained thus reads

14 $$\left[\begin{array}{c|c} T(s) & U(s) \\ \hline -V(s) & W(s) \end{array}\right].$$

Referring to the term "system matrix" mentioned above, we shall call the complete generator (6) for S_i' a "modified Rosenbrock system matrix" (MR-system matrix) for $(\mathscr{B}, C)$. The essential feature of an MR-system matrix as given by (6) is the special structure of the columns associated with the output variable y_0.

Many characteristic features of an R-representation $(\mathscr{B}, C)$ of the form described above can now be derived from the fact that any MR-system matrix for $(\mathscr{B}, C)$ of the form (6) is row equivalent to a generator of the form (9).

15 **Note.** **(i)** In Rosenbrock's studies there is an additional requirement concerning the size of the r by r-matrix $T(s)$ in the R-system matrix as given

by (14). It is namely required that r fulfil the condition (*cf.* Rosenbrock, 1970, p. 50)

16
$$r \geqslant n,$$

where $n = \partial(\det T(s)) = \partial(\det T(p))$ is the order of the R-representation. We shall here generally allow r to take other values also—the condition (16) will be explicitly stated only if this condition is essential in the context.

(ii) An R-representation $(\mathcal{B}, C)$ of the form shown in Fig. 7.9 can be interpreted either as a composition of a family of regular differential input-output relations, or as a decomposition of a given regular differential input-output relation (S_o above).

Now an R-representation is not very general as a composition—the most general composition has the structure discussed in section 7.2. As a decomposition, the R-representation has too little structure to be really interesting—note that an MR-system matrix of the form

17
$$\begin{array}{c} \begin{array}{ccc} x & y_0 & u_0 \end{array} \\ \left[\begin{array}{c:c::c} T(p) & 0 & -U(p) \\ \hdashline -I & I & 0 \end{array}\right] \end{array}$$

represents a trivial decomposition of the regular differential input-output relation generated by $[T(p) \vdots -U(p)]$.

In order to get really interesting and applicable representations additional structural requirements must be added. This then leads to the powerful state-space representation (*cf.* section 7.9), which thus is a special form of the R-representation.

(iii) It is important to note that there is a major conceptual difference between the matrix (13) and the corresponding R-system matrix (14). The former is based on the general theory of differential operators valid for any possible set of initial conditions, whereas the latter is formally based on the Laplace transform of the system equations with zero initial conditions (*cf.* Rosenbrock, 1970, p. 46). This rather loose derivation of the R-system matrix has in the past caused considerable confusion regarding the true significance of this matrix (*cf.* for instance the discussion in Rosenbrock, 1974a, c).

(iv) There is yet another crucial point which should be emphasized in this context. The MR-system matrix (6) for an R-representation $(\mathcal{B}, C)$ can, owing to the simple and fixed structure of the columns associated with y_0, be unambiguously represented by the corresponding R-system matrix (14). But the generator (9), which is row equivalent to (6) and which directly gives a generator for the overall input-output relation S_o determined by $(\mathcal{B}, C)$, can clearly not be represented in this way—the columns associated with y_0 are significant and they cannot be simply deleted. Thus only a proper subset of the set of all regular generators row equivalent to (6) can be represented by an R-system matrix.

This deficiency leads to serious consequences with regard to the applicability of the R-system matrix as a description of an R-representation. There is, for instance, no apparent way of determining a generator for S_o above on the basis of R-system matrices only (theorem 2.1 in Rosenbrock, 1970, chapter 3, section 2, illustrates the situation).

(v) The transfer matrix $\mathcal{G}_o(p)$ determined by S_o above is, according to (12), given by

18 $$\mathcal{G}_o(p) = A_0(p)^{-1}B_0(p) = A_0(p)^{-1}(P_3(p)U(p) + A_0(p)W(p)) = A_0(p)^{-1}P_3(p)U(p) + W(p).$$

From (10) it follows that $A_0(p)^{-1}P_3(p) = V(p)T(p)^{-1}$. Thus (18) can also be written as

19 $$\mathcal{G}_o(p) = V(p)T(p)^{-1}U(p) + W(p),$$

which expression can also be read directly from Fig. 7.9 (*cf.* Rosenbrock, 1970, chapter 2, section 2). Note that generally nothing can be concluded concerning the properness of $\mathcal{G}_o(p)$ on the basis of $W(p)$ only.

(vi) An R-representation ($\mathcal{B}$, C) as presented above is observable in the sense of section 7.2 if and only if $M(p)$ in (7) is unimodular, i.e. if and only if $T(p)$ and $V(p)$ in (1), (2) are right coprime. Now it is seen that a not unimodular $M(p)$ possibly appearing can be cancelled from the "output part" $V(p)T(p)^{-1} = V_1(p)T_1(p)^{-1}$ of $\mathcal{G}_o(p)$ as given by (19). In view of this Rosenbrock (Rosenbrock, 1970, chapter 2, section 5) calls the roots of $\det M(p)$ the "output-decoupling zeros" (o.d. zeros) of the R-representation under consideration. The set of roots of $\det M(p)$ is in Rosenbrock, 1970, denoted by $\{\gamma_i\}$ and multiple roots of $\det M(p)$ possibly appearing are appropriately repeated as members of this set. It turns out, however, that this interpretation leads to certain set theoretical difficulties. These difficulties can easily be avoided by letting $\{\gamma_i\}$ be represented by the corresponding generic polynomial $\det M(p)$. So, let $\gamma(p)$ denote the monic polynomial

20 $$\gamma(p) \propto \det M(p).$$

We shall say that $\gamma(p)$ represents $\{\gamma_i\}$. This notational convention will also be followed in the sequel.

To summarize:

21 Our R-representation ($\mathcal{B}$, C) is observable if and only if $T(p)$ and $V(p)$ are right coprime, i.e. if and only if $\gamma(p) = 1$ (there are no o.d. zeros).

Note that if the roots of det $T(p)$ are distinct, then they represent the "modes" of S_i' as generated by (6) (*cf.* Zadeh and Desoer, 1963, chapter 11), and further, the roots of $\gamma(p) \propto \det M(p)$ represent those modes of S_i' that are "not coupled" to the overall output y_0.

We shall discuss other kinds of zeros in a moment.

(vii) Next let us consider the controllability in the sense of section 6.4 of various input-output relations determined by an R-representation $(\mathcal{B}, C)$ of the form described above. To begin with it is noted that S_3 as generated by $[I \vdots -W(p)]$ (*cf.* Fig. 7.9) does not affect the controllability properties of S_i' and S_o—these properties depend solely on the series composition of S_1 and S_2. This means that the results (7.6.23) and (7.6.24) obtained in section 7.6 for a simple series composition with $A_2(p) = I$ are directly applicable in the present case. In terms of the present notations we thus have the following truths valid for *any* $W(p)$:

22 S_i' generated by (6) is controllable if and only if S_1 given by (1) is controllable, i.e. if and only if $T(p)$ and $U(p)$ are left coprime.

23 S_o generated by (12) is controllable if and only if $T_1(p)$ (*cf.* (7)) and $U(p)$ are left coprime.

A result of some use in the sequel and concerning the transfer matrix (19) can now be derived. Suppose that in the MR-system matrix (6) $T(p)$ and $V(p)$ are right coprime, and $T(p)$ and $U(p)$ left coprime. $M(p)$ in the generator (9) can then be chosen equal to I. From (23) it follows with $T(p) = T_1(p)$ (*cf.* (7)) that S_o is controllable, i.e. $A_0(p)$ and $B_0(p)$ as given by (12) are left coprime. Further (*cf.* (11))

24 $$\det T(p) \propto \det A_0(p).$$

This result clearly also implies the following more general result. Let (6) be an arbitrary MR-system matrix and consider the corresponding transfer matrix (*cf.* (18) and (19))

25 $$\mathcal{G}_o(p) = A_0(p)^{-1}B_0(p) = V(p)T(p)^{-1}U(p) + W(p).$$

Suppose that common divisors are cancelled from (25) yielding

$$\mathcal{G}_o(p) = A_{01}(p)^{-1}B_{01}(p) = V_1(p)T_1(p)^{-1}U_1(p) + W(p)$$

with $A_{01}(p)$ and $B_{01}(p)$ left coprime, $T_1(p)$ and $V_1(p)$ right coprime, and $T_1(p)$ and $U_1(p)$ left coprime. Then it follows that

26 $$\det T_1(p) \propto \det A_{01}(p).$$

(viii) In order to express some of the results obtained in the previous item in terms of decoupling zeros, and to derive the relationships between various sets of zeros, we proceed in the following way using the notations introduced above.

Consider a general MR-system matrix of the form (6), *cf.* also (7). Let $L(p)$ be a GCLD of $T(p)$ and $U(p)$ and write

27
$$T(p) = L(p)T_2(p),$$
$$U(p) = L(p)U_2(p).$$

Correspondingly, let $L_1(p)$ be a GCLD of $T_1(p)$ and $U(p)$ and write

28
$$T_1(p) = L_1(p)T_{11}(p),$$
$$U(p) = L_1(p)U_1(p).$$

Now we have $T(p) = T_1(p)M(p)$ (*cf.* (7)), and the first equality (28) gives

29
$$T(p) = L_1(p)T_{11}(p)M(p).$$

Next let $L_2(p)$ be a GCLD of $T_{11}(p)M(p)$ and $U_1(p)$. From (28) and (29) it then follows that $L_1(p)L_2(p)$ is a GCLD of $T(p)$ and $U(p)$. Comparing with (27) we see that $L_2(p)$ can be chosen so that

30
$$L(p) = L_1(p)L_2(p).$$

The same reasoning can be applied to the right divisors of $T(p)$ and $V(p)$. We thus start with (7). Next we determine a GCRD $M_1(p)$ of $T_2(p)$ and $V(p)$ and write

31
$$T_2(p) = T_{21}(p)M_1(p),$$
$$V(p) = V_2(p)M_1(p).$$

Here it holds that $T(p) = L(p)T_2(p)$ (*cf.* (27)), and the first equality (31) yields

32
$$T(p) = L(p)T_{21}(p)M_1(p).$$

Let $M_2(p)$ be a GCRD of $L(p)T_{21}(p)$ and $V_2(p)$. From (31) and (32) it follows that $M_2(p)M_1(p)$ is a GCRD of $T(p)$ and $V(p)$. Comparing with (7) we see that $M_2(p)$ can be chosen so that

33
$$M(p) = M_2(p)M_1(p).$$

The above results suggest the following self-explanatory factorizations of the MR-system matrix (6):

$$\begin{array}{c} \\ \end{array}\begin{array}{ccc} x & y_0 & u_0 \end{array}$$

$$\left[\begin{array}{c:c:c} T(p) & 0 & -U(p) \\ \hdashline -V(p) & I & -W(p) \end{array}\right] \tag{34}$$

$$= \left[\begin{array}{c:c:c} L_1(p)T_{11}(p)M_2(p)M_1(p) & 0 & -L_1(p)U_1(p) \\ \hdashline -V_1(p)M_2(p)M_1(p) & I & -W(p) \end{array}\right] \tag{35}$$

$$= \left[\begin{array}{c:c:c} L_1(p)L_2(p)T_{21}(p)M_1(p) & 0 & -L_1(p)L_2(p)U_2(p) \\ \hdashline -V_2(p)M_1(p) & I & -W(p) \end{array}\right]. \tag{36}$$

Finally let $L_0(p)$ be a GCLD of $A_0(p)$ and $B_0(p)$ (*cf.* (12)) and write

$$\begin{aligned} A_0(p) &= L_0(p)A_{01}(p) \quad \text{and} \\ B_0(p) &= L_0(p)B_{01}(p). \end{aligned} \tag{37}$$

The resulting transfer matrix $\mathscr{G}_o(p)$ is then (*cf.* (18) and (19)):

$$\mathscr{G}_o(p) = A_{01}(p)^{-1}B_{01}(p) \tag{38}$$

$$= V(p)T(p)^{-1}U(p) + W(p) \tag{39}$$

$$= V_1(p)T_{11}(p)^{-1}U_1(p) + W(p) \tag{40}$$

$$= V_2(p)T_{21}(p)^{-1}U_2(p) + W(p). \tag{41}$$

(40) is obtained from (39) so that $M(p) = M_2(p)M_1(p)$ is first cancelled from the "output part" $V(p)T(p)^{-1}$ of $\mathscr{G}_o(p)$ (*cf.* item (vi) above), and then $L_1(p)$ from the remaining part, whereas (41) is obtained from (39) so that $L(p) = L_1(p)L_2(p)$ is first cancelled from the "input part" $T(p)^{-1}U(p)$ of $\mathscr{G}_o(p)$, and then $M_1(p)$ from the remaining part.

It follows from (11) and (37) and from the various expressions for $T(p)$ contained in (34), (35), (36), that

$$\begin{aligned} \det T(p) &= \det L_1(p)T_{11}(p)M_2(p)M_1(p) \\ &= \det L_1(p)L_2(p)T_{21}(p)M_1(p) \\ &\propto \det M_2(p)M_1(p)L_0(p)A_{01}(p). \end{aligned} \tag{42}$$

Moreover, because of the coprimeness of the matrices appearing in (38), (40), and (41) above, it follows from the results obtained in the previous item (vii) that (note that all the matrices involved are square and nonsingular):

$$\det T_{11}(p) \propto \det T_{21}(p) \propto \det A_{01}(p). \tag{43}$$

Combining (42) and (43) we obtain:

44 $$\det L_1(p) \propto \det L_0(p),$$

45 $$\det L_2(p) \propto \det M_2(p).$$

All the results concerning various sets of zeros presented in Rosenbrock, 1970, chapter 2, section 5, now follow in a straightforward way from the above relationships. We shall give a brief presentation of these results.

Recall first that the roots of $\det M(p) = \det M_2(p)M_1(p)$ were called output-decoupling zeros because $M(p)$ could be cancelled from the "output part" $V(p)T(p)^{-1}$ of $\mathscr{G}_o(p)$. These o.d. zeros were denoted by $\{\gamma_i\}$ and represented by the monic polynomial (*cf.* (20))

46 $$\gamma(p) \propto \det M_1(p) \det M_2(p).$$

In the same way $L(p) = L_1(p)L_2(p)$ can be cancelled from the "input part" $T(p)^{-1}V(p)$ of $\mathscr{G}_o(p)$. The roots of $\det L(p)$ are therefore termed "input-decoupling zeros" (i.d. zeros). These i.d. zeros are denoted by $\{\beta_i\}$, and they are here represented by the monic polynomial

47 $$\beta(p) \propto \det L_1(p) \det L_2(p).$$

Because of (45), $\det M_2(p) \propto \det L_2(p)$ are common divisors of $\gamma(p)$ and $\beta(p)$. The roots of $\det M_2(p) \propto \det L_2(p)$ are called "input-output-decoupling zeros" (i.o.d. zeros). They are denoted by $\{\delta_i\}$ and they are here represented by the monic polynomial

48 $$\delta(p) \propto \det M_2(p) \propto \det L_2(p).$$

Other zeros used in Rosenbrock, 1970, are $\{\theta_i\}$ represented by the monic polynomial

49 $$\theta(p) = \gamma(p)/\delta(p) \propto \det M_1(p),$$

$\{\eta_i\}$ represented by the monic polynomial

50 $$\eta(p) \propto \det T(p),$$

and $\{\alpha_i\}$ represented by the monic polynomial (*cf.* (43))

51 $$\alpha(p) \propto \det T_{11}(p) \propto \det T_{21}(p) \propto \det A_{01}(p).$$

The members of $\{\alpha_i\}$ are, for obvious reasons, called the "poles" of $\mathscr{G}_o(p)$.

In terms of the above notations the contents of theorem 5.1 in Rosenbrock, 1970, chapter 2, section 5 now read:

52 $\theta(p)\,|\,\gamma(p)$ (i.e. $\theta(p)$ divides $\gamma(p)$), corresponds to $\{\gamma_i\} \supseteq \{\theta_i\}$,

53 $\delta(p)\,|\,\beta(p)$, corresponds to $\{\beta_i\} \supseteq \{\delta_i\}$,

54 $\beta(p)\gamma(p)/\delta(p)\,|\,\eta(p)$, corresponds to $\{\eta_i\} \supseteq \{\beta_i, \gamma_i\} - \{\delta_i\}$.

These relationships are evident consequences of the above results.
The results (22) and (23) in terms of the above polynomials now read:

55 S_i' generated by (6) is controllable if and only if $\beta(p) = 1$ (there are no i.d. zeros).

56 S_o generated by (12) is controllable if and only if $\delta(p) = \beta(p)$ (all the i.d. zeros are i.o.d. zeros).

It was pointed out in item (vi) that if the roots of $T(p)$ are distinct, then they represent the modes of S_i'. The roots of $\gamma(p) \propto \det M(p)$, i.e. the o.d. zeros, then represent those modes of S_i' that are not coupled to the overall output y_0. In the same way it is found that the roots of $\beta(p)$, i.e. the i.d. zeros, in this case represent those modes of S_i' that are not coupled to the overall input u_0, and the roots of $\delta(p)$, i.e. the i.o.d. zeros, those modes of S_i' that are coupled neither to the overall input u_0, nor to the overall output y_0 (for more details, see Zadeh and Desoer, 1963, chapter 11). Correspondingly, the roots of $\beta(p)/\delta(p) \propto \det L_0(p) \propto \det L_1(p)$ (*cf.* (44), (47), (48)) would represent those modes of S_o that are not coupled to the input u_0.

Many other kinds of zeros have been discussed in the literature. We shall not dwell on them in this context.

(ix) In section 6.2 we discussed at length the relationships between differential input-output relations and their generators. The results were summarized in theorem (6.2.1) and theorem (6.2.2). In the same way we may look for corresponding relationships between the various input-output relations determined by R-representations and the corresponding MR-system matrices. Here the situation is, however, more complicated than in section 6.2, because we have, for every R-representation, many interesting input-output relations.

In this item we shall present some preliminary comments on this matter. A more detailed presentation will be given in section 7.10.

Consider an R-representation $(\mathcal{B}, C)$ of the form depicted in Fig. 7.9, and let the notations introduced above be valid, except for the generator

(9) for S_1', which is renamed as (*cf.* (12) and (37))

57
$$\left[\begin{array}{c:c:c} M(p) & A_2(p) & -B_1(p) \\ \hdashline 0 & A_0(p) & -B_0(p) \end{array}\right]$$

58
$$= \left[\begin{array}{c:c:c} M(p) & A_2(p) & -B_1(p) \\ \hdashline 0 & L_0(p)A_{01}(p) & -L_0(p)B_{01}(p) \end{array}\right],$$

where $A_{01}(p)$ and $B_{01}(p)$ are left coprime. The corresponding MR-system matrix (6) can, according to (35), be written as

59
$$\left[\begin{array}{c:c:c} L_1(p)T_{11}(p)M(p) & 0 & -L_1(p)U_1(p) \\ \hdashline -V_1(p)M(p) & I & -W(p) \end{array}\right]$$

with (*cf.* (43), (44))

60
$$\det T_{11}(p) \propto \det A_{01}(p),$$
$$\det L_1(p) \propto \det L_0(p).$$

The construction of (59) from the given MR-system matrix (6) was explained in detail in the previous item (*cf.* (27), . . . , (35)).

Now suppose that a new R-representation $(\mathcal{B}_1, C)$ is constructed from $(\mathcal{B}, C)$ as follows. The MR-system matrix for $(\mathcal{B}_1, C)$ is formed from (6) by premultiplication by a suitable square polynomial matrix $R(p)$, $\det R(p) \neq 0$, of appropriate size according to

61
$$\underbrace{\left[\begin{array}{c:c} R_1(p) & 0 \\ \hdashline R_3(p) & I \end{array}\right]}_{R(p)} \left[\begin{array}{c:c:c} T(p) & 0 & -U(p) \\ \hdashline -V(p) & I & -W(p) \end{array}\right]$$

62
$$= \left[\begin{array}{c:c:c} R_1(p)T(p) & 0 & -R_1(p)U(p) \\ \hdashline R_3(p)T(p) - V(p) & I & -R_3(p)U(p) - W(p) \end{array}\right].$$

The order of $(\mathcal{B}_1, C)$ will thus be equal to the order of $(\mathcal{B}, C)$ + $\partial(\det R_1(p))$.

Note that $R(p)$ necessarily has the structure shown in (61).

It is easily seen that a row equivalent upper triangular form of (62) is obtained by premultiplication of (57) or (58) by a corresponding square polynomial matrix $S(p)$, $\det S(p) \neq 0$, according to

63 $$\underbrace{\left[\begin{array}{c:c} S_1(p) & S_2(p) \\ \hdashline 0 & S_4(p) \end{array}\right]}_{S(p)} \left[\begin{array}{c:c:c} M(p) & A_2(p) & -B_1(p) \\ \hdashline 0 & A_0(p) & -B_0(p) \end{array}\right]$$

64 $$= \left[\begin{array}{c:c:c} S_1(p)M(p) & A_{21}(p) & -B_{11}(p) \\ \hdashline 0 & S_4(p)L_0(p)A_{01}(p) & -S_4(p)L_0(p)B_{01}(p) \end{array}\right],$$

where $A_{21}(p)$ and $B_{11}(p)$ are determined by the multiplication indicated in (63). We have, of course,

65 $$\det R_1(p) \propto \det S_1(p) \det S_4(p).$$

Further, with S_o denoting the overall input-output relation determined by $(\mathcal{B}, C)$ and generated by $[A_0(p) \vdots -B_0(p)]$, and with S_{o1} denoting the overall input-output relation determined by $(\mathcal{B}_1, C)$ and generated by $[S_4(p)A_0(p) \vdots -S_4(p)B_0(p)]$, it generally holds that $S_o \subset S_{o1}$. More specifically, $S_o = S_{o1}$ if and only if $S_4(p)$ is unimodular.

This result can be read directly from (63), (64). Reading the corresponding result from (61), (62) is not that easy. To be able to do this we would need more information about the relationship between $R(p)$ in (61) and $S(p)$ in (63). We shall proceed in the following way.

The same procedure according to which (59) was obtained from (6) is applied in a obvious way to (62) yielding an expression for (62) of the form

66 $$\left[\begin{array}{c:c:c} L_{11}(p)T_{111}(p)M_{11}(p) & 0 & -L_{11}(p)U_{11}(p) \\ \hdashline -V_{11}(p)M_{11}(p) & I & -W_1(p) \end{array}\right],$$

where the interpretation of the different quantities should be clear on the basis of the construction employed. Comparing (64), (66) with (58), (59) and using (60) we immediately arrive at the result

67 $$\det T_{111}(p) \propto \det A_{01}(p),$$
$$\det L_{11}(p) \propto \det S_4(p) \det L_0(p) \propto \det S_4(p) \det L_1(p).$$

It can also be assumed that

68 $$M_{11}(p) = S_1(p)M(p).$$

Now it is concluded that the following statements are equivalent:

$$S_o = S_{o1},$$

$$S_4(p) \text{ is unimodular,}$$

69 $$\det L_{11}(p) \propto \det L_1(p),$$

$$\det S_1(p) \propto \det R_1(p).$$

Of course, if $R_1(p)$ is unimodular, then so are $S_1(p)$ and $S_4(p)$, and we have $S_o = S_{o1}$. But we can also have $S_o = S_{o1}$ even if $R_1(p)$ is not unimodular. This happens, according to the above results, whenever $S_1(p)M(p)$—a GCRD of $R_1(p)T(p)$ and $R_3(p)T(p) - V(p)$ in (62)—is such that $\det S_1(p) \propto \det R_1(p)$.

It can further be concluded that S_o is a proper subset of S_{o1} in particular if $S_1(p)$ is unimodular whereas $R_1(p)$ is not. This happens in particular if $R_1(p)T(p)$ and $V(p)$ are right coprime with $R_1(p)$ not unimodular. □

7.9 The state-space representation

Let $(\mathcal{B}, C)$ be an R-representation of the form discussed in the previous paragraph and shown in Fig. 7.9. $(\mathcal{B}, C)$ is said to be a "state-space representation" if the corresponding MR-system matrix (7.8.6) is of the form (2) below.

1 $$\begin{array}{c} \begin{array}{ccc} x & y_0 & u_0 \end{array} \\ \left[\begin{array}{c:c:c} T(p) & 0 & -U(p) \\ \hdashline -V(p) & I & -W(p) \end{array}\right] \end{array}$$

2 $$= \begin{array}{c} \begin{array}{ccc} x & y_0 & u_0 \end{array} \\ \left[\begin{array}{c:c:c} pI - F & 0 & -G \\ \hdashline -H & I & -K(p) \end{array}\right] \end{array},$$

where F, G, and H are constant matrices of appropriate sizes.

We shall, in connection with a state-space representation $(\mathcal{B}, C)$ as described above, use the following terminology (the notations introduced in the previous paragraph are used below without any further explanation).

An MR-system matrix of the form (2) representing a state-space representation is said to be a "state-space MR-system matrix".

The intermediate variable x is called the *state* (signal). The input-output relation S_1 generated by $[pI - F \,\vdots\, -G]$ is called the "input-state" relation, and the input-output relation S_2 (a mapping which can be identified with H) is called the "state-output" mapping.

Because the state-space representation is a special form of the R-representation, it is clear that the results derived in the previous paragraph for the R-representation also hold for the state-space representation. The additional structure does, however, also introduce some additional features in the state-space representation. We shall collect below a number of relevant results holding for a state-space representation $(\mathscr{B}, C)$ of the above kind. It is assumed that the reader is familiar with standard textbooks on the subject.

3 **Note.** **(i)** $T(p) = pI - F$ in (1), (2) is an r by r polynomial matrix operator with $r = n =$ the order of the state-space representation $=$ the dimension of the corresponding state-space (here $\mathbf{C}^n$, *cf.* Zadeh and Desoer, 1963, chapter 4).

(ii) The left-hand side of (2) is clearly of CRP-form (canonical row proper form, *cf.* appendix A2). It follows that the input-state relation S_1 (*cf.* Fig. 7.9) generated by $[pI - F \vdots -G]$ is always strictly proper, and also that the reduced internal input-output relation S_i' generated by (2), as well as the overall input-output relation S_o determined by $(\mathscr{B}, C)$, are proper (strictly proper) if and only if $K(p)$ is constant (zero). Further, $K(p)$ in (2) is uniquely determined by S_o (*cf.* section 6.3).

(iii) A state-space representation $(\mathscr{B}, C)$ as given above is, according to the statement (7.8.21), observable in the sense of section 7.2 if and only if $pI - F$ and H are right coprime ($\gamma(p) = 1$, i.e. there are no o.d. zeros). For a state-space representation our observability concept thus coincides with the ordinary observability concept (*cf.* Rosenbrock, 1970, chapter 5, section 3, theorem 3.1). The coprimeness of $pI - F$ and H can be tested in many ways (*cf.* Rosenbrock, 1970, chapter 2, section 6, theorem 6.2; appendix A2).

(iv) Concerning the controllability aspects in the sense of section 6.4 of a state-space representation $(\mathscr{B}, C)$ of the above kind we have the results stated in note (7.8.15), (vii). Thus the reduced internal input-output relation S_i' generated by (2) is controllable if and only if the input-state relation S_1 generated by $[pI - F \vdots -G]$ is controllable, i.e. if and only if $pI - F$ and G are left coprime ($\beta(p) = 1$, i.e. there are no i.d. zeros). Thus controllability in the sense of section 6.4 of the input-state relation S_1 coincides with the usual controllability concept for a state-space representation (*cf.* Rosenbrock, 1970, chapter 5, section 2, theorem 2.1). The coprimeness of $pI - F$ and G can again be tested in many ways (*cf.* Rosenbrock, 1970, chapter 2, section 6, theorem 6.2; appendix A2).

The controllability of the overall input-output relation S_o determined by a state-space representation is generally not considered as a separate concept in the classical state-space theory (*cf.* though the "output controllability" concept discussed in Ogata, 1967, section 7.1). □

7.10 The Rosenbrock representation and the state-space representation as decompositions of regular differential input-output relations. Equivalence relations

In this paragraph we shall consider a number of relevant properties of the R-representation and the state-space representation as decompositions. We shall follow the notational conventions introduced in sections 7.8 and 7.9.

Two basic facts

Let us start with two basic facts which are direct consequences of the results obtained in particular in section 7.8.

Let

1 $$S_o \subset \mathscr{X}^l \times \mathscr{X}^m$$

be a regular differential input-output relation, and let $(\mathscr{B}, C)$ be an R-representation as shown in Fig. 7.9 with the corresponding MR-system matrix given by (7.8.6) as

2 $$\begin{array}{ccc} x & y_0 & u_0 \end{array}$$
$$\left[\begin{array}{c:c:c} T(p) & 0 & -U(p) \\ \hdashline -V(p) & I & -W(p) \end{array}\right].$$

Note here that any polynomial matrix of the form (2) with $T(p)$ square and $\det T(p) \neq 0$ can be interpreted as an MR-system matrix determining a unique corresponding R-representation as shown in Fig. 7.9.

Let us then rewrite the equality (7.8.8) and (7.8.9) representing the transformation of (2) to a row equivalent upper triangular form as follows:

3 $$\underbrace{\left[\begin{array}{c:c} P_1(p) & A_2(p) \\ \hdashline P_3(p) & A_0(p) \end{array}\right]}_{P(p)} \overset{\begin{array}{ccc} x & y_0 & u_0 \end{array}}{\left[\begin{array}{c:c:c} T(p) & 0 & -U(p) \\ \hdashline -V(p) & I & -W(p) \end{array}\right]}$$

4 $$= \overset{\begin{array}{ccc} x & y_0 & u_0 \end{array}}{\left[\begin{array}{c:c:c} M(p) & A_2(p) & -B_1(p) \\ \hdashline 0 & A_0(p) & -B_0(p) \end{array}\right]}.$$

The matrix (4) is uniquely determined by (2) if it is made to be of CUT-form.

Note that $A_2(p)$ and $A_0(p)$ are right coprime because $P(p)$ is unimodular (*cf.* appendix A2). Recall also that (2) and (4) are regular generators for the reduced internal input-output relation S_i' determined by our R-representation $(\mathcal{B}, C)$.

The first of the two basic facts referred to above now reads: The R-representation $(\mathcal{B}, C)$ mentioned above is a decomposition of S_o as given by (1) if and only if $[A_0(p) \vdots -B_0(p)]$ as given by (4) is a generator for S_o.

Conversely, we may take any polynomial matrix of the form (4):

5 $$\left[\begin{array}{c:c:c} M(p) & A_2(p) & -B_1(p) \\ \hdashline 0 & A_0(p) & -B_0(p) \end{array}\right]$$

with $M(p)$ and $A_0(p)$ square, $\det M(p) \neq 0$, $\det A_0(p) \neq 0$, and with $A_2(p)$ and $A_0(p)$ right coprime. Then there is a unimodular polynomial matrix $Q(p)$ of appropriate size such that

6 $$\underbrace{\left[\begin{array}{c:c} T_1(p) & Q_2(p) \\ \hdashline -V_1(p) & Q_4(p) \end{array}\right]}_{Q(p)} \left[\begin{array}{c:c:c} M(p) & A_2(p) & -B_1(p) \\ \hdashline 0 & A_0(p) & -B_0(p) \end{array}\right]$$

7 $$= \left[\begin{array}{c:c:c} T_1(p)M(p) & 0 & -T_1(p)B_1(p) - Q_2(p)B_0(p) \\ \hdashline -V_1(p)M(p) & I & V_1(p)B_1(p) - Q_4(p)B_0(p) \end{array}\right].$$

The matrix (7) is again uniquely determined by (5) if the left-hand part (7) is made to be of CLT-form (canonical lower triangular form with respect to row equivalence).

Evidently, (7) qualifies as an MR-system matrix and it thus determines a unique corresponding R-representation $(\mathcal{B}, C)$.

Our second basic fact then simply reads: The R-representation $(\mathcal{B}, C)$ so obtained is a decomposition of S_o as given by (1) if and only if $[A_0(p) \vdots -B_0(p)]$ in (5) is a generator for S_o.

Input-output equivalence on the set $\mathcal{MR}$ of all MR-system matrices

The above considerations suggest the following formalization (*cf.* section 6.2).

As in section 6.2, let $\mathcal{R}$ denote the set of all regular differential input-output relations, and let, in addition, $\mathcal{MR}$ denote the set of all polynomial matrices which qualify as MR-system matrices, i.e.

8
$$\mathcal{MR} \triangleq \left\{ \left[\begin{array}{c:c:c} T(p) & 0 & -U(p) \\ \hdashline -V(p) & I & -W(p) \end{array} \right] \middle| T(p), V(p), \right.$$
$$\left. U(p), W(p) \text{ are polynomial matrices with } T(p) \text{ square and } \det T(p) \neq 0 \right\}.$$

Any $M \in \mathcal{MR}$ then determines a unique corresponding R-representation which in turn is a decomposition of a unique $S_o \in \mathcal{R}$. The assignment

9
$$M \mapsto S_o$$

thus defines a mapping

10
$$f : \mathcal{MR} \to \mathcal{R},$$

and this mapping is clearly surjective with respect to $\mathcal{R}$.

Two MR-system matrices M_1 and $M_2 \in \mathcal{MR}$ are now said to be *input-output equivalent* if the R-representations determined by them are input-output equivalent in the sense of section 7.2, i.e. if $f(M_1) = f(M_2)$.

Given $M_1, M_2 \in \mathcal{MR}$ it is an easy matter to test whether they are input-output equivalent or not.

11 **Remark.** An R-representation may arise as the result of forming a mathematical model of a real system, and such an R-representation may, of course, not be observable. However, if we are faced with the task of finding an R-representation which is a decomposition of a given regular differential input-output relation, then there is generally no reason to choose an R-representation that is not observable. Therefore, we could here have restricted the set (8) to comprise only elements with $T(p)$ and $V(p)$ right coprime, corresponding to observable R-representations only. We shall, however, not make this restriction in this context. □

Complete invariant for input-output equivalence on $\mathcal{MR}$

Input-output equivalence as defined above is of course an equivalence relation on $\mathcal{MR}$; it is, in fact, the natural equivalence determined by the mapping f as given by (9), (10). According to the terminology in MacLane and Birkhoff, 1967, chapter VIII, f can thus be called a *complete invariant* for input-output equivalence on $\mathcal{MR}$.

Canonical forms for input-output equivalence on $\mathcal{MR}$

A subset $\mathcal{MR}^* \subset \mathcal{MR}$ qualifying as a set of canonical forms for input-output equivalence on $\mathcal{MR}$ can be constructed in several ways (*cf.* sections 6.2 and 6.6). A simple way would be as follows.

Pick an arbitrary $M \in \mathcal{MR}$, and let this M have the form (2). Transform M to upper triangular form as shown in (3), (4) so that $[A_0(p) \vdots -B_0(p)]$ in (4) is a canonical form for row equivalence as presented in section 6.2, for instance a CUT-form. Then form the matrix (*cf.* (7.8.17))

12 $$M^* \triangleq \left[\begin{array}{c:c::c} A_0(p) & 0 & -B_0(p) \\ \hdashline -I & I & 0 \end{array}\right].$$

This matrix qualifies as an MR-system matrix representing a trivial observable R-representation, and it is clearly input-output equivalent to M. Moreover, M^* is uniquely determined by the input-output equivalence class in $\mathcal{MR}$ determined by M.

It follows that the set of all matrices of the form (12) which are contained in $\mathcal{MR}$ form a set $\mathcal{MR}^*$ of canonical forms for input-output equivalence on $\mathcal{MR}$.

The assignment

13 $$M \mapsto M^*$$

defines a mapping

14 $$g : \mathcal{MR} \to \mathcal{MR}^*,$$

which is surjective with respect to $\mathcal{MR}^*$, and g is then also qualified to be called a complete invariant for input-output equivalence on $\mathcal{MR}$.

The situation now corresponds to the situation depicted in Fig. 6.1 with $\mathcal{M}$ replaced by $\mathcal{MR}$ and $\mathcal{M}^*$ by $\mathcal{MR}^*$.

15 **Remark.** The results presented above strictly speaking contain all the information needed in this context concerning the properties of a general R-representation, and the equivalence relation introduced on the set $\mathcal{MR}$ of all MR-system matrices is a very natural one. More specific kinds of equivalences are needed only for the more specific state-space representation. However, a great deal has been written in the literature in particular about another equivalence relation on $\mathcal{MR}$ called "strict system equivalence" as presented in Rosenbrock, 1970, chapter 2, section 3. Because it was felt important to give this equivalence its proper place within the present framework, we shall continue our considerations concerning equivalence relations on $\mathcal{MR}$ and on certain subsets of $\mathcal{MR}$. □

Equivalent formulations of input-output equivalence on $\mathcal{MR}$

The input-output equivalence on $\mathcal{MR}$ introduced above can be formulated in many equivalent ways. Let us collect a number of them in the following theorem.

16 **Theorem.** *Let* $(\mathcal{B}_1, C_1)$ *and* $(\mathcal{B}_2, C_2)$ *be two R-representations as described in section 7.8 regarded as decompositions of the regular differential input-output relations* $S_{o1} \subset \mathcal{X}^{l_1} \times \mathcal{X}^{m_1}$ and $S_{o2} \subset \mathcal{X}^{l_2} \times \mathcal{X}^{m_2}$ *respectively, and let* S'_{i1} *and* S'_{i2} *be the corresponding reduced internal input-output relations determined by* $(\mathcal{B}_1, C_1)$ *and* $(\mathcal{B}_2, C_2)$ *and generated by the corresponding MR-system matrices*

17
$$\begin{array}{c} \quad x_1 \qquad\qquad\quad y_{01} \qquad u_{01} \\ \left[\begin{array}{c:c:c} T_{11}(p)M_1(p) & 0 & -U_1(p) \\ \hdashline -V_{11}(p)M_1(p) & I & -W_1(p) \end{array}\right], \end{array}$$

and

18
$$\begin{array}{c} \quad x_2 \qquad\qquad\quad y_{02} \qquad u_{02} \\ \left[\begin{array}{c:c:c} T_{21}(p)M_2(p) & 0 & -U_2(p) \\ \hdashline -V_{21}(p)M_2(p) & I & -W_2(p) \end{array}\right] \end{array}$$

respectively, where $M_1(p)$ *is a GCRD of* $T_{11}(p)M_1(p)$ *and* $V_{11}(p)M_1(p)$, *and* $M_2(p)$ *a GCRD of* $T_{21}(p)M_2(p)$ *and* $V_{21}(p)M_2(p)$. *There are then also row equivalent upper triangular forms* (20) *and* (22) *of* (17) *and* (18) *according to* (*cf.* (3), (4))

19
$$\underbrace{\left[\begin{array}{c:c} P_{11}(p) & A_{21}(p) \\ \hdashline P_{31}(p) & A_{01}(p) \end{array}\right]}_{P_1(p)} \begin{array}{c} \quad x_1 \qquad\qquad\quad y_{01} \qquad u_{01} \\ \left[\begin{array}{c:c:c} T_{11}(p)M_1(p) & 0 & -U_1(p) \\ \hdashline -V_{11}(p)M_1(p) & I & -W_1(p) \end{array}\right] \end{array}$$

20
$$= \begin{array}{c} x_1 \qquad\quad y_{01} \qquad\qquad u_{01} \\ \left[\begin{array}{c:c:c} M_1(p) & A_{21}(p) & -B_{11}(p) \\ \hdashline 0 & A_{01}(p) & -B_{01}(p) \end{array}\right] \end{array}$$

and

21
$$\underbrace{\left[\begin{array}{c:c} P_{12}(p) & A_{22}(p) \\ \hdashline P_{32}(p) & A_{02}(p) \end{array}\right]}_{P_2(p)} \begin{array}{c} \quad x_2 \qquad\qquad\quad y_{02} \qquad u_{02} \\ \left[\begin{array}{c:c:c} T_{21}(p)M_2(p) & 0 & -U_2(p) \\ \hdashline -V_{21}(p)M_2(p) & I & -W_2(p) \end{array}\right] \end{array}$$

22
$$
\begin{array}{c}
\begin{array}{ccc} x_2 & y_{02} & u_{02} \end{array}\\
= \left[\begin{array}{c:c:c} M_2(p) & A_{22}(p) & -B_{12}(p) \\ \hdashline 0 & A_{02}(p) & -B_{02}(p) \end{array}\right]
\end{array}
$$

with $P_1(p)$ and $P_2(p)$ unimodular, and with $A_{21}(p)$ and $A_{01}(p)$ right coprime and likewise with $A_{22}(p)$ and $A_{02}(p)$ right coprime. We may assume that $[A_{01}(p) \vdots -B_{01}(p)]$ and $[A_{02}(p) \vdots -B_{02}(p)]$ in (20) *and* (22) *are of CUT-form.*

Finally, putting $M_1(p) = I$ and $M_2(p) = I$ in (17), (18), (20), (22) *we obtain regular generators which we shall call the "observable parts" of* (17), (18), (20), *and* (22) *respectively. Let the regular differential input-output relations generated by the observable part of* (17), (20), *and* (18), (22) *be denoted by S'_{i11} and S'_{i21} respectively. Of course, S'_{i11} and S'_{i21} determine the same overall input-output relations as S'_{i1} and S'_{i2} respectively* (*cf.* section 7.6).

With the above notations the following statements are equivalent:

(i) *The MR-system matrices* (17) *and* (18) *are input-output equivalent, i.e.* $S_{o1} = S_{o2}$.
(ii) $[A_{01}(p) \vdots -B_{01}(p)] = [A_{02}(p) \vdots - B_{02}(p)]$.
(iii) *There exist polynomial matrices $S_1(p)$, $S_2(p)$, $X_1(p)$, $X_3(p)$ of appropriate sizes such that*

23
$$
\underbrace{\left[\begin{array}{c:c} S_1(p) & S_2(p) \\ \hdashline 0 & I \end{array}\right]}_{S(p)}
\left[\begin{array}{c:c:c} I & A_{21}(p) & -B_{11}(p) \\ \hdashline 0 & A_{01}(p) & -B_{01}(p) \end{array}\right]
$$

24
$$
= \left[\begin{array}{c:c:c} I & A_{22}(p) & -B_{12}(p) \\ \hdashline 0 & A_{02}(p) & -B_{02}(p) \end{array}\right]
\underbrace{\left[\begin{array}{c:c:c} X_1(p) & 0 & X_3(p) \\ \hdashline 0 & I & 0 \\ \hdashline 0 & 0 & I \end{array}\right]}_{X(p)}
$$

with $S_1(p) = X_1(p)$.
(iv) *There exist polynomial matrices $R_1(p)$, $R_3(p)$, $X_1(p)$, $X_3(p)$ of appropriate sizes such that*

25
$$
\underbrace{\left[\begin{array}{c:c} R_1(p) & 0 \\ \hdashline R_3(p) & I \end{array}\right]}_{R(p)}
\left[\begin{array}{c:c:c} T_{11}(p) & 0 & -U_1(p) \\ \hdashline -V_{11}(p) & I & -W_1(p) \end{array}\right]
$$

26
$$= \left[\begin{array}{c:c:c} T_{21}(p) & 0 & -U_2(p) \\ \hdashline -V_{21}(p) & I & -W_2(p) \end{array}\right] \underbrace{\left[\begin{array}{c:c:c} X_1(p) & 0 & X_3(p) \\ \hdashline 0 & I & 0 \\ \hdashline 0 & 0 & I \end{array}\right]}_{X(p)}$$

with $R_1(p)$ and $T_{21}(p)$ left coprime, and $X_1(p)$ and $T_{11}(p)$ right coprime.
(v) *There is a one-one correspondence (bijection) between the sets S'_{i11} and S'_{i21} of the form*

27
$$\begin{aligned} &\eta : S'_{i11} \to S'_{i21}, \\ &(u_0, (x_{11}, y_0)) \mapsto (u_0, (x_{21}, y_0)), \\ &x_{21} = Z_1(p)x_{11} + Z_3(p)u_0, \end{aligned}$$

where $Z_1(p)$ and $Z_3(p)$ are polynomial matrix operators of appropriate sizes. □

28 **Remark.** Note that $T_{11}(p)$ and $T_{21}(p)$ in (17), (18) are not required to be of the same size, i.e. $S_1(p)$, $X_1(p)$, $R_1(p)$, $Z_1(p)$ above may be nonsquare. Note also that we are, in items (iii), . . . , (v) of the above theorem, only concerned with the observable parts of the regular generators involved. □

The formulation of items (iii), . . . , (v) of the theorem was inspired by results presented in Fuhrmann, 1977, and Pernebo, 1977, see also Rosenbrock, 1977a, b.

29 **Proof.** The equivalence between items (i) and (ii) of theorem (16) follows directly from the fundamental facts referred to at the beginning of this paragraph.

Consider then item (iii) and suppose that the equality (23), (24) holds. It then trivially follows that $[A_{01}(p) \vdots -B_{01}(p)] = [A_{02}(p) \vdots -B_{02}(p)]$, i.e. (iii) implies (ii).

Conversely, suppose that (ii) holds and denote $[A_{01}(p) \vdots -B_{01}(p)] = [A_{02}(p) \vdots -B_{02}(p)] \triangleq [A_0(p) \vdots -B_0(p)]$. Construct the matrices $S_1(p)$, $S_2(p)$, $X_1(p)$, and $X_3(p)$ according to the following scheme.

$S_1(p)$ and $S_2(p)$ are chosen so that

30
$$S_1(p)A_{21}(p) + S_2(p)A_0(p) = A_{22}(p).$$

Such $S_1(p)$ and $S_2(p)$ can always be found because $A_{21}(p)$ and $A_0(p)$ were right coprime.

Then choose $X_1(p) = S_1(p)$, and $X_3(p)$ according to

31 $$X_3(p) = -S_1(p)B_{11}(p) - S_2(p)B_0(p) + B_{12}(p).$$

With $S_1(p)$, $S_2(p)$, $X_1(p)$ and $X_3(p)$ so chosen it is found that the equality (23), (24) is fulfilled. Thus (ii) implies (iii).

Next consider the equivalence between (iii) and (iv). The proof of this equivalence requires a great deal of rather tedious matrix manipulation. We shall therefore be content here with a brief outline of the proof. It is left to the reader to fill in the missing details.

Suppose that (iv) holds. Using the equalities (19), (20) and (21), (22), we obtain from (25), (26) the new equality

32 $$P_2(p)R(p)P_1(p)^{-1}\left[\begin{array}{c:c:c} I & A_{21}(p) & -B_{11}(p) \\ \hdashline 0 & A_{01}(p) & -B_{01}(p) \end{array}\right]$$

33 $$= \left[\begin{array}{c:c:c} I & A_{22}(p) & -B_{12}(p) \\ \hdashline 0 & A_{02}(p) & -B_{02}(p) \end{array}\right] X(p),$$

where $P_2(p)R(p)P_1(p)^{-1}$ is found to be equal to a polynomial matrix of the form

34 $$\left[\begin{array}{c:c} S_1(p) & S_2(p) \\ \hdashline 0 & S_4(p) \end{array}\right]$$

with $S_1(p) = X_1(p)$, $S_4(p)$ square and $\det S_4(p) \neq 0$. Finally it can be shown that the left coprimeness of $R_1(p)$ and $T_{21}(p)$ implies that $S_4(p)$ is unimodular—$S_4(p)$ is then, in fact, equal to I, because $[A_{01}(p) \vdots - B_{01}(p)]$ and $[A_{02}(p) \vdots -B_{02}(p)]$ were assumed to be of CUT-form. It follows that $P_2(p)R(p)P_1(p)^{-1}$ in (32) qualifies as $S(p)$ in (23), and $X(p)$ in (26) and (33) as $X(p)$ in (24). Thus (iv) implies (iii).

Now suppose that (iii) holds and that consequently $[A_{01}(p) \vdots -B_{01}(p)] = [A_{02}(p) \vdots -B_{02}(p)]$. Using the equalities (19), (20) and (21), (22), we obtain from (23), (24) the new equality

35 $$P_2(p)^{-1}S(p)P_1(p)\left[\begin{array}{c:c:c} T_{11}(p) & 0 & -U_1(p) \\ \hdashline -V_{11}(p) & I & -W_1(p) \end{array}\right]$$

36 $$= \left[\begin{array}{c:c:c} T_{21}(p) & 0 & -U_2(p) \\ \hdashline -V_{21}(p) & I & -W_2(p) \end{array}\right] X(p),$$

where $P_2(p)^{-1}S(p)P_1(p)$ is found to be equal to a polynomial matrix of the form

37 $$\left[\begin{array}{c:c} R_1(p) & 0 \\ \hdashline R_3(p) & I \end{array}\right].$$

Comparison of (35), (36) with (25), (26) shows that $X(p)$ in (24) and (36) qualifies as $X(p)$ in (26), and $P_2(p)^{-1}S(p)P_1(p)$ in (35) as $R(p)$ in (25) if it can be shown that $R_1(p)$ in (37) and $T_{21}(p)$ are left coprime, and that $X_1(p)$ in (24) and $T_{11}(p)$ are right coprime.

Now (35), (36), and (37) give the equality

38 $$R_3(p)T_{11}(p) - V_{11}(p) = -V_{21}(p)X_1(p).$$

It follows that if $N(p)$ is a common right divisor of $X_1(p)$ and $T_{11}(p)$, then $N(p)$ is necessarily also a right divisor of $V_{11}(p)$. However, $T_{11}(p)$ and $V_{11}(p)$ were assumed right coprime, i.e. $N(p)$ must be unimodular, and $X_1(p)$ in (24) and $T_{11}(p)$ consequently right coprime.

To show that $R_1(p)$ in (37) and $T_{21}(p)$ are left coprime requires a little more effort. The equality (35), (36) and the assumption that $R_1(p)$ and $T_{21}(p)$ are not left coprime can be shown to lead to the result that $[A_{01}(p) \vdots -B_{01}(p)]$ cannot be equal to $[A_{02}(p) \vdots -B_{02}(p)]$ (here we may apply the results presented in note (7.8.15), (ix), together with the fact that we have already shown that (iv) implies (iii) and thus also (i)). This is a contradiction and it is concluded that $R_1(p)$ in (37) and $T_{21}(p)$ must be left coprime. Consequently, (iii) implies (iv).

Finally, consider item (v) of the above theorem. It is immediately seen that (v) trivially implies (i). We shall show that (ii) implies (v).

So let us assume that $[A_{01}(p) \vdots -B_{01}(p)] = [A_{02}(p) \vdots -B_{02}(p)] \triangleq [A_0(p) \vdots -B_0(p)]$, and let us form the relation S'_{i121} defined as the set of all pairs of the form $(u_0, (x_{11}, x_{21}, y_0))$ with $(u_0, (x_{11}, y_0)) \in S'_{i11}$ and $(u_0, (x_{21}, y_0)) \in S'_{i21}$. S'_{i121} is clearly a regular differential input-output relation generated by the regular generator

39 $$\begin{array}{cccc} x_{11} & x_{21} & y_0 & u_0 \end{array}$$
$$\left[\begin{array}{c:c:c:c} I & 0 & A_{21}(p) & -B_{11}(p) \\ \hdashline 0 & I & A_{22}(p) & -B_{12}(p) \\ \hdashline 0 & 0 & A_0(p) & -B_0(p) \end{array}\right]$$

which is obtained from (20) and (22) in an obvious way.

Referring to the equalities (19), (20) and (21), (22), there are regular generators, row equivalent to (39), given by

40
$$\begin{array}{c} \\ \\ \\ \end{array}\begin{matrix} x_{11} & x_{21} & y_0 & u_0 \end{matrix}$$
$$\left[\begin{array}{c:c:c::c} T_{11}(p) & 0 & 0 & -U_1(p) \\ \hdashline 0 & I & A_{22}(p) & -B_{12}(p) \\ \hdashline -V_{11}(p) & 0 & I & -W_1(p) \end{array}\right]$$

and also by

41
$$\begin{matrix} x_{11} & x_{21} & y_0 & u_0 \end{matrix}$$
$$\left[\begin{array}{c:c:c::c} I & 0 & A_{21}(p) & -B_{11}(p) \\ \hdashline 0 & T_{21}(p) & 0 & -U_2(p) \\ \hdashline 0 & -V_{21}(p) & I & -W_2(p) \end{array}\right].$$

These generators can be further transformed to row equivalent forms according to

42
$$\begin{matrix} x_{11} & x_{21} & y_0 & u_0 \end{matrix}$$
$$\left[\begin{array}{c:c:c::c} T_{11}(p) & 0 & 0 & -U_1(p) \\ \hdashline A_{22}(p)V_{11}(p) & I & 0 & -B_{12}(p) + A_{22}(p)W_1(p) \\ \hdashline -V_{11}(p) & 0 & I & -W_1(p) \end{array}\right]$$

and

43
$$\begin{matrix} x_{11} & x_{21} & y_0 & u_0 \end{matrix}$$
$$\left[\begin{array}{c:c:c::c} I & A_{21}(p)V_{21}(p) & 0 & -B_{11}(p) + A_{21}(p)W_2(p) \\ \hdashline 0 & T_{21}(p) & 0 & -U_2(p) \\ \hdashline 0 & -V_{21}(p) & I & -W_2(p) \end{array}\right]$$

respectively.

Now take any $(u_0, (x_{11}, y_0)) \in S'_{i11}$ and let x_{21} be such that $(u_0, (x_{21}, y_0)) \in S'_{i21}$. Then $(u_0, (x_{11}, x_{21}, y_0))$ is a member of S'_{i121} and we have, according to (42), (43), the equalities

44
$$x_{21} = -A_{22}(p)V_{11}(p)x_{11} + (B_{12}(p) - A_{22}(p)W_1(p))u_0$$

and

45
$$x_{11} = -A_{21}(p)V_{21}(p)x_{21} + (B_{11}(p) - A_{21}(p)W_2(p))u_0.$$

It is now easily concluded that (44) defines the bijection η (*cf.* (27)), and (45) its inverse η^{-1}. The theorem has thus been proved. □

46 **Remark.** **(i)** The equivalence between items (ii) and (iii) of theorem (16) above was very easy to prove, whereas the equivalence between items (iii) and (iv) of the same theorem turned out to be rather hard to verify. The reason for this is that the upper triangular forms (20) and (22) of the given MR-system matrices (17) and (18) are much more suitable for considerations of this kind than the original MR-system matrices themselves. This proves once again the great applicability of the upper triangular forms.
(ii) The usefulness of the equivalent formulations of the input-output equivalence on $\mathcal{MR}$ contained in theorem (16) can be questioned. This holds in particular for items (iii), . . . , (v). Are these items useful, for instance, for testing whether two given MR-system matrices (17) and (18) are input-output equivalent or not, or for generating a new MR-system matrix (18) from an original one (17) so that (18) is input-output equivalent to (17). The answer to both these questions seems to be no! Both the testing of given MR-system matrices for input-output equivalence as well as the generation of new input-output equivalent MR-system matrices from original ones are most naturally performed over the row equivalent upper triangular forms of the MR-system matrices involved, without the use of the additional matrices appearing in items (iii), . . . , (v) of the theorem. These items are included in the theorem because they are closely related to results presented in the literature (*cf.* Fuhrmann, 1977, and Pernebo, 1977). □

The input-output equivalence on $\mathcal{MR}$ as defined above does not impose any specific requirements on the o.d. zeros (*cf.* note (7.8.15), (vi)) of the R-representations involved, i.e. the input-output equivalence of two MR-system matrices (17) and (18) does not depend on the divisors $M_1(p)$ and $M_2(p)$. It is now possible to introduce conditions on the o.d. zeros also in a number of ways. We shall do this by using a slightly generalized version of Rosenbrock's "strict system equivalence" as defined in Rosenbrock, 1970, chapter 2, section 3. The generalization is performed by including a preliminary "compatibility" transformation in the definition of the equivalence.

The Rosenbrock strict system equivalence slightly generalized

Let the notations introduced in the introductory part of theorem (16) above be valid, and consider two MR-system matrices as given by (17) and (18) respectively with the corresponding row equivalent upper triangular forms as given by (20) and (22). Let $T_{11}(p)M_1(p)$ and $M_1(p)$ be of size r_1 by r_1 and $T_{21}(p)M_2(p)$ and $M_2(p)$ of size r_2 by r_2. Further let the orders of $(\mathcal{B}_1, C_1)$ and $(\mathcal{B}_2, C_2)$ be n_1 and n_2 respectively, i.e.

$\partial(\det T_{11}(p)M_1(p)) = n_1$ and $\partial(\det T_{21}(p)M_2(p)) = n_2$. The preliminary compatibility transformation comprises the following rather trivial augmentation (*cf.* Rosenbrock, 1970, chapter 2, section 2) of (17), (18), (20), and (22). Denote (17) by M_1, (20) by $\tilde{M}_1$, (18) by M_2, and (22) by $\tilde{M}_2$. Then choose any integer $r \geqslant \max\{r_1, r_2, n_1, n_2\}$ and form the new MR-system matrices M_1' and M_2' according to

47 $$M_1' = \left[\begin{array}{c|c} I_{r-r_1} & 0 \\ \hline 0 & M_1 \end{array}\right],$$

and

48 $$M_2' = \left[\begin{array}{c|c} I_{r-r_2} & 0 \\ \hline 0 & M_2 \end{array}\right],$$

where I_{r-r_1} and I_{r-r_2} are identity matrices of sizes $(r - r_1)$ by $(r - r_1)$ and $(r - r_2)$ by $(r - r_2)$ respectively. The row equivalent upper triangular forms $\tilde{M}_1'$ and $\tilde{M}_2'$ of (47) and (48) are then obviously given by

49 $$\tilde{M}_1' = \left[\begin{array}{c|c} I_{r-r_1} & 0 \\ \hline 0 & \tilde{M}_1 \end{array}\right],$$

and

50 $$\tilde{M}_2' = \left[\begin{array}{c|c} I_{r-r_2} & 0 \\ \hline 0 & \tilde{M}_2 \end{array}\right]$$

respectively. It is found that the new R-representations $(\mathcal{B}_1', C_1')$ and $(\mathcal{B}_2', C_2')$ defined by (47) and (48) respectively possess the same relevant properties as $(\mathcal{B}_1, C_1)$ and $(\mathcal{B}_2, C_2)$ respectively (the reader is advised to consider the details).

Now let us rename $M_1', M_2', \tilde{M}_1'$ and $\tilde{M}_2'$ in an obvious way according to (*cf.* (17), (18), (20), (22))

51 $$M_1' \triangleq \left[\begin{array}{c|c|c} T_{11}(p)M_1(p) & 0 & -U_1(p) \\ \hline -V_{11}(p)M_1(p) & I & -W_1(p) \end{array}\right],$$

52 $$M_2' \triangleq \left[\begin{array}{c|c|c} T_{21}(p)M_2(p) & 0 & -U_2(p) \\ \hline -V_{21}(p)M_2(p) & I & -W_2(p) \end{array}\right],$$

53 $$\tilde{M}_1' \triangleq \left[\begin{array}{c:c:c} M_1(p) & A_{21}(p) & -B_{11}(p) \\ \hdashline 0 & A_{01}(p) & -B_{01}(p) \end{array}\right],$$

54 $$\tilde{M}_2' \triangleq \left[\begin{array}{c:c:c} M_2(p) & A_{22}(p) & -B_{12}(p) \\ \hdashline 0 & A_{02}(p) & -B_{02}(p) \end{array}\right].$$

It may be assumed that $[A_{01}(p) \vdots -B_{01}(p)]$ and $[A_{02}(p) \vdots -B_{02}(p)]$ above are of CUT-form. Note that $T_{11}(p)M_1(p)$, $T_{21}(p)M_2(p)$, $M_1(p)$, and $M_2(p)$ in (51), . . . , (54) are now of the same size r by r.

The original MR-system matrices (17) and (18) are now said to be "strictly system equivalent" (in our slightly generalized sense) if the corresponding augmented MR-system matrices M_1' and M_2' as given by (51) and (52) are related by a "transformation of strict system equivalence", i.e. if there are polynomial matrices $R_1(p), R_3(p), Y_1(p)$, and $Y_3(p)$ of appropriate sizes with $R_1(p)$ and $Y_1(p)$ unimodular such that

55 $$\underbrace{\left[\begin{array}{c:c} R_1(p) & 0 \\ \hdashline R_3(p) & I \end{array}\right]}_{R(p)} \left[\begin{array}{c:c:c} T_{11}(p)M_1(p) & 0 & -U_1(p) \\ \hdashline -V_{11}(p)M_1(p) & I & -W_1(p) \end{array}\right] \underbrace{\left[\begin{array}{c:c:c} Y_1(p) & 0 & Y_3(p) \\ \hdashline 0 & I & 0 \\ \hdashline 0 & 0 & I \end{array}\right]}_{Y(p)}$$

56 $$= \left[\begin{array}{c:c:c} T_{21}(p)M_2(p) & 0 & -U_2(p) \\ \hdashline -V_{21}(p)M_2(p) & I & -W_2(p) \end{array}\right].$$

In view of the results obtained in note (7.8.15), (ix) it is clear that the original MR-system matrices (17) and (18) are strictly system equivalent in the above sense if and only if there are polynomial matrices $S_1(p)$, $S_2(p)$, $Y_1(p)$, and $Y_3(p)$ of appropriate sizes and with $S_1(p)$ and $Y_1(p)$ unimodular such that M_1' and M_2' as given by (53) and (54) are related according to

57 $$\underbrace{\left[\begin{array}{c:c} S_1(p) & S_2(p) \\ \hdashline 0 & I \end{array}\right]}_{S(p)} \left[\begin{array}{c:c:c} M_1(p) & A_{21}(p) & -B_{11}(p) \\ \hdashline 0 & A_{01}(p) & -B_{01}(p) \end{array}\right] \underbrace{\left[\begin{array}{c:c:c} Y_1(p) & 0 & Y_3(p) \\ \hdashline 0 & I & 0 \\ \hdashline 0 & 0 & I \end{array}\right]}_{Y(p)}$$

58
$$= \left[\begin{array}{c:c:c} M_2(p) & A_{22}(p) & -B_{12}(p) \\ \hdashline 0 & A_{02}(p) & -B_{02}(p) \end{array}\right]$$

implying that $[A_{01}(p) \vdots -B_{01}(p)] = [A_{02}(p) \vdots -B_{02}(p)]$ (they were assumed to be of CUT-form).

In what follows the term "strict system equivalence" will be used in the above generalized sense.

Strict system equivalence—an equivalence relation on $\mathcal{MR}$

It turns out that strict system equivalence as defined above is indeed an equivalence relation on the set $\mathcal{MR}$ of all MR-system matrices, because it possesses the necessary reflexivity, symmetricity and transitivity properties (*cf.* MacLane and Birkhoff, 1967, chapter I, § 4). That the reflexivity and symmetricity conditions are fulfilled is immediately clear (recall that $R(p)$, $S(p)$ and $Y(p)$ in (55), . . . , (58) are unimodular), whereas it requires some effort to show that the transitivity condition is also satisfied (this property is a direct consequence of theorem 5 in Pernebo, 1977 and theorem 4 in Rosenbrock 1977*a*). Note that it is essential for the transitivity property that r in (47), (48) fulfil the condition $r \geqslant \max\{r_1, r_2, n_1, n_2\}$ (*cf.* Rosenbrock, 1970, chapter 3, section 4, example 3.1 and 3.2; Pernebo, 1977, example 1).

Strictly system equivalent R-representations

If two MR-system matrices are strictly system equivalent, then we shall also say that the corresponding R-representations are strictly system equivalent.

59 **Note.** **(i)** In view of the equality (57), (58) it is clear that strict system equivalence implies input-output equivalence. The converse turns out to be generally true only in the observable case, i.e. two observable R-representations as well as their corresponding MR-system matrices are strictly system equivalent if and only if they are input-output equivalent. That input-output equivalence in the observable case implies strict system equivalence follows from the considerations contained in the next item.
(ii) Consider an observable R-representation $(\mathcal{B}, C)$ determined by a corresponding MR-system matrix

60
$$\left[\begin{array}{c:c:c} T(p) & 0 & -U(p) \\ \hdashline -V(p) & I_m & -W(p) \end{array}\right],$$

where the matrix sizes are as in (7.8.6), and where $T(p)$ and $V(p)$ are right coprime.

Let a row equivalent upper triangular form of (60) be (*cf.* (7.8.9))

61
$$\left[\begin{array}{c:c:c} I_r & A_2(p) & -B_1(p) \\ \hdashline 0 & A_0(p) & -B_0(p) \end{array}\right].$$

Then (60) is strictly system equivalent to the following MR-system matrix:

62
$$\left[\begin{array}{c:c:c} A_0(p) & 0 & -B_0(p) \\ \hdashline -I_m & I_m & 0 \end{array}\right]$$

which represents a trivial R-representation (*cf.* note (7.8.15), (ii) and (12) above). This can be demonstrated by showing that the augmented MR-system matrices

63
$$\left[\begin{array}{c:c:c:c} I_m & 0 & 0 & 0 \\ \hdashline 0 & T(p) & 0 & -U(p) \\ \hdashline 0 & -V(p) & I_m & -W(p) \end{array}\right]$$

and

64
$$\left[\begin{array}{c:c:c:c} I_r & 0 & 0 & 0 \\ \hdashline 0 & A_0(p) & 0 & -B_0(p) \\ \hdashline 0 & -I_m & I_m & 0 \end{array}\right]$$

are related by a transformation of strict system equivalence as given by (55), (56) (exercise for the reader).

The above result clearly shows that input-output equivalence implies strict system equivalence in the case of observable R-representations.

Note further that (63) and (64) are related by a transformation of strict system equivalence even though it may happen that the augmentation performed does not satisfy the condition stated in (47), (48).

(iii) It was already noted above that two R-representations which are strictly system equivalent determine the same overall input-output relation, i.e. the overall input-output relation is invariant under strict system equivalence. Many other properties of R-representations are also invariant under this equivalence, e.g. (*cf.* Rosenbrock, 1970, chapter 2):

The order of an R-representation.
The overall transfer matrix determined by an R-representation.
65 All the various sets of decoupling zeros associated with an R-representation, as well as the monic polynomials representing these sets (*cf.* note (7.8.15)).

These results are easy consequences of the defining equalities given above.
(iv) On the basis of the previous results we arrive at the following conclusions:
Given an MR-system matrix $M_1 \in \mathcal{MR}$ it is easy to construct a new MR-system matrix $M_2 \in \mathcal{MR}$ such that M_1 and M_2 are either input-output equivalent or strictly system equivalent. Conversely, given M_1 and $M_2 \in \mathcal{MR}$, it is easy to determine whether M_1 and M_2 are input-output equivalent or not. In the observable case it is equally easy to perform a test for strict system equivalence, because in this case strict system equivalence coincides with input-output equivalence. If, however, the R-representations determined by M_1 and $M_2 \in \mathcal{MR}$ are not observable, then there is in general no known easy way of determining whether they are strictly system equivalent or not (the test can be performed via the state-space representation, *cf.* Rosenbrock, 1970, chapter 2, section 3). This difficulty reduces the usefulness of the concept of strict system equivalence in the general case.
(v) Let $M \in \mathcal{MR}$ be an MR-system matrix and let $\mathscr{C}_M \subset \mathcal{MR}$ denote the equivalence class consisting of all MR-system matrices strictly system equivalent to M. $\mathscr{C}_M$ then also contains state-space MR-system matrices of the form (7.9.2) corresponding to state-space representations. Let $\mathscr{C}_{M\text{st}} \subset \mathscr{C}_M$ denote the set of all state-space MR-system matrices contained in $\mathscr{C}_M$. Then it follows from theorem 3.3 in Rosenbrock, 1970, chapter 2, section 3 that the elements of $\mathscr{C}_{M\text{st}}$ are related by a similarity transformation—a transformation well-known from the state-space theory.

If only observable R-representations are considered, then the strict system equivalence appearing above can be replaced by the input-output equivalence.
(vi) Let the notations introduced in the previous item hold and consider the problem of how to generate the set $\mathscr{C}_{M\text{st}}$ on the basis of a general element $M \in \mathscr{C}_M$, $M \notin \mathscr{C}_{M\text{st}}$, i.e. how to construct the set of all state-space MR-system matrices strictly system equivalent to a general MR-system matrix.

This problem can be solved on the basis of theorem 3.2 in Rosenbrock, 1970, chapter 2, section 3, which shows—at least in principle—how a state-space MR-system matrix—call it M_{st}—strictly system equivalent to the given MR-system matrix M can be found. Then $M_{st} \in \mathscr{C}_{Mst}$, and the remaining elements of $\mathscr{C}_{Mst}$ can be generated from M_{st} by means of suitable similarity transformations.

The above problem is closely related to the following "realization" problem:

Let S_o be a regular differential input-output relation generated by the regular generator $[A_0(p) \vdots -B_0(p)]$. Construct a state-space MR-system matrix M_{st} such that the state-space representation determined by M_{st} is an observable decomposition of S_o.

The above problem can be solved in several ways.

One possibility is to start with a trivial R-representation as a decomposition of S_o (*cf.* (12)), and then apply Rosenbrock's construction mentioned above.

Other, more direct methods have also been described in the literature (*cf.* for instance Polak, 1969, Wolovich, 1973 and 1974, Guidorzi, 1975, Wolovich and Guidorzi, 1977). Some additional comments concerning this matter can be found in note (8.5.35), (iv) below.

Generally speaking it is a cumbersome procedure to construct a state-space representation that is a decomposition (observable or not) of a given regular differential input-output relation S_o. Such a construction should therefore be performed only when absolutely necessary. One of the main objectives of the polynomial systems theory presented in this book is even to provide a means of dealing efficiently with a certain class of important problems without using the cumbersome state-space representation.

(vii) Fuhrmann, 1977, and Pernebo, 1977, have suggested generalizations of the original strict system equivalence as defined by Rosenbrock. It is found that the generalized strict system equivalences suggested by these authors actually coincide with our definition given above.

Other closely related definitions of system equivalence can be found in Wolovich, 1974, Morf, 1975, Pugh and Shelton, 1978. □

7.11 Observer synthesis problem

In sections 7.1 and 7.2 we discussed the elimination of superfluous output components and observability of compositions and presented sufficient and necessary conditions for observability. Now we shall return to this question and show how the developed methodology applies to synthesis and practical

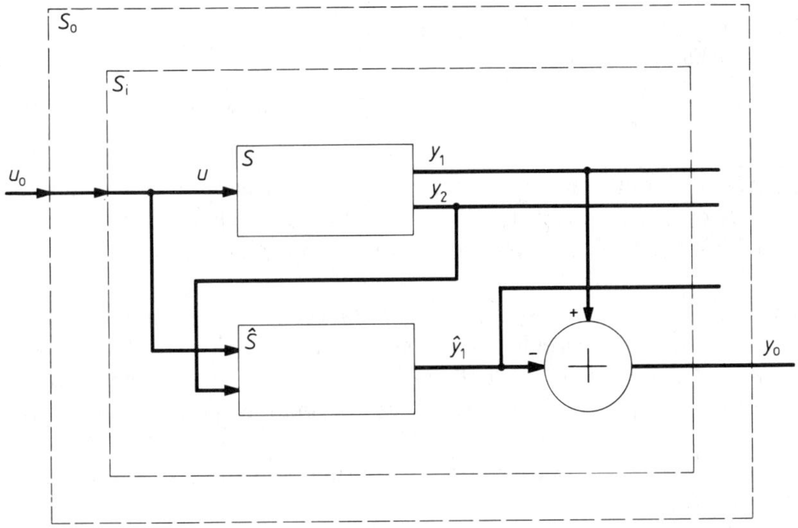

Fig. 7.10. The observer synthesis problem.

realization of systems for observing unmeasurable output components.

The problem can be depicted by Fig. 7.10.

$S \subset \mathscr{X}^r \times \mathscr{X}^q$ is a regular differential input-output relation with two outputs $y_1 \in \mathscr{X}^{q_1}$, $y_2 \in \mathscr{X}^{q_2}$. Suppose that S is generated by the equation

1
$$\underbrace{\begin{bmatrix} A_1(p) & A_2(p) \\ A_3(p) & A_4(p) \end{bmatrix}}_{A(p)} \begin{bmatrix} y_1 \\ y_2 \end{bmatrix} = \underbrace{\begin{bmatrix} B_1(p) \\ B_2(p) \end{bmatrix}}_{B(p)} u.$$

Only the output y_2 can be observed and used together with u as input of a regular differential input-output relation $\hat{S}$ called an observer. The overall input-output relation S_o is constructed only for the problem formulation because its output $y_0 = y_1 - \hat{y}_1$ cannot be observed.

2 **Problem.** Consider the composition shown in Fig. 7.10 and suppose that S is given and regular. The problem is to find a regular differential input-output relation $\hat{S}$ such that the internal input-output relation S_i is regular and at least the following conditions are fulfilled.

3 If $(u, (y_1, y_2)) \in S$ then $((u, y_2), y_1) \in \hat{S}$.
This guarantees that if

4 $$\hat{y}_1|(-\infty, t) \cap T = y_1|(-\infty, t) \cap T \quad \text{for some } t \in T$$

then $\hat{y}_1 = y_1$.

5 The overall input-output relation S_o should be asymptotically stable and the roots of its characteristic polynomial should be located in a satisfactory way with regard to the resulting dynamic properties. □

6 **Remark.** The observer should also fulfil certain other conditions. For instance $\hat{S}$ should be realizable as simply as possible, which in practice means that it has to be controllable and of low order. Furthermore, the properties of S_o should be insensitive to small variations in parameter values. □

The first step towards a solution is to bring the generator for S to upper triangular form

7 $$\left[\begin{array}{c:c:c} \tilde{A}_1(p) & \tilde{A}_2(p) & -\tilde{B}_1(p) \\ \hdashline 0 & \tilde{A}_4(p) & -\tilde{B}_2(p) \end{array}\right].$$

Let $[C(p) :: -D_1(p) \ \vdots \ -D_2(p)]$ be a generator for $\hat{S}$. The internal input-output relation S_i is then regular and generated by

8 $$\left[\begin{array}{c:c:c:c} \tilde{A}_1(p) & \tilde{A}_2(p) & 0 & 0 \\ \hdashline 0 & \tilde{A}_4(p) & 0 & 0 \\ \hdashline 0 & -D_2(p) & C(p) & 0 \\ \hdashline -I & 0 & I & I \end{array}\right] \left[\begin{array}{c} y_1 \\ \hdashline y_2 \\ \hdashline \hat{y}_1 \\ \hdashline y_0 \end{array}\right] = \left[\begin{array}{c} \tilde{B}_1(p) \\ \hdashline \tilde{B}_2(p) \\ \hdashline D_1(p) \\ \hdashline 0 \end{array}\right] u_0,$$

which can be brought to the form

9 $$\left[\begin{array}{c:c:c:c} -I & 0 & I & I \\ \hdashline 0 & \tilde{A}_2(p) & \tilde{A}_1(p) & \tilde{A}_1(p) \\ \hdashline 0 & \tilde{A}_4(p) & 0 & 0 \\ \hdashline 0 & -D_2(p) & C(p) & 0 \end{array}\right] \left[\begin{array}{c} y_1 \\ \hdashline y_2 \\ \hdashline \hat{y}_1 \\ \hdashline y_0 \end{array}\right] = \left[\begin{array}{c} 0 \\ \hdashline \tilde{B}_1(p) \\ \hdashline \tilde{B}_2(p) \\ \hdashline D_1(p) \end{array}\right] u_0.$$

Thus y_1 can be eliminated. Further, there exists a unimodular matrix

10 $$P(p) \triangleq \left[\begin{array}{c:c:c} P_1(p) & P_2(p) & P_3(p) \\ \hdashline P_4(p) & P_5(p) & P_6(p) \\ \hdashline P_7(p) & P_8(p) & P_9(p) \end{array}\right]$$

with appropriate dimensions such that

11
$$[P_7(p) \,\vdots\, P_8(p) \,\vdots\, P_9(p)] \left[\begin{array}{c|c|c||c} \tilde{A}_2(p) & \tilde{A}_1(p) & \tilde{A}_1(p) & -\tilde{B}_1(p) \\ \hline \tilde{A}_4(p) & 0 & 0 & -\tilde{B}_2(p) \\ \hline -D_2(p) & C(p) & 0 & -D_1(p) \end{array}\right]$$

$$= [0 \,\vdots\, 0 \,\vdots\, P_7(p)\tilde{A}_1(p) \,\vdots\vdots\, -P_7(p)\tilde{B}_1(p) - P_8(p)\tilde{B}_2(p) - P_9(p)D_1(p)].$$

Hence S_o is generated by

12
$$P_7(p)\tilde{A}_1(p)y_0 = (P_7(p)\tilde{B}_1(p) + P_8(p)\tilde{B}_2(p) + P_9(p)D_1(p))u_0.$$

The first candidate

Because $[\tilde{A}_1(p) \,\vdots\vdots\, -\tilde{B}_1(p) \,\vdots\, \tilde{A}_2(p)]$ is regular it can be used as a first candidate for an observer. Taking

13
$$P(p) = \left[\begin{array}{c|c|c} I & 0 & 0 \\ \hline 0 & I & 0 \\ \hline I & 0 & -I \end{array}\right]$$

gives that S_i' ($\triangleq S_i$ with y_1 eliminated) is generated by

14
$$\left[\begin{array}{c|c|c||c} \tilde{A}_2(p) & \tilde{A}_1(p) & \tilde{A}_1(p) & -\tilde{B}_1(p) \\ \hline \tilde{A}_4(p) & 0 & 0 & -\tilde{B}_2(p) \\ \hline 0 & 0 & \tilde{A}_1(p) & 0 \end{array}\right].$$

Hence S_o is generated by

15
$$\tilde{A}_1(p)y_0 = 0.$$

This observer satisfies the condition (3) but not necessarily the other conditions, so that new candidates must be found.

Generation of new candidates

Let S' be the set of pairs $((u_0, y_2), y_1)$ such that $(u_0, (y_1, y_2)) \in S$ and $\hat{S}'$ the set of pairs $(u_0, (\hat{y}_1, y_2))$ such that $((u_0, y_2), \hat{y}_1) \in \hat{S}$. Then the condition (3) can be written as $S' \subset \hat{S}$, which is equivalent to $S \subset \hat{S}'$. Let S_2 denote the set of pairs $(u_0, (y_1, y_2))$ generated by the equation

16
$$\tilde{A}_4(p)y_2 = \tilde{B}_2(p)u_0.$$

Because $S \subset S_2$, $S \subset \hat{S}'$ is equivalent to $S \subset \hat{S}' \cap S_2$. $\hat{S}' \cap S_2$ is generated by

17 $$\left[\begin{array}{c|c||c} C(p) & -D_2(p) & -D_1(p) \\ \hline 0 & \tilde{A}_4(p) & -\tilde{B}_2(p) \end{array}\right].$$

Now, the condition $S \subset \hat{S}' \cap S_2$ is fulfilled if and only if there exists a matrix $T(p) \in \mathbf{C}[p]^{q \times q}$ such that

18 $$\left[\begin{array}{c|c||c} C(p) & -D_2(p) & -D_1(p) \\ \hline 0 & \tilde{A}_4(p) & -\tilde{B}_2(p) \end{array}\right] = \underbrace{\left[\begin{array}{c|c} T_1(p) & T_2(p) \\ \hline T_3(p) & T_4(p) \end{array}\right]}_{T(p)} \left[\begin{array}{c|c||c} \tilde{A}_1(p) & \tilde{A}_2(p) & -\tilde{B}_1(p) \\ \hline 0 & \tilde{A}_4(p) & -\tilde{B}_2(p) \end{array}\right].$$

In addition, $T_3(p) = 0$ and $T_4(p) = I$ by regularity.

Furthermore, if $[C(p) \,\vdots\vdots\, -D_1(p) \,\vdots\, -D_2(p)]$ is chosen as

19 $$[C(p) \,\vdots\vdots\, -D_1(p) \,\vdots\, -D_2(p)] = [T_1(p) \,\vdots\, T_2(p)] \left[\begin{array}{c||c|c} \tilde{A}_1(p) & -\tilde{B}_1(p) & \tilde{A}_2(p) \\ \hline 0 & -\tilde{B}_2(p) & \tilde{A}_4(p) \end{array}\right]$$

and

20 $$P(p) = \left[\begin{array}{c|c|c} I & 0 & 0 \\ \hline 0 & I & 0 \\ \hline T_1(p) & T_2(p) & -I \end{array}\right],$$

we obtain the generator

21 $$\left[\begin{array}{c|c|c||c} \tilde{A}_2(p) & \tilde{A}_1(p) & \tilde{A}_1(p) & -\tilde{B}_1(p) \\ \hline \tilde{A}_4(p) & 0 & 0 & -\tilde{B}_2(p) \\ \hline 0 & 0 & T_1(p)\tilde{A}_1(p) & 0 \end{array}\right]$$

for S_1'. Hence S_o is generated by

22 $$T_1(p)\tilde{A}_1(p)y_0 = 0.$$

The desired properties may be found by choosing suitable $T_1(p)$ and $T_2(p)$. Nevertheless it is easily seen that if the input-output relation generated by (15) is unstable, S_o is unstable irrespective of $T_1(p)$.

Another reasoning

There is also another course of reasoning which results in the same kind of structure for observers. Suppose that S_o is required to be a subset of a regular $\tilde{S}$ generated by $[E(p) \,\vdots\, 0]$. Then it must hold that

23 $$P_7(p)\tilde{B}_1(p) + P_8(p)\tilde{B}_2(p) + P_9(p)D_1(p) = 0.$$

Furthermore, by (11)

24 $$P_7(p)\tilde{A}_2(p) + P_8(p)\tilde{A}_4(p) - P_9(p)D_2(p) = 0,$$

25 $$P_7(p)\tilde{A}_1(p) + P_9(p)C(p) = 0.$$

(23), . . . , (25) together are equivalent to

26 $$P_9(p)[C(p) \,\vdots\vdots\, -D_1(p) \,\vdots\, -D_2(p)] = -[P_7(p) \,\vdots\, P_8(p)] \left[\begin{array}{c|c|c} \tilde{A}_1(p) & -\tilde{B}_1(p) & \tilde{A}_2(p) \\ \hline 0 & -\tilde{B}_2(p) & \tilde{A}_4(p) \end{array}\right].$$

If $P_9(p)C(p)$ is further assumed to be a right divisor of $E(p)$, i.e. there exists $L(p) \in \mathbf{C}[p]^{q_1 \times q_1}$ such that

27 $$E(p) = L(p)P_9(p)C(p) = U(p)C(p)$$

with $U(p) \triangleq L(p)P_9(p)$, the generator for the observer $\hat{S}$ must satisfy

28 $$U(p)[C(p) \,\vdots\vdots\, -D_1(p) \,\vdots\, -D_2(p)] = [T_1(p) \,\vdots\, T_2(p)] \left[\begin{array}{c|c|c} \tilde{A}_1(p) & -\tilde{B}_1(p) & \tilde{A}_2(p) \\ \hline 0 & -\tilde{B}_2(p) & \tilde{A}_4(p) \end{array}\right]$$

and

29 $$T_1(p)\tilde{A}_1(p) = E(p),$$

where

30 $$[T_1(p) \,\vdots\, T_2(p)] \triangleq -L(p)[P_7(p) \,\vdots\, P_8(p)].$$

Conversely, suppose that (28) and (29) hold and let $V(p) \in \mathbf{C}[p]^{q_1 \times q_1}$ be a greatest common left divisor of $U(p)$ and $[T_1(p) \,\vdots\, T_2(p)]$ such that $U(p) = V(p)U_1(p)$, $[T_1(p) \,\vdots\, T_2(p)] = V(p)[T_{11}(p) \,\vdots\, T_{21}(p)]$ with $U_1(p)$, $[T_{11}(p) \,\vdots\, T_{21}(p)]$ left coprime. Choosing $P(p)$ so that $P_7(p) = -T_{11}(p)$, $P_8(p) = -T_{21}(p)$ and $P_9(p) = U_1(p)$ gives that S_o is generated by

31 $$T_{11}(p)\tilde{A}_1(p)y_0 = 0.$$

Hence $S_o \subset \tilde{S}$, i.e. every generator satisfying (28) and (29) for some $U(p)$ and $[T_1(p) \,\vdots\, T_2(p)]$ fulfils our requirement. This leads to many interesting problems concerning, for instance, the minimality of the observer. We shall, however, not pursue this problem further in this context.

Properness

Properness is an important property from the realization point of view.

It can be shown that there always exists a proper observer satisfying (19). The transfer matrix determined by S is

32 $$\mathscr{G}(p) \triangleq \begin{bmatrix} \mathscr{G}_1(p) \\ \hdashline \mathscr{G}_2(p) \end{bmatrix} \triangleq \begin{bmatrix} \tilde{A}_1(p)^{-1}(\tilde{B}_1(p) - \tilde{A}_2(p)\tilde{A}_4(p)^{-1}\tilde{B}_2(p)) \\ \hdashline \tilde{A}_4(p)^{-1}\tilde{B}_2(p) \end{bmatrix}$$

and the transfer matrix determined by $\hat{S}$

33 $$\begin{aligned} \hat{\mathscr{G}}(p) &\triangleq [\hat{\mathscr{G}}_1(p) \,\vdots\, \hat{\mathscr{G}}_2(p)] \\ &\triangleq [\tilde{A}_1(p)^{-1}(\tilde{B}_1(p) + T_1(p)^{-1}T_2(p)\tilde{B}_2(p)) \,\vdots\, -\tilde{A}_1(p)^{-1}(\tilde{A}_2(p) \\ &\quad + T_1(p)^{-1}T_2(p)\tilde{A}_4(p))]. \end{aligned}$$

Further, $\hat{\mathscr{G}}_1(p)$ can be written as

34 $$\hat{\mathscr{G}}_1(p) = \mathscr{G}_1(p) - \hat{\mathscr{G}}_2(p)\mathscr{G}_2(p).$$

Hence $\hat{\mathscr{G}}_1(p)$ is proper if $\mathscr{G}(p)$ and $\hat{\mathscr{G}}_2(p)$ are proper. So suppose that S is proper and consider only the part

35 $$[C(p) \,\vdots\vdots\, -D_2(p)] = [T_1(p) \,\vdots\, T_2(p)] \begin{bmatrix} \tilde{A}_1(p) & \tilde{A}_2(p) \\ 0 & \tilde{A}_4(p) \end{bmatrix}$$

of the generator for $\hat{S}$.

Let $T_1(p)$ be such that $T_1(p)\tilde{A}_1(p)$ is row proper with row degrees $r_j \geq \partial(\tilde{A}_4(p)) - 1$, $j \in \mathbf{q}_1$. By note (A2.94) there exist $T_2(p)$ and $D_2(p)$ such that

36 $$\begin{aligned} T_1(p)\tilde{A}_2(p) &= -T_2(p)\tilde{A}_4(p) - D_2(p), \\ \partial(D_2(p)) &< \partial(\tilde{A}_4(p)). \end{aligned}$$

The transfer matrix

37 $$C(p)^{-1}D_2(p) = -\tilde{A}_1(p)^{-1}T_1(p)^{-1}(T_1(p)\tilde{A}_2(p) + T_2(p)\tilde{A}_4(p))$$

is then proper.

If the row degrees of $T_1(p)\tilde{A}_1(p)$ are higher than $\partial(\tilde{A}_4(p)) - 1$ and $\mathscr{G}_1(p)$ is strictly proper, a strictly proper observer results. Note that $\partial(\tilde{A}_4(p))$ determines only an upper limit for the row degrees of $C(p)$ and in many cases it is possible to achieve lower degrees, particularly if $T_1(p)$ is not required to be arbitrary. On the other hand, if S is not proper, it may be necessary to take $C(p)$ of higher row degrees.

Algebraic equations

The dimensions of S can be reduced if there are some algebraic equations between the signals of S. Suppose that the signals satisfy the algebraic equation

38
$$\underbrace{[K_1 \vdots K_2]}_{K}\begin{bmatrix} y_1 \\ \hdashline y_2 \end{bmatrix} = Lu_0,$$

where K and L are matrices of appropriate dimensions over **C**. This equation can be brought by elementary row operations and permutations to the form

39
$$\left[\begin{array}{c:c:c} K_{11} & K_{12} & K_{21} \\ \hdashline 0 & 0 & K_{22} \end{array}\right]\begin{bmatrix} P_1y_1 \\ \hdashline P_2y_1 \\ \hdashline y_2 \end{bmatrix} = \begin{bmatrix} L_1 \\ \hdashline L_2 \end{bmatrix} u_0.$$

where K_{11} is **C**-unimodular and

$$P \triangleq \begin{bmatrix} P_1 \\ \hdashline P_2 \end{bmatrix}$$

is a permutation matrix.

Let S' be the set of pairs $(u_0, (y_1, y_2))$ satisfying (39). Because all pairs of S are supposed to satisfy (39), $S \cap S'$ must equal S. Using this and elementary row operations gives that S is generated by an equation of the form

40
$$\left[\begin{array}{c:c:c} K_{11} & K_{12} & K_{21} \\ \hdashline 0 & \tilde{A}'_1(p) & \tilde{A}'_2(p) \\ \hdashline 0 & 0 & \tilde{A}'_4(p) \end{array}\right]\begin{bmatrix} P_1y_1 \\ \hdashline P_2y_1 \\ \hdashline y_2 \end{bmatrix} = \begin{bmatrix} L_1 \\ \hdashline \tilde{B}'_1(p) \\ \hdashline \tilde{B}'_2(p) \end{bmatrix} u_0.$$

Because

41
$$K_{11}z_1 + K_{12}z_2 + K_{21}y_2 = L_1u_0$$

has a unique solution z_1 for each u_0, y_2, z_2, it is possible to construct an observer for estimating P_2y_1 only.

The algebraic equations can be found from row proper generators for S. This follows from the fact that the sets of row degrees of row proper matrices are invariants for the row equivalence (*cf.* appendix A2) and since one row proper generator for S is obtained by bringing the submatrix

42
$$\left[\begin{array}{c:c:c} \tilde{A}'_1(p) & \tilde{A}'_2(p) & -\tilde{B}'_1(p) \\ \hdashline 0 & \tilde{A}'_4(p) & -\tilde{B}'_2(p) \end{array}\right]$$

in (40) to row proper form.

Algorithms for constructing observers

The observers can be constructed using the analogous algorithm to the algorithm in section 1.5 for constructing feedback controllers. Because the modification to the present case is quite obvious we do not consider the detailed algorithm here. The idea, however, is the sequential use of (18).

7.12 Feedback compensator synthesis

In section 1.5 we already discussed the application of polynomial systems theory to feedback compensation and controller synthesis. In this section we shall continue this discussion and present some results concerning the solution of feedback compensator synthesis problems.

Feedback composition

Consider the feedback composition depicted by Fig. 7.11.

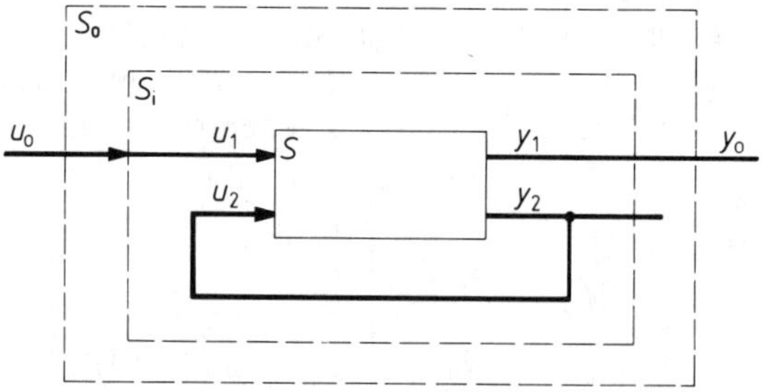

Fig. 7.11. A basic feedback composition.

In Fig. 7.11 $S \subset \mathscr{X}^r \times \mathscr{X}^q$ is a regular differential input-output relation with two inputs $u_1 \in \mathscr{X}^{r_1}$, $u_2 \in \mathscr{X}^{r_2}$ and two outputs $y_1 \in \mathscr{X}^{q_1}$, $y_2 \in \mathscr{X}^{q_2}$, where $q_2 = r_2$. Let S be generated by

1
$$\begin{array}{c} \begin{array}{cccc} y_1 & y_2 & u_1 & u_2 \end{array} \\ \underbrace{\begin{bmatrix} A_1(p) & A_2(p) \\ A_3(p) & A_4(p) \end{bmatrix}}_{A(p)} \underbrace{\begin{bmatrix} -B_1(p) & -B_2(p) \\ -B_3(p) & -B_4(p) \end{bmatrix}}_{-B(p)}, \end{array}$$

which without loss of generality can be assumed to be in an upper right block triangular form with $A_3(p) = 0$. Furthermore, let $L(p)$ be a GCLD of $A_4(p)$ and $B_4(p)$. Then we get the generator

2
$$\begin{array}{c} \begin{matrix} y_1 & y_2 & u_1 & u_2 \end{matrix} \\ \left[\begin{array}{c:c::c:c} A_1(p) & A_2(p) & -B_1(p) & -B_2(p) \\ \hdashline 0 & L(p)A_{41}(p) & -B_3(p) & -L(p)B_{41}(p) \end{array}\right]. \end{array}$$

The characteristic polynomial of S is the monic polynomial corresponding to

$$\det A_1(p) \det L(p) \det A_{41}(p) \neq 0.$$

The internal input-output relation S_i determined by the composition is generated by

3
$$\underbrace{\left[\begin{array}{c:c} A_1(p) & A_2(p) - B_2(p) \\ \hdashline 0 & L(p)(A_{41}(p) - B_{41}(p)) \end{array}\right.}_{A_i(p)} \underbrace{\left.\begin{array}{c} -B_1(p) \\ \hdashline -B_3(p) \end{array}\right]}_{-B_i(p)}.$$

(columns: y_1, y_2, u_0)

Now, if $\det(A_{41}(p) - B_{41}(p)) \neq 0$ then the composition is regular and the characteristic polynomial of S_i is the monic polynomial corresponding to

4
$$\det A_1(p) \det L(p) \det(A_{41}(p) - B_{41}(p)).$$

Thus we see that the feedback composition does not change the divisor $\det A_1(p) \det L(p)$ of the characteristic polynomial. Furthermore, if $[A_{41}(p) \vdots -B_{41}(p)]$ is strictly proper then the order of the composition is equal to the order of S (*cf.* note (7.2.24) (v)).

The internal-overall input-output relation S_{io} determined by the composition is generated by

5
$$\begin{array}{c} \begin{matrix} y_1 & y_2 & y_0 & u_0 \end{matrix} \\ \left[\begin{array}{c:c:c::c} -I & 0 & I & 0 \\ \hdashline A_1(p) & A_2(p) - B_2(p) & 0 & -B_1(p) \\ \hdashline 0 & L(p)(A_{41}(p) - B_{41}(p)) & 0 & -B_3(p) \end{array}\right], \end{array}$$

which is row equivalent to

6

$$
\begin{array}{cccc}
y_1 & y_2 & y_0 & u_0
\end{array}
$$
$$
\left[
\begin{array}{c:c:c:c}
-I & 0 & I & 0 \\
\hdashline
0 & A_2(p) - B_2(p) & A_1(p) & -B_1(p) \\
\hdashline
0 & L(p)(A_{41}(p) - B_{41}(p)) & 0 & -B_3(p)
\end{array}
\right].
$$

Hence the composition is observable if and only if $A_2(p) - B_2(p)$ and $L(p)(A_{41}(p) - B_{41}(p))$ are right coprime.

Let $P(p)$ be a unimodular matrix of the form

7

$$
P(p) = \left[
\begin{array}{c:c:c}
I & 0 & 0 \\
\hdashline
0 & P_1(p) & P_2(p) \\
\hdashline
0 & P_3(p) & P_4(p)
\end{array}
\right]
$$

bringing the generator (6) to a right upper block triangular form

8

$$
\begin{array}{cccc}
y_1 & y_2 & y_0 & u_0
\end{array}
$$
$$
\left[
\begin{array}{c:c:c:c}
-I & 0 & I & 0 \\
\hdashline
0 & M(p) & P_1(p)A_1(p) & -P_1(p)B_1(p) - P_2(p)B_3(p) \\
\hdashline
0 & 0 & P_3(p)A_1(p) & -P_3(p)B_1(p) - P_4(p)B_3(p)
\end{array}
\right],
$$

where $M(p)$ is a GCRD of $A_2(p) - B_2(p)$ and $L(p)(A_{41}(p) - B_{41}(p))$. Thus the overall input-output relation S_0 determined by the composition is generated by

9

$$
[P_3(p)A_1(p) \;\vdots\; -(P_3(p)B_1(p) + P_4(p)B_3(p))],
$$

which is regular if the composition is regular.

The transfer matrix determined by S_0 is

10

$$
\mathcal{G}_0(p) = A_1(p)^{-1}P_3(p)^{-1}(P_3(p)B_1(p) + P_4(p)B_3(p)).
$$

Because

11

$$
P_3(p)(A_2(p) - B_2(p)) + P_4(p)L(p)(A_{41}(p) - B_{41}(p)) = 0
$$

we obtain

12

$$
\mathcal{G}_0(p) = A_1(p)^{-1}(B_1(p) - (A_2(p) - B_2(p))(A_{41}(p)
$$
$$
- B_{41}(p))^{-1}L(p)^{-1}B_3(p)).
$$

The transfer matrix determined by S is

13 $$\mathcal{G}(p) = \left[\begin{array}{c|c} \mathcal{G}_1(p) & \mathcal{G}_2(p) \\ \hline \mathcal{G}_3(p) & \mathcal{G}_4(p) \end{array}\right]$$

$$= \left[\begin{array}{c|c} A_1(p)^{-1} & -A_1(p)^{-1}A_2(p)A_{41}(p)^{-1}L(p)^{-1} \\ \hline 0 & A_{41}(p)^{-1}L(p)^{-1} \end{array}\right] \left[\begin{array}{c|c} B_1(p) & B_2(p) \\ \hline B_3(p) & L(p)B_{41}(p) \end{array}\right].$$

Substituting (13) into (12) after some calculations we obtain

14 $$\mathcal{G}_o(p) = \mathcal{G}_1(p) + \mathcal{G}_2(p)(I - \mathcal{G}_4(p))^{-1}\mathcal{G}_3(p)$$

where the existence of the inverse $(I - \mathcal{G}_4(p))^{-1}$ is guaranteed by the regularity of the composition.

15 **Note.** The effect of a feedback on an input-output relation has an interesting interpretation. Let S be the differential input-output relation considered above and define a differential input-output relation S_2 as follows

16 $$S_2 = \{((u_1, u_2), (y_1, y_2)) \,|\, u_1 \in \mathcal{X}^{r_1}, u_2 \in \mathcal{X}^{r_2}, y_1 \in \mathcal{X}^{q_1}, y_2 \in \mathcal{X}^{r_2} \text{ and } y_2 = u_2\}.$$

Now the feedback composition can be regarded as the intersection $S \cap S_2$, i.e. the composition only rejects some input-output pairs of the original input-output relation. This property can, of course, be generalized to all compositions. □

More general feedback composition

The feedback composition considered above is a basic feedback composition and every feedback composition can be reduced to it. However, it is often more useful to divide the composition to a feedforward part and a feedback part. Then we obtain a feedback composition depicted by Fig. 7.12.

Let $S_1 \subset \mathcal{X}^{r_1} \times \mathcal{X}^{q_1}$ be a differential input-output relation with two inputs $u_{11} \in \mathcal{X}^{r_{11}}$, $u_{12} \in \mathcal{X}^{r_{12}}$ and two outputs $y_{11} \in \mathcal{X}^{q_{11}}$, $y_{12} \in \mathcal{X}^{q_{12}}$, and $S_{12} \subset \mathcal{X}^{r_2} \times \mathcal{X}^{q_2}$ be a differential input-output relation with two inputs $u_{21} \in \mathcal{X}^{r_{21}}$, $u_{22} \in \mathcal{X}^{r_{22}}$ and one output $y_2 \in \mathcal{X}^{q_2}$. Furthermore, let $r_{12} = q_2$ and $q_{12} = r_{22}$.

Suppose that S_1 is generated by a regular upper right block triangular generator

17 $$\begin{array}{cccc} y_{11} & y_{12} & u_{11} & u_{12} \end{array}$$
$$\left[\begin{array}{c|c||c|c} A_1(p) & A_2(p) & -B_1(p) & -B_2(p) \\ \hline 0 & L(p)A_{41}(p) & -B_3(p) & -L(p)B_{41}(p) \end{array}\right],$$

where $L(p)$ is a GCLD of $L(p)A_{41}(p)$ and $L(p)B_{41}(p)$, and let S_2 be generated by a regular generator

18 $$\begin{matrix} y_2 & & u_{21} & & u_{22} \\ [C(p) & \vdots\vdots & -D_1(p) & \vdots & -D_2(p)]. \end{matrix}$$

Then the internal input-output relation S_i determined by the composition

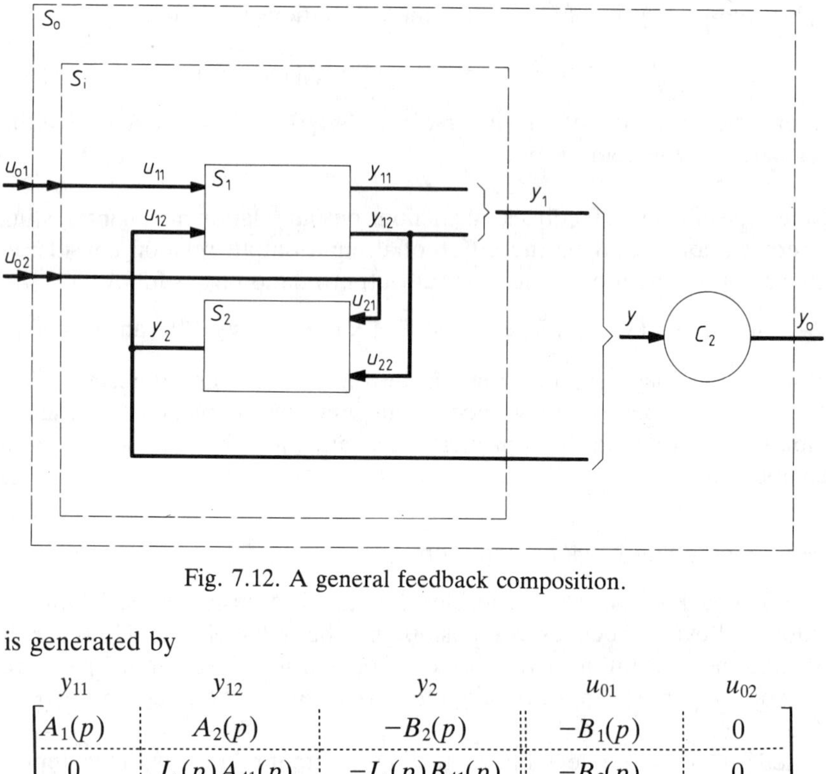

Fig. 7.12. A general feedback composition.

is generated by

19 $$\begin{matrix} y_{11} & y_{12} & y_2 & u_{01} & u_{02} \end{matrix}$$
$$\left[\begin{array}{c:c:c::c:c} A_1(p) & A_2(p) & -B_2(p) & -B_1(p) & 0 \\ \hdashline 0 & L(p)A_{41}(p) & -L(p)B_{41}(p) & -B_3(p) & 0 \\ \hdashline 0 & -D_2(p) & C(p) & 0 & -D_1(p) \end{array}\right].$$

Some authors, for instance Wolovich (Wolovich, 1974) consider a little different feedback composition depicted in Fig. 7.13, which seems to be more general than ours. However, it is easy to show that the compositions are in a sense equivalent.

So let S_2' be generated by a regular generator

20 $$\begin{matrix} y_2 & & u_{21} & & u_{22} & & u_{23} \\ [C'(p) & \vdots\vdots & -D_1'(p) & \vdots & -D_2'(p) & \vdots & -D_3'(p)]. \end{matrix}$$

If we take into account the feedback connection $u_{23} = y_2$ and replace S_2' by the overall input-output relation determined by this feedback composition, we obtain a feedback composition of the type of Fig. 7.12. Its feedback part S_2 is generated by

21 $$\begin{matrix} y_2 & u_{21} & u_{22} \\ [C'(p) - D_3'(p) & -D_1'(p) & -D_2'(p)]. \end{matrix}$$

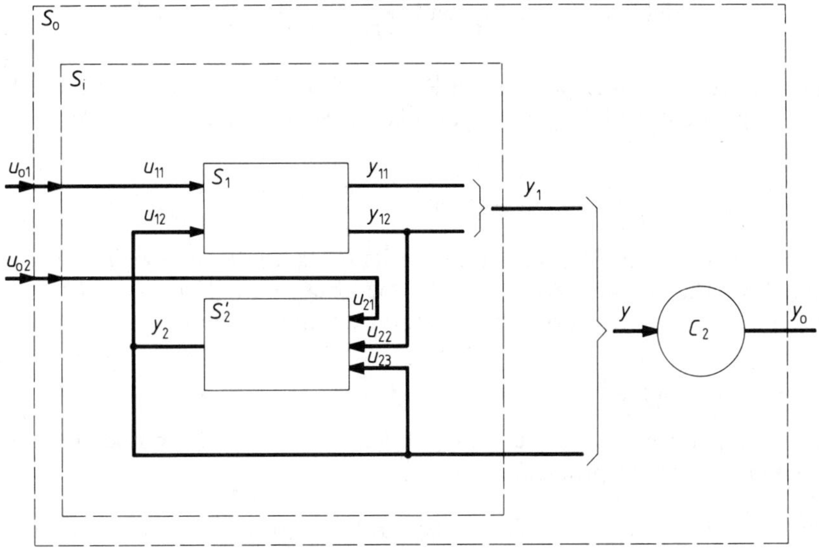

Fig. 7.13. The feedback composition of Wolovich.

Comparing (21) to (18) gives the equations

22 $$\begin{aligned} C'(p) - D_3'(p) &= C(p), \\ D_1'(p) &= D_1(p), \\ D_2'(p) &= D_2(p) \end{aligned}$$

for transforming our composition to a composition of Wolovich and vice versa.

Because the composition in Fig. 7.13 does not offer any advantages we shall return to our original composition in Fig. 7.12. The internal input-output relation S_i determined by that composition is generated by

23
$$\begin{array}{c} \quad y_{11} \qquad\qquad y_{12} \qquad\qquad\qquad y_2 \qquad\qquad\qquad u_{01} \qquad\qquad u_{02} \\ \left[\begin{array}{c|c|c||c|c} A_1(p) & A_2(p) & -B_2(p) & -B_1(p) & 0 \\ \hline 0 & L(p)A_{41}(p) & -L(p)B_{41}(p) & -B_3(p) & 0 \\ \hline 0 & -D_2(p) & C(p) & 0 & -D_1(p) \end{array}\right]. \end{array}$$

The characteristic polynomial of S_i is the monic polynomial corresponding to

24
$$\det A_1(p) \det L(p) \det \left[\begin{array}{c|c} A_{41}(p) & -B_{41}(p) \\ \hline -D_2(p) & C(p) \end{array}\right],$$

i.e. the characteristic polynomial has a divisor $\det A_1(p) \det L(p)$ which is independent of the input-output relation S_2. However, the divisor

25
$$\det \left[\begin{array}{c|c} A_{41}(p) & -B_{41}(p) \\ \hline -D_2(p) & C(p) \end{array}\right]$$

depends on S_2, in fact, we can locate its roots arbitrarily by choosing S_2 suitably. That is the reason why the feedback composition is commonly used for compensation of unsatisfactory dynamic systems.

Feedback synthesis

In section 1.5 we already considered feedback synthesis and defined the feedback synthesis problem.

26 **Problem.** Consider the feedback composition shown in Fig. 7.12 and suppose that S_1 is given, regular and strictly proper. The problem is to find a regular and proper (alternatively strictly proper) feedback compensator S_2 such that S_i is regular and the roots of the characteristic polynomial of S_i, i.e. the roots of the polynomial (24), are located in a satisfactory way with regard to the resulting dynamic properties of the feedback composition under consideration—in particular it is required that all these roots have negative real parts, implying an asymptotically stable S_i.

In order to make the problem nontrivial let it be assumed that the characteristic polynomial of S_1 corresponding to

27
$$\det A_1(p) \det L(p) \det A_{41}(p)$$

has at least one root which is not satisfactorily located. □

28 **Remark.** There are often many other qualifications that the input-output relation S_2 has to fulfil. The most important points are that S_2 is controllable

and realizable as simply as possible. Furthermore, the properties of S_i must be insensitive to parameter variations. This is to some degree guaranteed when S_1 and S_2 are proper and one of them strictly proper (*cf.* note (7.2.24) (v)). There are usually some requirements for the overall input-output relation S_o, too. For instance, the overall transfer function has to be suitable and so on, but these requirements are strongly dependent on the situation. □

29 **Remark.** If we have a given domain $C \subset \mathbf{C}$ where the roots of the characteristic polynomial are located in a satisfactory way, we could extend our ring of scalars $\mathbf{C}[p]$ to the ring $\mathbf{C}[p]_{S_C}$ of quotients, where S_C is the set of monic polynomials having all roots in C (*cf.* appendices A1 and A2, Pernebo, 1978). This would also require a corresponding modification of our signal space. Because this does not seem to offer any advantages in solving our problem but all calculations have to be performed in the polynomial structure, we do not use this possibility here. □

Now, it is easily seen that our problem has a solution if and only if the roots of $\det A_1(p) \det L(p)$ are satisfactorily located. Hence in trying to solve the problem the first thing to do would be to determine a generator (17) for different orderings and different groupings of the components of $y_1 = (y_{11}, y_{12})$, and then to choose an ordering and grouping—if it exists—yielding a satisfactory location of the roots of $\det A_1(p) \det L(p)$. Of course, a corresponding analysis could also be performed with respect to the choice of u_{12}—it may happen that a satisfactory result can be obtained using only some of the control inputs available as components of u_{12}.

Suppose then that the roots of $\det A_1(p) \det L(p)$ are satisfactorily located. Hence the problem is to find the feedback compensator S_2 such that the roots of (25) are satisfactorily located. For simplifying the notations rename (25) as

30
$$\det \left[\begin{array}{c:c} A(p) & -B(p) \\ \hdashline -D(p) & C(p) \end{array}\right]$$

where $A(p)$ and $B(p)$ are left coprime q by q- and q by r-matrices, respectively, and $[A(p) \vdots -B(p)]$ is strictly proper.

The first candidate for $[C(p) \vdots -D(p)]$

Because $A(p)$ and $B(p)$ are assumed to be left coprime, there then exists a unimodular matrix $Q(p)$ of the form

31
$$Q(p) = \left[\begin{array}{c:c} A(p) & -B(p) \\ \hdashline Q_3(p) & Q_4(p) \end{array}\right]$$

(*cf*. appendix A1). Recall here the construction of $Q(p)$. First the matrix $[A(p) \vdots -B(p)]$ is brought by the elementary column operations, i.e. post-multiplying it by a unimodular matrix $P(p)$, to the form

32 $$[L(p) \vdots 0] = [A(p) \vdots -B(p)] \underbrace{\begin{bmatrix} P_1(p) & P_2(p) \\ P_3(p) & P_4(p) \end{bmatrix}}_{P(p)},$$

where $L(p)$ is a GCLD of $A(p)$ and $B(p)$. Without loss of generality $L(p)$ can be taken as equal to I. Let $Q(p)$ be the inverse of $P(p)$. Then we obtain

33 $$[A(p) \vdots -B(p)] = [I \vdots 0] \underbrace{\begin{bmatrix} Q_1(p) & Q_2(p) \\ Q_3(p) & Q_4(p) \end{bmatrix}}_{Q(p)}$$

$$= [Q_1(p) \vdots Q_2(p)],$$

which proves our assertion.

Now taking

34 $$[C(p) \vdots -D(p)] = [Q_4(p) \vdots Q_3(p)]$$

gives

35 $$\det \begin{bmatrix} A(p) & -B(p) \\ -D(p) & C(p) \end{bmatrix} = \text{constant} \neq 0,$$

whose roots (the empty set!) are satisfactorily located. However, it is easy to show that the roots are very sensitive to small variations in parameters. In fact, we can locate the roots to arbitrary places in the complex plane making only small variations in parameters.

In order to generate other candidates for $[C(p) \vdots -D(p)]$ we can use the fundamental result presented below.

36 **A fundamental theorem.** *Suppose that the matrix* $Q(p)$ *of the form* (31) *is given. Then it holds*:
(i) *For any regular and proper* (*or strictly proper as desired*) *generator* $[C(p) \vdots -D(p)]$ *the degree of* (30) *is equal to the sum of the degrees of* $\det A(p)$ *and* $\det C(p)$ *implying that* (30) *is nonzero. Furthermore there exist corresponding polynomial matrices* $T_3(p)$ *and* $T_4(p)$ *of appropriate sizes* ($T_4(p)$ *square*, $\det T_4(p) \neq 0$) *such that*

37 $$\begin{bmatrix} A(p) & -B(p) \\ -D(p) & C(p) \end{bmatrix} = \begin{bmatrix} I & 0 \\ T_3(p) & T_4(p) \end{bmatrix} \begin{bmatrix} A(p) & -B(p) \\ Q_3(p) & Q_4(p) \end{bmatrix},$$

with

38
$$\det\left[\begin{array}{c|c} A(p) & -B(p) \\ \hline -D(p) & C(p) \end{array}\right] \propto \det T_4(p).$$

(ii) *For any (nonempty) set C of complex numbers there is always a regular and proper (or strictly proper as desired) generator* $[C(p) \vdots -D(p)]$ *of appropriate size such that* (30) *is nonzero and the roots of* (30) *belong to C. Moreover,* $C(p)$ *and* $D(p)$ *can always be chosen left coprime.* □

39 **Proof.** **(i)** Bring first the generators $[A(p) \vdots -B(p)]$ and $[C(p) \vdots -D(p)]$ to a row proper form. Then the degree of det $A(p)$ is equal to the sum of the row degrees of $[A(p) \vdots -B(p)]$ and the degree of det $C(p)$ is equal to the sum of the row degrees of $[C(p) \vdots -D(p)]$. Furthermore, then the matrix

40
$$\left[\begin{array}{c|c} A(p) & -B(p) \\ \hline -D(p) & C(p) \end{array}\right]$$

is also row proper and the degree of (30) is equal to the sum of the row degrees of (40), hence equal to the sum of the degrees of det $A(p)$ and det $C(p)$.

The second part of item (i) is proved as follows. Using the identity $P(p)\,Q(p) = I$ we obtain

41
$$\left[\begin{array}{c|c} A(p) & -B(p) \\ \hline -D(p) & C(p) \end{array}\right]$$

$$= \left[\begin{array}{c|c} A(p) & -B(p) \\ \hline -D(p) & C(p) \end{array}\right] \underbrace{\left[\begin{array}{c|c} P_1(p) & P_2(p) \\ \hline P_3(p) & P_4(p) \end{array}\right] \left[\begin{array}{c|c} A(p) & -B(p) \\ \hline Q_3(p) & Q_4(p) \end{array}\right]}_{I}$$

$$= \left[\begin{array}{c|c} I & 0 \\ \hline -D(p)P_1(p) + C(p)P_3(p) & -D(p)P_2(p) + C(p)P_4(p) \end{array}\right] \cdot \left[\begin{array}{c|c} A(p) & -B(p) \\ \hline Q_3(p) & Q_4(p) \end{array}\right]$$

$$= \left[\begin{array}{c|c} I & 0 \\ \hline T_3(p) & T_4(p) \end{array}\right] \left[\begin{array}{c|c} A(p) & -B(p) \\ \hline Q_3(p) & Q_4(p) \end{array}\right],$$

where

42 $$T_3(p) = -D(p)P_1(p) + C(p)P_3(p),$$

43 $$T_4(p) = -D(p)P_2(p) + C(p)P_4(p).$$

(ii) According to appendix A1 the transfer matrix $A(p)^{-1}B(p)$ can be presented in the form

44 $$A(p)^{-1}B(p) = B_1(p)A_1(p)^{-1}$$

with $A_1(p)$ and $B_1(p)$ right coprime. This was seen as follows. Postmultiplying the matrix

45 $$\left[\begin{array}{c|c} A(p) & -B(p) \\ \hline 0 & I \end{array}\right]$$

by $P(p)$ (*cf.* (32)) gives

46 $$\left[\begin{array}{c|c} A(p) & -B(p) \\ \hline 0 & I \end{array}\right]\left[\begin{array}{c|c} P_1(p) & P_2(p) \\ \hline P_3(p) & P_4(p) \end{array}\right] = \left[\begin{array}{c|c} I & 0 \\ \hline P_3(p) & P_4(p) \end{array}\right],$$

47 $$\det A(p) \det P(p) = \det P_4(p) \neq 0,$$

which implies that $\det P_4(p) \neq 0$. Because

48 $$A(p)\, P_2(p) - B(p)P_4(p) = 0$$

we obtain

49 $$A(p)^{-1}B(p) = P_2(p)P_4(p)^{-1}$$

with $P_2(p)$ and $P_4(p)$ right coprime. Hence $P_4(p)$ and $P_2(p)$ can be taken as $A_1(p)$ and $B_1(p)$, respectively. Furthermore, without loss of generality $P_4(p)$ can be assumed to be column proper with column degrees $c_1, c_2, \ldots, c_r$. It turns out that $c_1, c_2, \ldots, c_r$ in fact coincide with the controllability indices (suitably ordered, *cf.* Wolovich, 1974, section 3.6) of a controllable and observable state-space representation of the transfer matrix (49).

Thus we have

50 $$\left[\begin{array}{c|c} A(p) & -B(p) \\ \hline Q_3(p) & Q_4(p) \end{array}\right]\left[\begin{array}{c} P_2(p) \\ \hline P_4(p) \end{array}\right] = \left[\begin{array}{c} 0 \\ \hline I \end{array}\right].$$

Suppose now, that $T_4(p)$ is an arbitrary r by r-matrix and multiply equation (50) by the matrix

51
$$\left[\begin{array}{c:c} I & 0 \\ \hdashline 0 & T_4(p) \end{array}\right].$$

We obtain

52
$$\left[\begin{array}{c:c} A(p) & -B(p) \\ \hdashline T_4(p)Q_3(p) & T_4(p)Q_4(p) \end{array}\right]\left[\begin{array}{c} P_2(p) \\ \hdashline P_4(p) \end{array}\right] = \left[\begin{array}{c} 0 \\ \hdashline T_4(p) \end{array}\right].$$

By note (A2.94), (i) there exist unique matrices $T_3(p)$ and $D(p)$ such that

53
$$T_4(p)\,Q_3(p) = T_3(p)A(p) - D(p),$$
$$\partial(D(p)) < \partial(A(p)).$$

Premultiplying (52) by

54
$$\left[\begin{array}{c:c} I & 0 \\ \hdashline -T_3(p) & I \end{array}\right]$$

gives

55
$$\left[\begin{array}{c:c} A(p) & -B(p) \\ \hdashline -D(p) & \underbrace{T_3(p)B(p) + T_4(p)Q_4(p)}_{C(p)} \end{array}\right]\left[\begin{array}{c} P_2(p) \\ \hdashline P_4(p) \end{array}\right] = \left[\begin{array}{c} 0 \\ \hdashline T_4(p) \end{array}\right].$$

Denote

56
$$C(p) \triangleq T_3(p)\,B(p) + T_4(p)\,Q_4(p)$$

and let the column degrees of $T_4(p)$ be $d_1, d_2, \ldots, d_r$. Write

57
$$C(p) = C_s p^s + C_{s-1}p^{s-1} + \ldots + C_0 \quad \text{with}$$
$$C_0, C_1, \ldots, C_s \in \mathbf{C}^{r\times r}, s = \partial(C(p)),$$

58
$$P_4(p) = \underbrace{C_{P_4}\begin{bmatrix} p^{c_1} & 0 & \ldots 0 \\ 0 & p^{c_2} & \ldots 0 \\ \vdots & \vdots & \vdots \\ 0 & 0 & \ldots p^{c_r} \end{bmatrix}}_{(P_4(p))^c} + \underbrace{\text{terms of lower degree}}_{(P_4(p))^l}$$

with $C_{P_4} \in \mathbf{C}^{r\times r}$, det $C_{P_4} \neq 0$,

59 $$T_4(p) = \underbrace{C_{T_4}\begin{bmatrix} p^{d_1} & 0 & \dots & 0 \\ 0 & p^{d_2} & \dots & 0 \\ \vdots & \vdots & & \vdots \\ 0 & 0 & \dots & p^{d_r} \end{bmatrix}}_{(T_4(p))^c} + \underbrace{\text{terms of lower degree}}_{(T_4(p))^l}.$$

with $C_{T_4} \in \mathbf{C}^{r\times r}$.

Consider the equation

60 $$-D(p)\,P_2(p) + C(p)\,P_4(p) = T_4(p).$$

Because $P_2(p)P_4(p)^{-1}$ was assumed to be strictly proper, the column degrees of $P_2(p)$ are lower than the corresponding column degrees of $P_4(p)$ (*cf.* appendix A2). Hence from (60) it is easily seen that it must hold

61 $$\partial(C(p)) = s \leqslant \max\{d_1 - c_1, d_2 - c_2, \dots, d_r - c_r, \partial(A(p)) - 2\}.$$

Suppose then that $T_4(p)$ is chosen column proper with column degrees satisfying

62 $$d_j = \partial(A(p)) + c_j - 1, j = 1, 2, \dots, r.$$

This implies that

63 $$\partial(C(p)) = s = \partial(A(p)) - 1.$$

From (60) we obtain

64 $$-D(p)\,P_2(p) + C(p)\,P_4(p)$$
$$= C_s\,C_{P_4}\begin{bmatrix} p^{s+c_1} & 0 & \dots & 0 \\ 0 & p^{s+c_2} & \dots & 0 \\ \vdots & \vdots & & \vdots \\ 0 & 0 & \dots & p^{s+c_r} \end{bmatrix} + \text{terms of lower degree}$$
$$= C_{T_4}\begin{bmatrix} p^{d_1} & 0 & \dots & 0 \\ 0 & p^{d_2} & \dots & 0 \\ \vdots & \vdots & & \vdots \\ 0 & 0 & \dots & p^{d_r} \end{bmatrix} + \text{terms of lower degree}.$$

By column properness det $C_{P_4} \neq 0$ and det $C_{T_4} \neq 0$, which implies that

C_s has also to satisfy $\det C_s \neq 0$. Hence $[C(p) \vdots -D(p)]$ is row proper and the row degrees of $D(p)$ are not higher than the row degrees of $C(p)$, which are all equal to $\partial(A(p)) - 1$. Thus $[C(p) \vdots -D(p)]$ is proper.

Now we have shown that a proper feedback compensator always exists. Because $T_4(p)$ was chosen arbitrarily we can take a $T_4(p)$ which is for instance in CUT-form with the diagonal entries having their roots in a given domain C of the complex plane. □

65 **Remark.** Our proof gives also the upper limits $r(\partial(A(p)) - 1)$ and $r(\partial(A(p)) - 1) + \partial(\det A(p))$ for the orders of S_2 and S_i, respectively, where r is the number of columns of $B(p)$. In order to get as low orders as possible, it is advantageous to bring $A(p)$ first to row proper form. It should, however, be noted that the orders given are only upper limits and it is often possible to obtain lower orders. □

Algorithms for choosing candidates

In section 1.5 we already discussed algorithms for constructing candidates for a feedback compensator, so that we shall omit them in this connection. However, these algorithms can often lead to much simpler feedback compensators than the construction above.

Dual approach

In section 7.7 we showed that the differential input-output relation generated by the regular $[A(p) \vdots -B(p)]$ with $A(p)$ and $B(p)$ left coprime can be decomposed into an observable series composition comprising the differential input-output relation generated by $[A_1(p) \vdots -I]$ followed by the differential input-output relation generated by $[I \vdots -B_1(p)]$, where $A_1(p)$ and $B_1(p)$ are right coprime and satisfy

66
$$A(p)^{-1}B(p) = B_1(p)A_1(p)^{-1}.$$

The internal input-output relation determined by this composition is generated by

67
$$\begin{array}{ccc} y_1 & y_0 & u_0 \end{array}$$
$$\left[\begin{array}{c:c:c} A_1(p) & 0 & -I \\ \hdashline -B_1(p) & I & 0 \end{array}\right].$$

The differential input-output relation generated by the regular

$[C(p) \vdots -D(p)]$ can be decomposed in the same way using the factorization

68 $$C(p)^{-1}D(p) = D_1(p)C_1(p)^{-1}$$

with $C_1(p)$ and $D_1(p)$ right coprime.

The decompositions above change the part

69 $$\left[\begin{array}{c:c} A(p) & -B(p) \\ \hdashline -D(p) & C(p) \end{array}\right]$$

of the generator of our feedback composition to

70 $$\left[\begin{array}{c:c:c:c} A_1(p) & 0 & 0 & -I \\ \hdashline -B_1(p) & I & 0 & 0 \\ \hdashline 0 & -I & C_1(p) & 0 \\ \hdashline 0 & 0 & -D_1(p) & I \end{array}\right].$$

Furthermore, using elementary row operations and column interchanges this can be brought to the form

71 $$\left[\begin{array}{c:c:c:c} I & 0 & 0 & -B_1(p) \\ \hdashline 0 & I & -D_1(p) & 0 \\ \hdashline 0 & 0 & -C_1(p) & B_1(p) \\ \hdashline 0 & 0 & -D_1(p) & A_1(p) \end{array}\right].$$

Thus we obtain

72 $$\det\left[\begin{array}{c:c} A(p) & -B(p) \\ \hdashline -D(p) & C(p) \end{array}\right] \propto \det\left[\begin{array}{c:c} C_1(p) & B_1(p) \\ \hdashline D_1(p) & A_1(p) \end{array}\right].$$

The problem is now to find right coprime $C_1(p)$ and $D_1(p)$ such that the roots of (72) are satisfactorily located.

Let $P(p)$ and $Q(p)$ be the unimodular matrices

73 $$P(p) = \left[\begin{array}{c:c} P_1(p) & B_1(p) \\ \hdashline P_3(p) & A_1(p) \end{array}\right],$$

74 $$Q(p) = \left[\begin{array}{c:c} A(p) & -B(p) \\ \hdashline Q_3(p) & Q_4(p) \end{array}\right]$$

satisfying

75 $$P(p)Q(p) = I$$

(*cf.* (32), (33), (46), . . . , (49)). Then we obtain

76 $$\begin{bmatrix} C_1(p) & B_1(p) \\ D_1(p) & A_1(p) \end{bmatrix} = \begin{bmatrix} P_1(p) & B_1(p) \\ P_3(p) & A_1(p) \end{bmatrix} \begin{bmatrix} A(p) & -B(p) \\ Q_3(p) & Q_4(p) \end{bmatrix} \begin{bmatrix} C_1(p) & B_1(p) \\ D_1(p) & A_1(p) \end{bmatrix}$$
$$= \begin{bmatrix} P_1(p) & B_1(p) \\ P_3(p) & A_1(p) \end{bmatrix} \begin{bmatrix} U_1(p) & 0 \\ U_3(p) & I \end{bmatrix},$$

where

77 $$U_1(p) = A(p)C_1(p) - B(p)D_1(p),$$

78 $$U_3(p) = Q_3(p)C_1(p) + Q_4(p)D_1(p).$$

Thus we have a fundamental result for constructing matrices $C_1(p)$ and $D_1(p)$. If we transpose (76) we get an equation which is analogous to (37). Therefore we can apply proof (39) to show that if $B_1(p)A_1(p)^{-1}$ is strictly proper, there always exist $C_1(p)$ and $D_1(p)$ such that $D_1(p)C_1(p)^{-1}$ is proper and the roots of (72) can be located arbitrarily. The upper limits for the degrees of $C_1(p)$ and (72) are $q(\partial(A_1(p)) - 1)$ and $q(\partial(A_1(p)) - 1) + \partial(A_1(p))$, respectively, where q is the number of rows in $B_1(p)$, i.e. the number of columns in $A(p)$.

Part III

Differential Systems. The Vector Space Structure

The basic concepts of systems theory as presented in part I of this book were decisively based on the central concept of a "system" regarded as a family of input-output mappings (*cf.* (2.1.2)). This system concept then led to a number of additional concepts and attributes, e.g. the "state" concept, the concepts of a "dynamic system", of a "linear system", and of an "interconnection of a family of systems" (see chapters 2 and 3). Any system determines a corresponding unique input-output relation (*cf.* (2.1.3)), whereas a given input-output relation can generally be determined by several systems. Accordingly, problems in which systems are involved must generally be treated by means of methods founded on the basic system concept. It is then natural that various state sets appear in an explicit way in this context (*cf.* section 3.1).

In part II of the book we chose, nevertheless, another approach. We discussed a number of problems concerning systems governed by ordinary linear time-invariant differential equations on the basis of the input-output relations generated by such equations—the input-output mappings which constitute the underlying systems, and the associated state sets never entered into the process. A special mathematical machinery was devised for this purpose. This course of approach was very briefly motivated by asserting that, under certain circumstances, natural suitable systems could always be assigned to the various input-output relations involved (*cf.* chapter 4).

We must now return to this matter. First of all we must give an explicit method for assigning a unique suitable system to a given input-output relation. At the same time we would like to be able to develop sufficiently strong mathematical structures in order to make calculations easy to perform. Secondly we must investigate the relationship between compositions of input-output relations of the form discussed in chapter 7, and the corresponding underlying interconnections of systems. In addition there is the concept of controllability as defined in a rather formal way in section 6.4. This definition requires a more detailed and illuminating motivation. The present part III of the book will be devoted to the above subjects.

Realizability conditions

As was already pointed out at the beginning of part II, it is assumed, throughout the rest of the book, that the set of realizability conditions (*cf.* section 2.2) comprises the causality property only.

8

The projection method

In this chapter we shall devise a constructive method for assigning a unique finite dimensional linear dynamic system to any regular differential input-output relation of the form discussed in chapters 6 and 7—provided that the underlying signal space possesses suitable properties. The method utilizes a projection mapping, and we shall therefore call it the *projection method*.

In order to achieve the intended objective we have, unfortunately, to choose a space of generalized functions (distributions) as our basic signal space. Now the general theory of generalized functions (*cf.* Gelfand and Shilov, 1964), is rather hard to digest, and many readers may therefore feel ill at ease with it. There is, however, one bright spot. The space needed is namely a rather well-behaved space of generalized functions (called the space of "piecewise infinitely regularly differentiable complex-valued generalized functions on T", with $T \subset \mathbf{R}$, an open interval, *cf.* appendix A4), and many concepts used in the context of ordinary pointwise defined complex-valued functions on $T \subset \mathbf{R}$ can be straightforwardly generalized to make sense also with respect to this space of generalized functions. With some care we may therefore safely use most of the familiar terminology and notions of ordinary functions also in this generalized case. This should help the reader to feel comfortable. It may be pointed out that almost the same space is, in fact, tacitly used also in engineering applications of the well-known Laplace transform method. Usually this fact does not cause any bother—because it is never noticed!

The projection method mentioned above is based on an idea which was briefly outlined in the well-known book by Zadeh and Desoer, 1963. The idea was later developed and formalized in Blomberg and Salovaara, 1968, Blomberg *et al.*, 1969, and Blomberg, 1975. The projection method leads to a vector space structure that is closely related to the field structures of the Laplace transform method and the Mikusinski operational calculus (Mikusinski, 1959). It forms a true unifying bridge between various kinds of representations of systems governed by ordinary linear time-invariant differential equations. It should be mentioned that the vector space structure obtained in this context with the aid of the projection method is also closely related to the more general structures discussed in Kamen, 1975, Sontag, 1976, and Ylinen, 1975.

8.1 Reason for choosing a space of generalized functions as signal space

In section 4.2 we discussed the problem of how to assign a system to a given regular differential input-output relation $S \subset \mathscr{X}^r \times \mathscr{X}^q$ generated by the regular generator $[A(p) \vdots -B(p)]$, where $\mathscr{X}$ was a suitable signal space consisting of complex-valued functions defined on some open time interval $T \subset \mathbf{R}$. We chose a $t_0 \in T$ and a pair $(u_1, y_1) \in S$, and followed the time evolution of (u_1, y_1) on the interval $T \cap (-\infty, t_0)$. The observed initial segments of u_1 and y_1 on this interval were denoted by u_1^- and y_1^- respectively. Then we formed the set $S_{(u_1^-, y_1^-)}$ consisting of all pairs (u, y) belonging to S and coinciding with (u_1^-, y_1^-) on $T \cap (-\infty, t_0)$. Formally (*cf.* (4.2.3))

1 $$S_{(u_1^-, y_1^-)} \triangleq \{(u, y) \mid (u, y) \in S \quad \text{and} \quad u \mid T \cap (-\infty, t_0) = u_1^- \quad \text{and} \quad y \mid T \cap (-\infty, t_0) = y_1^-\}.$$

Further it was noted that (1) is a mapping for every possible (u_1^-, y_1^-), and that the really interesting part of $S_{(u_1^-, y_1^-)}$ is given by the final segments from t_0 onwards of the input-output pairs contained in this mapping.

A system $\mathscr{S}$ with $S = \cup \mathscr{S}$ could thus be formed as the family of all possible mappings of the form (1), i.e. as (*cf.* (4.2.4))

2 $$\mathscr{S} \triangleq \{S_{(u_1^-, y_1^-)} \mid u_1^- = u \mid T \cap (-\infty, t_0) \quad \text{and } y_1^- = y \mid T \cap (-\infty, t_0) \quad \text{for some } (u, y) \in S\}.$$

With $t_0 \in T$ fixed, the system (2) is clearly uniquely determined by S.

Now our primal goal was to construct a unique linear dynamic (causal) system corresponding to any given regular differential input-output relation S of the form discussed in chapter 6 and 7, where the signal space $\mathscr{X}$ was assumed to be regular and to possess the richness property (5.1.5). So what can be said about system (2) above?

First of all it is concluded that system (2)—with the signal space $\mathscr{X}$ so chosen—certainly cannot be a linear system in the sense of the definition given in section 2.3, because the domain of $S_{(u_I^-, y_I^-)}$ for $u_I^- \neq 0$ is clearly not a vector space. Hence system (2) does not satisfy our conditions.

The difficulty caused here by nonzero initial segments can, however, easily be avoided by replacing the mappings $S_{(u_I^-, y_I^-)}$ above by mappings $S^+_{(u_I^-, y_I^-)}$, of the form

3

$$S^+_{(u_I^-, y_I^-)} | T \cap (-\infty, t_0) \triangleq 0,$$

$$S^+_{(u_I^-, y_I^-)} | T \cap [t_0, \infty) \triangleq S_{(u_I^-, y_I^-)} | T \cap [t_0, \infty).$$

Note that $S^+_{(u_I^-, y_I^-)}$ and $S_{(u_I^-, y_I^-)}$ represent the same input-output behaviour on $T \cap [t_0, \infty)$.

Now it can be seen that the domain of $S^+_{(u_I^-, y_I^-)}$ can be a vector space for any pair (u_I^-, y_I^-) of initial segments only if $\mathscr{X}$ allows step functions at t_0 so that the final input segment from t_0 onwards can be chosen independently of the initial input segment u_I^-. But then $\mathscr{X}$ must also contain the derivatives of every order of step functions at t_0 (*cf.* section 4.1), and so $\mathscr{X}$ must necessarily be a space of generalized functions.

Accordingly, we shall next choose a suitable space of generalized functions as our signal space $\mathscr{X}$, and then we shall devise a construction assigning a unique finite dimensional linear dynamic system to any regular differential input-output relation S of the form discussed in chapters 6 and 7. Our present goal has thereby been achieved.

8.2 The basic signal space $\mathscr{D}$ of generalized functions. Projection mappings. Subspaces of $\mathscr{D}$. Generalized causality

Motivated by the considerations in the previous paragraph, let us, in this part of the book choose the space $\mathscr{D}$ of "piecewise infinitely regularly differentiable complex-valued generalized functions on T", with $T \subset \mathbf{R}$ an open interval, as our basic signal space $\mathscr{X}$. This space fulfils all the conditions imposed on a regular signal space along with the special conditions discussed in the previous paragraph for *any* $t_0 \in T$ (and not just for a fixed $t_0 \in T$). It also possesses the richness property (5.1.5).

The space $\mathscr{D}$ of generalized functions

A precise presentation of the space $\mathscr{D}$ of generalized functions on T is given in appendix A4. We shall in the following note collect a number of relevant properties possessed by $\mathscr{D}$ and its elements.

1 **Note.** **(i)** $\mathscr{D}$ is a vector space over $\mathbf{C}$ with respect to the natural operations. It is a subspace of the space of all complex-valued generalized functions on T.
(ii) Any element of $\mathscr{D}$, and any kth-order derivative of such an element can be expressed as (*cf.* (A4.19))

2 $$f + \sum_{i \in I} c_i \delta_{t_i}^{(k_i)}.$$

f here denotes the "regular" part of (2). It is a generalized function which can and will be identified with an ordinary piecewise continuous function $T \to \mathbf{C}$. $\delta_{t_i}^{(k_i)}$, $t_i, \in T$, $k_i \in \{0, 1, 2, \ldots\}$, denotes a delta function of order k_i at t_i. The c_i's are complex numbers, and the sum

$$\sum_{i \in I, t_i \in \theta} c_i \delta_{t_i}^{(k_i)}$$

contains, for every bounded interval $\theta \subset T$, at most a finite number of nonzero terms. The zero of $\mathscr{D}$ is identified with the ordinary zero function $T \to \mathbf{C}$.

From now on we shall assume that f in an expression of the form (2) is always left continuous, i.e. that

3 $$f(t-) = f(t)$$

for any $t \in T$.
(iii) The elements of $\mathscr{D}$ are differentiable (in the generalized sense) any number of times according to the rule that the derivative of the unit step U_t at $t \in T$ yields the delta function $\delta_t^{(0)} \triangleq \delta_t$ of order zero at t, and that the derivative of $\delta_t^{(k)}$ is $\delta_t^{(k+1)}$ for $k \in \{0, 1, 2, \ldots\}$. With p denoting the differentiation operator and with $x \in \mathscr{D}$ it thus holds that $p^k x$—the kth-order derivative of x—is again an element of $\mathscr{D}$ for any $k \in \{0, 1, 2, \ldots\}$. p can, in fact, be regarded as a linear mapping $\mathscr{D} \to \mathscr{D}$.
(iv) The space C^∞ of infinitely continuously differentiable ordinary functions $T \to \mathbf{C}$ is regarded as a subspace of $\mathscr{D}$.
(v) $\mathscr{D}$ qualifies—in a generalized sense—as a regular signal space possessing the richness property (5.1.5). □

Of course, $\mathscr{D}$ is not really a space of ordinary time functions defined on the time set $T \subset \mathbf{R}$. Many concepts used in the context of ordinary time functions can, nevertheless, be generalized to make sense also with respect to $\mathscr{D}$. One such important concept is "the value of an element of $\mathscr{D}$ at $t \in T$".

The value of an element of $\mathscr{D}$ at a point $t \in T$

To begin with we note the following point.

Let x be an arbitrary element of $\mathscr{D}$ represented in the form (2), and let

$t \in T$. Then there exists an $\varepsilon > 0$ so that the restrictions of x to $(t - \varepsilon, t)$ and $(t, t + \varepsilon)$ can be identified with the corresponding restrictions of f.

Accordingly, we can define the left-hand limit $x(t-)$ and the right-hand limit $x(t+)$ of x at t as the corresponding limits of f, i.e. as $f(t-)$ and $f(t+)$ respectively. Finally, referring to (3) above, we shall define $x(t-) = f(t-)$ as "the value of $x \in \mathscr{D}$ at $t \in T$", and this value will be denoted by $x(t)$. Heuristically speaking this means that if x contains a delta function or a step at $t \in T$, then the corresponding "jump" appears "immediately to the right of t".

Projection mappings

Next we shall introduce two important *projection* mappings $\mathscr{D} \to \mathscr{D}$.

Let x again be an arbitrary element of $\mathscr{D}$ represented in the form (2), and let t be an arbitrary element of T. The projection mappings $\lceil_t$ and $\rceil_t$ associated with t are defined as follows.

$$\lceil_t : \mathscr{D} \to \mathscr{D},$$

$$x \mapsto \lceil_t f + \sum_{i \in I} c_i \lceil_t \delta_{t_i}^{(k_i)} \quad \text{with}$$

$$(\lceil_t f)(\tau) = \begin{cases} 0 \text{ for } \tau \in T,\ \tau \leqslant t, \\ f(\tau) \text{ for } \tau \in T,\ \tau > t, \end{cases}$$

4 $$\lceil_t \delta_{t_i}^{(k_i)} = \begin{cases} 0 \text{ (the zero function on } T) \text{ for } t_i < t, \\ \delta_{t_i}^{(k_i)} \text{ for } t_i \geqslant t, \end{cases}$$

$$\rceil_t : \mathscr{D} \to \mathscr{D} \quad \text{with}$$

$$\rceil_t + \lceil_t = I \quad \text{(the identity on } \mathscr{D}).$$

The mappings $\lceil_t$ and $\rceil_t$ defined above are indeed clearly projections, i.e. $\lceil_t \circ \lceil_t = \lceil_t$ and $\rceil_t \circ \rceil_t = \rceil_t$.

Note that $\rceil_t x$ and $\lceil_t x$ above can, roughly speaking, be interpreted as the products of x and the left-hand unit step $(1 - U_t)$ and the ordinary right hand unit step U_t (*cf.* (A4.3)) at t respectively so that $\rceil_t x$ does not contain any delta functions possibly contained in x at t, whereas $\lceil_t x$ does contain such delta functions.

Extensions of $\lceil_t$ and $\rceil_t$

In what follows we shall, without further notice, feel free to extend the mappings $\lceil_t$ and $\rceil_t$ as defined above to suitable sets, lists, etc. comprising elements of $\mathcal{D}$ with the natural interpretation that the projections are applied pointwise, componentwise, etc.

Useful subspaces of $\mathcal{D}$

We can now form a number of useful subspaces of $\mathcal{D}$ with the aid of mappings of the form $\lceil_t$ and $\rceil_t$ as given above. We shall introduce the following subspaces for an arbitrary $t \in T$.

$$\mathcal{D}^+ \triangleq \{x | x \in \mathcal{D} \quad \text{and there exists a } t \in T \text{ such that } \rceil_t x = 0 \text{ (the zero function on } T)\},$$

5 $$\mathcal{D}_t^+ \triangleq \lceil_t \mathcal{D} \subset \mathcal{D}^+,$$

$$\mathcal{D}^- \triangleq \{x | \in \mathcal{D} \text{ and there exists a } t \in T \text{ such that } \lceil_t x = 0\},$$

$$\mathcal{D}_t^- \triangleq \rceil_t \mathcal{D} \subset \mathcal{D}^-.$$

All the sets given above are obviously subspaces of $\mathcal{D}$. In addition, $\mathcal{D}_t^+$ is a subspace of $\mathcal{D}^+$ and $\mathcal{D}_t^-$ a subspace of $\mathcal{D}^-$. $\mathcal{D}^+$ ($\mathcal{D}^-$) consists of elements which are zero to the left (right) of some $t \in T$—the t possibly depending on the element considered. $\mathcal{D}_t^+(\mathcal{D}_t^-)$ consists of elements which are zero to the left (right) of a common $t \in T$. Let us note the following important properties possessed by the subspaces (5).

6 **Note.** **(i)** The elements of $\mathcal{D}_t^-$ and $\mathcal{D}_t^+$ possess the concatenation property:

7 $$\mathcal{D} = \mathcal{D}_t^- \oplus \mathcal{D}_t^+.$$

Thus let $x \in \mathcal{D}$ be arbitrary and form $x^- \triangleq \rceil_t x$ and $x^+ \triangleq \lceil_t x$. Then we have $x^- \in \mathcal{D}_t^-$, $x^+ \in \mathcal{D}_t^+$ and $x = x^- + x^+$. Conversely, take any $x^- \in \mathcal{D}_t^-$ and $x^+ \in \mathcal{D}_t^+$. Then there is an $x \in \mathcal{D}$ such that $x = x^- + x^+$, $\rceil_t x = x^-$, and $\lceil_t x = x^+$. The elements x^- and x^+ appearing here are thus "non-overlapping" in a certain sense (*cf.* appendix A4).
(ii) The spaces $\mathcal{D}^+$, $\mathcal{D}^-$, and $\mathcal{D}_t^+$ are—in a generalized sense—regular signal spaces, but they do not possess the richness property (5.1.5). The space $\mathcal{D}_t^-$ is not a signal space.

That $\mathcal{D}^+$, $\mathcal{D}^-$, and $\mathcal{D}_t^+$ are signal spaces, whereas $\mathcal{D}_t^-$ is not, should be evident (note that $\mathcal{D}_t^-$ is not invariant under the differentiation operator p—consider for instance the derivative of the left-hand unit step $(1 - U_t)$ at t). Moreover, the proof presented in appendix A4 of the regularity of $\mathcal{D}$ can be directly applied to show that $\mathcal{D}^+$ and $\mathcal{D}_t^+$ are regular (additional details relating to this point are given in note (8.3.7) below). An obvious modification of the proof shows that $\mathcal{D}^-$ is also regular. Finally, none of the signal spaces mentioned here possess the richness property (5.1.5), because the ordinary exponential functions are not elements of these spaces (*cf.* note (5.1.6), (i)).

(iii) Now let $\mathscr{X}$ be any one of the regular signal spaces $\mathscr{D}$, $\mathscr{D}^+$, $\mathscr{D}^-$, and $\mathscr{D}_t^+$, and let $\mathbf{C}[p]$ as before denote the set of all polynomial operators of the form (4.1.1). According to previous results, presented in chapters 4 and 5, $\mathscr{X}$ is then a (left) $\mathbf{C}[p]$-module. Moreover, p is an indeterminate over $\mathbf{C}$, $\mathbf{C}[p]$ is a commutative ring with unity, and also an integral domain. □

The causality property

Causality plays an important role in our considerations. The definition of this fundamental concept was given in definition (2.2.2). This definition is, however, applicable only in cases where the time functions are ordinary pointwise defined functions. If generalized functions are involved, a generalized form of causality must be introduced. We shall do this in the following way.

To begin with, let $\mathscr{P}$ denote the set of all piecewise continuous functions on our time interval T contained in $\mathscr{D}$ ($\mathscr{P}$ is then a subspace of $\mathscr{D}$), and let $\rceil_t$ be the projection mapping (4) for some $t \in T$. Then $\mathscr{P}$ is invariant under $\rceil_t$, i.e. $\rceil_t \mathscr{P} \subset \mathscr{P}$. Next consider an input-output mapping $s : \mathscr{P} \to \mathscr{P}$ (*cf.* (2.2.1)). It is then easily confirmed that the causality of s in the sense of definition (2.2.2) is equivalent to the statement that the equality

8 $$\rceil_t(s(\rceil_t u)) = \rceil_t(s(u))$$

holds for any $u \in \mathscr{P}$ and any $t \in T$ (*cf.* Willems, 1971, 2.4).

The generalized causality

We are now in a position to generalize the concept of causality as follows.

Let $\mathscr{X}$ be a subspace of $\mathscr{D}$ such that $\mathscr{X}$ is invariant under $\rceil_t$ for any $t \in T$, and let $s : \mathscr{X}^r \to \mathscr{X}^q$ be an input-output mapping. We shall say that s is causal in a generalized sense if there is an equality of the form (8) holding for any $u \in \mathscr{X}^r$ and any $t \in T$.

Note that sums and compositions of causal input-output mappings are again causal.

9 **Note.** Let $\mathscr{X}$ be a signal space considered as a subspace of $\mathscr{D}$ and suppose that $\mathscr{X}$ is invariant under the projection mapping $\rceil_t$ as given by (4) for any $t \in T$. Let $a(p) : \mathscr{X} \to \mathscr{X}$ be a polynomial operator of the form (4.1.1), and let $A(p) : \mathscr{X}^r \to \mathscr{X}^q$ be a polynomial matrix operator of the form (4.1.3)

both regarded as input-output mappings. Using expressions of the form (A4.20) for the elements of $\mathscr{X}$ it is easily seen that both $a(p)$ and $A(p)$ are causal in our generalized sense. □

10 **Remark.** It was previously stated (see the introduction to part II) that our set of realizability conditions in this context comprises the causality property only. According to the above results (note (9)), differential operators are considered causal, and thus also realizable. This is strictly speaking not in agreement with the common view that pure differentiators cannot be technically realized. We shall regard this discrepancy as a purely formal one—it is, after all, technically possible to realize differentiators with great accuracy.

It is, nevertheless, in practical applications often desirable to avoid technical difficulties, which always arise in realizing differentiators. In connection with the feedback and observer problems discussed in sections 7.11 and 7.12 emphasis was therefore laid on the (strict) properness of the input-output relations involved. □

8.3 The vector space $\mathscr{X}$ over $\mathbf{C}(p)$

In this paragraph we shall discuss the consequences of choosing one of the spaces $\mathscr{D}^+$, $\mathscr{D}^-$, or $\mathscr{D}_t^+$ for some $t \in T$ (*cf.* (8.2.5)) as the signal space.

The situation so far

To begin with, recall some of the basic facts presented in section 5.2; *cf.* also appendix A3.

So, let $\mathscr{X}$ be an arbitrary signal space, and let $L(\mathscr{X}, \mathscr{X})$ denote the algebra of all linear mappings $\mathscr{X} \to \mathscr{X}$ with addition and scalar multiples defined pointwise, and with a composition corresponding to multiplication. Further, let $\mathbf{C}[p]$ denote the set of polynomial operators of the form (4.1.1). Then $\mathbf{C}[p]$ forms a commutative subalgebra of $L(\mathscr{X}, \mathscr{X})$, and $\mathscr{X}$ can be regarded as a (left) $\mathbf{C}[p]$-module.

Now, if $\mathscr{X}$ happens to be regular, then the differentiation operator $p: \mathscr{X} \to \mathscr{X}$ is an indeterminate over $\mathbf{C}$. It follows that in this case $\mathbf{C}[p]$ is not only a commutative ring with unity, but also an integral domain. Consequently, the field of quotients of $\mathbf{C}[p]$, denoted by $\mathbf{C}(p)$, can be formed. The elements of $\mathbf{C}(p)$ are equivalence classes on $\mathbf{C}[p] \times (\mathbf{C}[p] - \{0\})$, and each class is representable by quotients of the form $b(p)/a(p)$ with $a(p), b(p) \in \mathbf{C}[p]$, $a(p) \neq 0$. $\mathbf{C}[p]$ is ring isomorphic in a natural way to the subring of $\mathbf{C}(p)$ consisting of equivalence classes representable by quotients of the form $b(p)/1$ for some $b(p) \in \mathbf{C}[p]$. The field operations in $\mathbf{C}(p)$ correspond to the usual operations for quotients. It should be emphasized that there is, at this stage, no identification between the elements of $\mathbf{C}(p)$ and the elements of some subset of $L(\mathscr{X}, \mathscr{X})$.

Now the following question arises. Is it possible to strengthen the $\mathbf{C}[p]$-module $\mathscr{X}$ above in such a way that $\mathscr{X}$ becomes a vector space over $\mathbf{C}(p)$ with $\mathbf{C}[p]$ regarded as a subring of $\mathbf{C}(p)$?

This question has been discussed in detail elsewhere (*cf.* appendix A3, Blomberg *et al.*, 1969, Blomberg and Salovaara, 1968, Ylinen, 1975). Here we shall be content with giving only the results that are relevant in this context.

Existence of the vector space $\mathscr{X}$ over $\mathbf{C}(p)$

1 **Theorem.** *Let $\mathscr{X}$ be a regular signal space regarded as a (left) $\mathbf{C}[p]$-module, with $\mathbf{C}[p]$ interpreted as a subring of the algrebra $L(\mathscr{X}, \mathscr{X})$ of all linear mappings $\mathscr{X} \rightarrow \mathscr{X}$ (cf. section 5.2), and let $\mathbf{C}(p)$ denote the field of quotients of $\mathbf{C}[p]$ as described previously. In addition, let every nonzero element of $\mathbf{C}[p]$ be invertible, i.e. an automorphism of $\mathscr{X}$. Then the $\mathbf{C}[p]$-module $\mathscr{X}$ can be strengthened to a vector space $\mathscr{X}$ over $\mathbf{C}(p)$ with every quotient $b(p)/a(p)$ of $\mathbf{C}(p)$ ($b(p) \in \mathbf{C}[p]$, $a(p) \in \mathbf{C}[p] - \{0\}$) identified with the corresponding element $a(p)^{-1}b(p)$ of $L(\mathscr{X}, \mathscr{X})$, and with the operations of the $\mathbf{C}[p]$-module $\mathscr{X}$ extended accordingly.* □

The essential thing here is that every nonzero element of $\mathbf{C}[p]$ is assumed to be invertible. It was observed previously (note (5.2.1), (iii)) that if the signal space $\mathscr{X}$ also possesses the richness property (5.1.5), then the nonconstant elements of $\mathbf{C}[p]$ are not automorphisms, and the above vector space structure cannot be obtained.

For the proof of the above theorem, see appendix A3. The proof is, surprisingly enough, far from trivial.

Rational matrix operators

Consider a (left) $\mathbf{C}[p]$-module $\mathscr{X}$ and suppose that the conditions of theorem (1) are fulfilled so that this module can be strengthened in a natural way to the vector space $\mathscr{X}$ over the field of quotients $\mathbf{C}(p)$, with $\mathbf{C}(p)$ regarded as a subset of $L(\mathscr{X}, \mathscr{X})$ as explained above.

A q by r-matrix $\mathscr{G}(p)$ with entries from the field of quotients $\mathbf{C}(p)$ as given by $(q, r \in \{1, 2, 3, \ldots\})$

2 $$\mathcal{G}(p) \triangleq \begin{bmatrix} g_{11}(p) & g_{12}(p) & \ldots & g_{1r}(p) \\ g_{21}(p) & g_{22}(p) & \ldots & g_{2r}(p) \\ \vdots & \vdots & & \vdots \\ g_{q1}(p) & g_{q2}(p) & \ldots & g_{qr}(p) \end{bmatrix}$$

with the $g_{ij}(p) \in \mathbf{C}(p)$, then clearly represents a linear mapping $\mathcal{X}^r \rightarrow \mathcal{X}^q$. For any $u = (u_1, u_2, \ldots, u_r) \in \mathcal{X}^r$ the corresponding $y = (y_1, y_2, \ldots, y_q) = \mathcal{G}(p)u \in \mathcal{X}^q$ is evaluated by interpreting $y = \mathcal{G}(p)u$ as a matrix equality with u and y regarded as column matrices.

We shall usually identify a mapping represented by a matrix of the form (2) with the matrix itself, and call it a "rational matrix operator". If we look at (2) just as a matrix with entries from the field of quotients $\mathbf{C}(p)$, then the term "rational matrix" will be used. This is in agreement with the convention explained in section 5.3. Statements holding for general matrices over a field (concerning rank, inverses, etc.) naturally also hold for rational matrices.

3 **Note.** Consider a rational matrix operator $\mathcal{G}(p): \mathcal{X}^r \rightarrow \mathcal{X}^q$ as given by (2). Then $\mathcal{G}(p): \mathcal{X}^r \rightarrow \mathcal{X}^q$ is a surjection with respect to $\mathcal{X}^q$ if and only if $\mathcal{G}(p)$ regarded as a rational matrix is of rank q (implies that $q \leq r$). Further, $\mathcal{G}(p): \mathcal{X}^r \rightarrow \mathcal{X}^q$ has an inverse $\mathcal{G}(p)^{-1}: \mathcal{X}^q \rightarrow \mathcal{X}^r$ if and only if $r = q$ and det $\mathcal{G}(p) \neq 0$. The matrix representing $\mathcal{G}(p)^{-1}: \mathcal{X}^q \rightarrow \mathcal{X}^r$ is in this case the rational matrix obtained as the inverse of the rational matrix representing $\mathcal{G}(p): \mathcal{X}^r \rightarrow \mathcal{X}^q$. □

Rational matrices and transfer matrices

Rational matrices, i.e. matrices with entries from the field of quotients $\mathbf{C}(p)$, were already encountered in section 6.3, where we defined the transfer matrix determined by a regular generator $[A(p) \,\vdots\, -B(p)]$ as the rational matrix (*cf.* (6.3.1))

4 $$\mathcal{G}(p) = A(p)^{-1}B(p).$$

Conversely it is known (*cf.* appendices A1, A2, and (6.4.7)) that every rational matrix $\mathcal{G}(p)$ of the form (2) can be factored as $\mathcal{G}(p) = A(p)^{-1} B(p)$, where $A(p)$ and $B(p)$ are suitable polynomial matrices. In this context we shall therefore regard the terms "rational matrix" and "rational matrix operator" as synonymous to the term "transfer matrix" and "transfer matrix operator" respectively.

Choosing one of the spaces $\mathscr{D}^+$, $\mathscr{D}^-$, or $\mathscr{D}_t^+$ as the signal space $\mathscr{X}$

Now, let the signal space $\mathscr{X}$ be one of the nontrivial regular signal spaces $\mathscr{D}^+$, $\mathscr{D}^-$, or $\mathscr{D}_t^+$ for some $t \in T$, as given by (8.2.5), and let

5 $$a(p) \triangleq a_0 + a_1 p + \ldots + a_n p^n, a_n \neq 0$$

be a polynomial operator $\mathscr{X} \to \mathscr{X}$ as given by (4.1.1). The regularity of $\mathscr{X}$ implies that $a(p)$ is a surjection with respect to $\mathscr{X}$.

Next, consider the equation $a(p)y = 0$. The set of all solutions $y \in \mathscr{D}$ to this equation is known to form an n-dimensional subspace of C^∞ spanned by functions of exponential type (*cf.* note (5.1.6), (i), and appendix A4). But then $a(p)y = 0$ can have only one solution $y \in \mathscr{X}$, namely $y = 0$. Consequently, $a(p)$ is an injection. It is thus concluded that $a(p): \mathscr{X} \to \mathscr{X}$ as given by (5) has an inverse $a(p)^{-1}: \mathscr{X} \to \mathscr{X}$.

We have thus proved the following corollary to theorem (1) above.

6 **Corollary.** *Let the signal space $\mathscr{X}$ be one of the regular signal spaces $\mathscr{D}^+$, $\mathscr{D}^-$, or $\mathscr{D}_t^+$ for some $t \in T$ as given by* (8.2.5), *and let $\mathscr{X}$ be regarded as a* $\mathbf{C}[p]$*-module (cf. section* 5.2). *Then the* $\mathbf{C}[p]$*-module $\mathscr{X}$ can be strengthened to a vector space $\mathscr{X}$ over* $\mathbf{C}(p)$*—the field of quotients of* $\mathbf{C}[p]$*—as described in theorem* (1). □

7 **Note.** Let the signal space $\mathscr{X}$ throughout items (i), . . . , (iii) below be equal to either of the regular signal spaces $\mathscr{D}^+$ or $\mathscr{D}_{t_0}^+$ for some $t_0 \in T$. The space $\mathscr{D}^-$ is not considered here, because it is not a suitable signal space as far as dynamical (causal) systems are concerned.

(i) Consider a differential input-output relation $S \subset \mathscr{X}^r \times \mathscr{X}^q$ generated by a matrix differential equation of the form (*cf.* (4.2.1))

8 $$A(p)y = B(p)u,$$

where $u \in \mathscr{X}^r$, $y \in \mathscr{X}^q$, and where $A(p)$ and $B(p)$ are polynomial matrix operators of sizes q by q and q by r respectively with $\det A(p) \neq 0$. For any given $u \in \mathscr{X}^r$ there is then, according to the above results, a unique corresponding $y \in \mathscr{X}^q$ given by

9 $$y = A(p)^{-1}B(p)u = \mathscr{G}(p)u$$

with $\mathscr{G}(p): \mathscr{X}^r \to \mathscr{X}^q$ a rational matrix operator. We thus have the situation (*cf.* note (4.2.6), (iii)), that the relation S generated by (8) really is an input-output mapping $\mathscr{X}^r \to \mathscr{X}^q$, in fact, $S = \mathscr{G}(p)$.

It should be emphasized that we are here calculating in the vector space $\mathscr{X}$ over the field of quotients $\mathbf{C}(p)$. The polynomial matrices $A(p)$ and $B(p)$ appearing above should therefore be interpreted as special kinds of rational matrices. Further, in order to solve y from (8) for a given u, all the usual operations for solving a set of vector space equations can be applied. In particular it is possible to premultiply (8) by any q by q nonsingular rational matrix without influencing the solution. This means that (8) can always be replaced by an equivalent equation, where the polynomial matrix $A(p)$ is, say, upper triangular, or even diagonal.

It turns out that $\mathcal{G}(p)$ as given above is a causal input-output mapping in the generalized sense discussed in section 8.2. In order to show this we need an explicit expression for the y determined by (9) for a given u. Such an expression can be constructed as, for instance, outlined in the next item.

(ii) Consider the situation described in the previous item and in particular equation (9). Let the entries of $\mathcal{G}(p)$ in (9) be denoted as in (2).

Next let $u \in \mathcal{X}^r$ be arbitrary but fixed. This means that there is a $t_0 \in T$ such that $u \in \mathcal{D}_{t_0}^{+r}$. The y given by (9) is then necessarily a member of $\mathcal{D}_{t_0}^{+q}$, and an explicit expression for this y can be constructed as follows.

Let $y \triangleq (y_1, y_2, \ldots, y_q)$ and $u \triangleq (u_1, u_2, \ldots, u_r)$, and consider, for instance, y_1. $y = \mathcal{G}(p)u$ gives the following equation determining y_1:

10 $$y_1 = g_{11}(p)u_1 + g_{12}(p)u_2 + \ldots + g_{1r}(p)u_r.$$

Let $a(p)$ be a common polynomial multiple (preferably a least common multiple) of the polynomial denominators of $g_{11}(p), g_{12}(p), \ldots, g_{1r}(p)$. Premultiplication of (10) by $a(p)$ then gives an equivalent equation for y_1 of the form

11 $$a(p)y_1 = x$$

with

12 $$x \triangleq a(p)g_{11}(p)u_1 + a(p)g_{12}(p)u_2 + \ldots + a(p)g_{1r}(p)u_r.$$

$a(p)g_{11}(p)$ etc. in (12) are of course now polynomials in p.

Next write the components $u_1, u_2, \ldots, u_r$ of u in the form (A4.24), and compute x as given by (12). Bring x to the form (A4.24). Of course, $x \in \mathcal{D}_{t_0}^{+}$. Write x as

13 $$x = f + \Sigma_u,$$

where f is a piecewise continuous function $T \to \mathbf{C}^q$ with $f(t) = 0$ for all $t \leqslant t_0$, and where every component of Σ_u consists of a sum of delta functions of various orders at various points t_i of T with $t_i \geqslant t_0$ for all t_i. Clearly, both f and Σ_u are members of $\mathcal{D}_{t_0}^{+}$.

We have now arrived at a situation which is closely related to the situation represented by (A4.25). The difference is merely that here we are interested in solutions y_1 to (11) with $y_1 \in \mathcal{D}_{t_0}^{+}$ rather than with $y_1 \in \mathcal{D}$ as in (A4.25).

So write

14 $$y_1 = y_0 + \Sigma_y,$$

where $y_0 \in \mathcal{D}_{t_0}^{+}$ is chosen so that

15 $$a(p)y_0 = f$$

and $\Sigma_y \in \mathcal{D}_{t_0}^{+}$ so that

16 $$a(p)\,\Sigma_y = \Sigma_u.$$

The y_0 and Σ_y as given by (15) and (16) always exist and are unique.

Σ_y as given by (16) can be found by means of the method outlined in (A4.24), . . . , (A4.29) with t_0 above identified with t_0 in (A4.29).

y_0 as given by (15) is a piecewise continuous function $T \to \mathbf{C}$, and it is well known that it can be expressed as

17 $$y_0(t) = \int_{t_0}^{t} g(t-\tau)\, f(\tau)\, d\tau$$

for every $t \in T$, where $g: T \to \mathbf{C}$, $g(t) = 0$ for all $t \leqslant 0$, is the "weighting function" (also called "impulse response function" or "Green's function") associated with $a(p)$ (see e.g. Zadeh and Desoer, 1963, section 5.2). g is obtained as the unique solution to

18 $$a(p)g = \delta_0, \quad g(t) = 0 \quad \text{for all } t \leqslant 0.$$

It is then easily found that y_1 as given by (14) is the unique solution in $\mathcal{D}_{t_0}^{+}$ to (11).

Finally the remaining components $y_2, \ldots, y_q$ of $y = \mathcal{G}(p)u \in \mathcal{D}_{t_0}^{+q}$ as given by (9) for the chosen $u \in \mathcal{D}_{t_0}^{+r}$ are determined analogously.

(iii) Let $u \in \mathcal{X}^r$ be chosen as in item (ii) above, i.e. $u \in \mathcal{D}_{t_0}^{+r}$ for some $t_0 \in T$, and let $y \in \mathcal{D}_{t_0}^{+q}$ given by $y = \mathcal{G}(p)u$ according to (9) be determined as described. Then take an arbitrary $t_1 \in T$. It follows that $\rceil_{t_1} u \in \mathcal{D}_{t_0}^{+r}$, and that there is a unique $z \in \mathcal{D}_{t_0}^{+q}$ given by

19 $$z = \mathcal{G}(p)\, \rceil_{t_1} u.$$

This z can again be explicitly constructed as described in item (ii). It is readily found that the following equality then holds:

20 $$\begin{aligned} \rceil_{t_1} z &= \rceil_{t_1}(\mathcal{G}(p)\, \rceil_{t_1} u) \\ &= \rceil_{t_1} y = \rceil_{t_1}(\mathcal{G}(p)u). \end{aligned}$$

Because $u \in \mathcal{X}^r$ and $t_1 \in T$ were arbitrary, it follows according to our definition given in section 8.2 (*cf.* in particular (8.2.8)) that $\mathcal{G}(p) =$

$A(p)^{-1}B(p):\mathscr{X}^r \to \mathscr{X}^q$ as given by (9) is a causal input-output mapping.
(iv) In many cases the Laplace transform method, including tables over function-transform pairs, could be conveniently used to determine some of the functions appearing above. This holds in particular if the space of Laplace transformable functions is extended to include also delta functions. Such an extension is possible on the basis of the general theory for generalized functions. For additional material relating to this matter, see the book by Zadeh and Desoer, 1963, in particular chapters 4 and 5, and note (8.4.20), (vi) below. □

8.4 Compositions of projections and differential operators. Initial condition mappings

Compositions of projections of the kind introduced in section 8.2 and differential operators prove to be a useful tool in the sequel. The main question is: What happens when the order of the mappings in a composition of a projection and a differential operator is reversed?

A fundamental relationship

To begin with let us derive a fundamental relationship.

Let $A(p):\mathscr{D}^q \to \mathscr{D}^s$ be a polynomial matrix operator of the form (4.1.3), let $\rceil_t$ and $\lceil_t$ for some $t \in T$ be projections of the form (8.2.4) appropriately extended, and let $x \in \mathscr{D}^q$ be arbitrary. Note that $\mathscr{D}^q$ and $\mathscr{D}^s$ can be written as the direct sums $\mathscr{D}_t^{-q} \oplus \mathscr{D}_t^{+q}$ and $\mathscr{D}_t^{-s} \oplus \mathscr{D}_t^{+s}$ respectively (*cf.* note (8.2.6), (i)), and that x can thus be written in a unique way as

1 $$x = x^- + x^+$$

with

2 $$x^- \triangleq \rceil_t x \in \mathscr{D}_t^{-q},$$
$$x^+ \triangleq \lceil_t x \in \mathscr{D}_t^{+q}.$$

It follows that $A(p)x$ can be written as

3 $$A(p)(x^- + x^+) = A(p)x^- + A(p)x^+$$

with $A(p)x^+ \in \mathscr{D}_t^{+s}$.

On taking the image under $\lceil_t$ of both sides of (3) and observing that $\lceil_t A(p)x^+ = A(p)x^+$ we arrive at

4 $$\lceil_t(A(p)(x^- + x^+)) = \lceil_t(A(p)x^-) + A(p)x^+,$$

or equivalently

5 $$\lceil_t(A(p)x) = \lceil_t(A(p)\rceil_t x) + A(p)\lceil_t x.$$

Now the left-hand side of (5) is just the image of x under $\lceil_t \circ A(p):\mathfrak{D}^q \to \mathfrak{D}_t^{+s}$, whereas the last term on the right is the image of x under $A(p) \circ \lceil_t:\mathfrak{D}^q \to \mathfrak{D}_t^{+s}$. The difference between these images is given by the image of x under $\lceil_t \circ A(p) \circ \rceil_t:\mathfrak{D}^q \to \mathfrak{D}_t^{+s}$ or equivalently, by the image of $x^- \triangleq \rceil_t x$ under $\lceil_t \circ A(p):\mathfrak{D}^q \to \mathfrak{D}_t^{+s}$.

A corresponding relationship concerning compositions of $\rceil_t$ and $A(p)$ is found in the following way. From $A(p)(x^- + x^+) = (\rceil_t + \lceil_t)(A(p)(x^- + x^+))$ it follows that

6 $$\rceil_t(A(p)(x^- + x^+)) = -\lceil_t(A(p)(x^- + x^+)) + A(p)x^- + A(p)x^+.$$

Substitution of (4) in (6) then yields

7 $$\rceil_t(A(p)(x^- + x^+)) = -\lceil_t(A(p)x^-) + A(p)x^-,$$

or equivalently

8 $$\rceil_t(A(p)x) = -\lceil_t(A(p)\rceil_t x) + A(p)\rceil_t x.$$

Again it is thus concluded that the difference between the images of x under $\rceil_t \circ A(p):\mathfrak{D}^q \to \mathfrak{D}_t^{-s}$ and $A(p) \circ \rceil_t:\mathfrak{D}^q \to \mathfrak{D}^s$ respectively is given by the image of x under $\lceil_t \circ A(p) \circ \rceil_t:\mathfrak{D}^q \to \mathfrak{D}_t^{+s}$.

Consider the difference $\lceil_t(A(p)\rceil_t x) = \lceil_t(A(p)x^-)$ appearing above. The components of x^- and of its derivatives px^-, p^2x^-, . . . contain jumps at t of magnitudes given by $-x^-(t)$, $-(px^-)(t)$, . . . It is thus clear that the components of $A(p)x^-$ generally contain delta functions of various orders at t caused by these jumps and their derivatives. From t to the right $A(p)x^-$ is zero. Finally it is concluded that $\lceil_t A(p)x^-$ must be zero except for the delta functions contained in $A(p)x^-$ at t. Note in passing that the values of x^-, px^-, . . . at t with $x^- \triangleq \rceil_t x$ coincide with the corresponding values of x, px, . . . This fact will occasionally be used later on.

Projections and polynomial matrix operators do not generally commute

As a result it is thus concluded that projections of the kind introduced in section 8.2 and polynomial matrix operators do not generally commute.

Initial condition mappings

We shall, in the sequel, mostly use expressions of the form (4) and (7). In order to simplify the notations we shall therefore denote the restriction of $\lceil_t \circ A(p):\mathcal{D}^q \to \mathcal{D}_t^{+s}$ to $\mathcal{D}_t^{-q}$ by A_t^0, i.e.

9 $$A_t^0 \triangleq \lceil_t \circ A(p)\,|\,\mathcal{D}_t^{-q}:\mathcal{D}_t^{-q} \to \mathcal{D}_t^{+s}.$$

A_t^0 will be called the *initial condition mapping* at t associated with the polynomial matrix operator $A(p):\mathcal{D}^q \to \mathcal{D}^s$ (note the obvious notational convention). The initial condition mapping is closely related to the effect caused by the initial conditions appearing in the theory of differential equations (*cf.* section 8.5)—hence the name. If t is fixed and clear from the context, then the simpler notation A^0 is used for A_t^0. Using the initial condition mapping (9) we can thus write (4) and (7) as

10 $$\lceil_t(A(p)\,(x^- + x^+)) = A_t^0 x^- + A(p)x^+$$

and

11 $$\rceil_t(A(p)\,(x^- + x^+)) = -A_t^0 x^- + A(p)x^-$$

respectively with

12 $$A_t^0 x^- = \lceil_t(A(p)x^-)$$

for every $x^- \in \mathcal{D}_t^{-q}$ and every $x^+ \in \mathcal{D}_t^{+q}$.

Note that initial condition mappings as defined above are clearly linear mappings.

In what follows we shall need more explicit expressions for (12).

A fundamental lemma

The following lemma gives a result that is fundamental in this context (*cf.* Blomberg *et al.*, 1969, section 5.2).

13 **Lemma.** Let $a(p):\mathcal{D} \to \mathcal{D}$ *be a polynomial operator* (*cf.* (4.1.1)) *of the form* $a(p) = p^n$ *for some* $n \in \{0, 1, 2, \ldots\}$, *and let* $t \in T$. *Furthermore, let* $a_t^0:\mathcal{D}_t^- \to \mathcal{D}_t^+$ *be the initial condition mapping at* t *associated with* $a(p) = p^n$ *given by* (*cf.* (12))

14 $$a_t^0 x^- = \lceil_t(p^n x^-)$$

for every $x^- \in \mathcal{D}_t^-$. *Then* $a_t^0 x^-$ *in* (14) *can be expressed as*

15 $$a_t^0 x^- = -(p^{n-1}x^-)(t)\,\delta_t - (p^{n-2}x^-)(t)\,\delta_t^{(1)} - \ldots - x^-(t)\,\delta_t^{(n-1)},$$

where $(p^i x^-)(t)$ *denotes the value of* $p^i x^- \in \mathcal{D}_t^-$ *at* t (*cf.* section 8.2). *For* $n = 0$ (15) *is interpreted as* $a_t^0 x^- = 0$, *i.e.* a_t^0 *is in this case the zero mapping.* □

16 **Proof.** The proof of the above lemma is easily performed by induction on

n and noting that for $n = 1$ we have $a_t^0 x^- = \lceil_t (px^-) = -x^-(t)\delta_t$, i.e. $a_t^0 x^-$ comprises just the derivative $-x^-(t)\delta_t$ of the jump $-x^-(t)$ appearing in x^- at t. □

The initial condition mapping associated with a polynomial operator

On the basis of the result stated in lemma (13) above it is now an easy matter to determine the initial condition mapping $a_t^0 : \mathcal{D}_t^- \to \mathcal{D}_t^+$ at t associated with an arbitrary polynomial operator $a(p) : \mathcal{D} \to \mathcal{D}$ of the form (4.1.1). Thus let $a(p)$ be given as

17 $$a(p) \triangleq a_0 + a_1 p + \ldots + a_n p^n.$$

It is then found that

18 $$a_t^0 x^- = \lceil_t (a(p)x^-)$$

$$= -\left[\left(\sum_{i=1}^{n} a_i (p^{i-1}x^-)(t)\right)\delta_t + \left(\sum_{i=2}^{n} a_i (p^{i-2}x^-)(t)\right)\delta_t^{(1)} + \ldots + a_n x^-(t)\,\delta_t^{(n-1)}\right]$$

holds for every $x^- \in \mathcal{D}_t^-$. For $\partial(a(p)) < 1$, a_t^0 is interpreted as the zero mapping.

If $x \in \mathcal{D}$ is such that $x^- = \rceil_t x$, then the values $x^-(t)$, $(px^-)(t), \ldots,$ $(p^{n-1}x^-)(t)$ appearing in (15) and (18) are equal to—and can be replaced by—the values $x(t)$, $(px)(t), \ldots, (p^{n-1}x)(t)$ respectively.

The initial condition mapping associated with a polynomial matrix operator

Next consider a polynomial matrix operator $A(p) : \mathcal{D}^q \to \mathcal{D}^s$ of the form (4.1.3), and let $A_t^0 : \mathcal{D}_t^{-q} \to \mathcal{D}_t^{+s}$ be the initial condition mapping at t associated with $A(p)$. Then, using obvious properties and interpretations (in particular, $\lceil_t \circ a_{ij}(p)$ and $\lceil_t \circ A(p)$ are written simply as $\lceil_t a_{ij}(p)$ and $\lceil_t A(p)$ respectively).

19 $$A_t^0 x^- = \lceil_t (A(p)x^-)$$

$$= \underbrace{\begin{bmatrix} \lceil_t a_{11}(p) & \lceil_t a_{12}(p) & \ldots & \lceil_t a_{1q}(p) \\ \lceil_t a_{21}(p) & \lceil_t a_{22}(p) & \ldots & \lceil_t a_{2q}(p) \\ \vdots & \vdots & & \vdots \\ \lceil_t a_{s1}(p) & \lceil_t a_{s2}(p) & \ldots & \lceil_t a_{sq}(p) \end{bmatrix}}_{\lceil_t A(p)} \underbrace{\begin{bmatrix} x_1^- \\ x_2^- \\ \vdots \\ x_q^- \end{bmatrix}}_{x^-}$$

$$= \underbrace{\begin{bmatrix} a^0_{11t} & a^0_{12t} & \dots & a^0_{1qt} \\ a^0_{21t} & a^0_{22t} & \dots & a^0_{2qt} \\ \vdots & \vdots & & \vdots \\ a^0_{s1t} & a^0_{s2t} & \dots & a^0_{sqt} \end{bmatrix}}_{A^0_t} \underbrace{\begin{bmatrix} x_1^- \\ x_2^- \\ \vdots \\ x_q^- \end{bmatrix}}_{x^-}$$

holds for every $x^- \triangleq (x_1^-, x_2^-, \dots, x_q^-) \in \mathcal{D}_t^{-q}$.

20 **Note.** **(i)** It was concluded above that projections and polynomial matrix operators do not generally commute. They do commute if the polynomial matrix operator is a constant matrix operator. This follows readily from (19) with $a^0_{ijt} = 0$ for all i, j. Projections and unimodular polynomial matrix operators would probably also be expected to commute. It can, however, easily be shown that this does not generally hold for unimodular polynomial matrix operators of size q by q with $q > 1$ (consider, for instance, $A(p) = \begin{bmatrix} 1 & p \\ 0 & 1 \end{bmatrix}$ and determine $\lceil_t(A(p)\rceil_t x) \neq 0$ for $x = \begin{bmatrix} 0 \\ 1 \end{bmatrix}$, where 1 is a constant signal).

(ii) Let $A(p)$, $B(p)$, $C(p)$, and $D(p)$ be suitable polynomial matrix operators and let L and M be suitable constant matrices such that

21 $$C(p) = L\,A(p)M$$

and

22 $$D(p) = A(p) + B(p).$$

Furthermore, let A^0, B^0, C^0, D^0 be the initial condition mappings at some fixed t associated with $A(p)$, $B(p)$, $C(p)$, and $D(p)$ respectively as given by (19). Then the following equalities hold

23 $$C^0 = L\,A^0 M,$$

and

24 $$D^0 = A^0 + B^0.$$

These results follow readily on the basis of the defining equality (19). We shall use the above results later on.

(iii) Next we shall give an alternative formulation of $a^0 x^-$ (t fixed) as given by (18). For this purpose we shall introduce a number of notational conventions.

For any $n \in \{0, 1, 2, \dots\}$, let $\mathbf{C}_n[p]$ denote the subset of $\mathbf{C}[p]$ consisting of the set of all polynomial operators $\mathscr{D} \to \mathscr{D}$ of the form (4.1.1) of degree $\leqslant n$. $\mathbf{C}_n[p]$ constitutes an $(n+1)$-dimensional vector space over $\mathbf{C}$.

To every $a(p) = a_0 + a_1 p + \ldots + a_n p^n \in \mathbf{C}_n[p]$ we assign a corresponding coefficient list $a \triangleq (a_0, a_1, \ldots, a_n) \in \mathbf{C}^{n+1}$. This assignment determines a one-one correspondence between the elements of $\mathbf{C}_n[p]$ and $\mathbf{C}^{n+1}$.

To every $a(p) = a_0 + a_1 p + \ldots + a_n p^n \in \mathbf{C}[p]$ we assign a corresponding d by d coefficient matrix $\underline{a} \in \mathbf{C}^{d \times d}$ as follows.

If $\partial(a(p)) < 1$, then $d = 1$ and $\underline{a}$ is the 1 by 1 zero matrix.

If $\partial(a(p)) \geqslant 1$, then $d = \partial(a(p))$ and $\underline{a}$ is formed according to

25
$$\underline{a} = \begin{bmatrix} a_1 & a_2 & \ldots & a_{d-1} & a_d \\ a_2 & a_3 & \ldots & a_d & 0 \\ \vdots & \vdots & & \vdots & \vdots \\ a_{d-1} & a_d & \ldots & 0 & 0 \\ a_d & 0 & \ldots & 0 & 0 \end{bmatrix}$$

with $\det \underline{a} = a_d^d \neq 0$. $\underline{a}$ is thus invertible as a constant matrix for $\partial(a(p)) \geqslant 1$.

Later we shall also use augmented coefficient matrices which are obtained from the coefficient matrices as defined above by adding a suitable number of zero rows at the bottom and zero columns on the right (*cf.* item (iv) below).

To every $x^- \in \mathscr{D}_t^-$ we assign a corresponding initial condition list with respect to $\mathbf{C}^n$ (and t) given by $x^0 \triangleq (x^-(t), (px^-)(t), \ldots, (p^{n-1}x^-)(t)) \in \mathbf{C}^n$. The actual value of n is not indicated in x^0—this value will be clear from the context. If $x \in \mathscr{D}$ is such that $x^- = \rceil_t x$, then we shall also say that x^0 above is the initial condition list corresponding to x.

Finally it is noted that $\delta_t^{(i)}$ for every $i \in \{0, 1, 2, \ldots\}$ can be written as $p^i \delta_t$.

Now consider (17) and (18) and let the $a(p)$ in (17) be an arbitrary element of $\mathbf{C}[p]$. Determine $d \geqslant 1$ and form the d by d coefficient matrix $\underline{a} \in \mathbf{C}^{d \times d}$ associated with $a(p)$ as described above. Choose $x^- \in \mathscr{D}^-$ and form the corresponding initial condition list $x^0 \triangleq (x^-(t), (px^-)(t), \ldots, (p^{d-1}x^-)(t)) \in \mathbf{C}^d$. Compute the corresponding coefficient list $c \triangleq (c_0, c_1, \ldots, c_{d-1}) \in \mathbf{C}^d$ from the matrix equation

26
$$c = \underline{a} x^0.$$

Form the $c(p) \in \mathbf{C}_{d-1}[p]$ corresponding to the coefficient list c found above. Then (18) can be written as

27
$$a^0 x^- = -c(p)\delta_t.$$

In view of the results obtained we arrive at the following conclusion.

When x^- is varied over the whole of $\mathscr{D}_t^-$, the corresponding initial condition list x^0 in (26) varies over the whole of $\mathbf{C}^d$. If $\partial(a(p))<1$, we have $d=1$ and $\underline{a}=0$. Thus c and $c(p)$ are in this case zero for every $x^0 \in \mathbf{C}^d$. If $\partial(a(p)) \geqslant 1$, we have $d=\partial(a(p))$ and $\underline{a}$ is invertible. Thus c and $c(p)$ vary in this case over the whole of $\mathbf{C}^d$ and $\mathbf{C}_{d-1}[p]$ respectively when x^0 is varied over $\mathbf{C}^d$.

Finally the above results lead to the conclusion that for $\partial(a(p)) \geqslant 1$, the range of a^0 is a $d=\partial(a(p))$-dimensional subspace of $\mathscr{D}_t^+$ spanned by the set of linearly independent delta functions $\{\delta_t, \delta_t^{(1)}, \ldots, \delta_t^{(d-1)}\}$, i.e. this set forms a basis for the subspace considered.

(iv) A formulation corresponding to (26) and (27) above can also be given in the multivariable case for A^0x^- (t fixed) as given by (19).

Thus let $A(p):\mathscr{D}^q \to \mathscr{D}^s$ be an arbitrary polynomial matrix operator of the form (4.1.3). Determine the d_{ij} by d_{ij} coefficient matrices $\underline{a}_{ij}$ associated with the corresponding entries $a_{ij}(p)$ of $A(p)$ as described above (*cf.* (25)). Every $\underline{a}_{ij}$ is then augmented as described in the previous item by adding (if necessary) zero rows and columns to a d_{ri} by d_{cj}-matrix $\underline{a}'_{ij}$, where $d_{ri} \triangleq \max\{d_{i1}, d_{i2}, \ldots, d_{iq}\}$ and $d_{cj} \triangleq \max\{d_{1j}, d_{2j}, \ldots, d_{sj}\}$. The augmented coefficient matrices thus obtained are then used to form the d_1 by d_2 coefficient matrix $\underline{A} \in \mathbf{C}^{d_1 \times d_2}$, $d_1 \triangleq d_{r1}+d_{r2}+\ldots+d_{rs}$, $d_2 \triangleq d_{c1}+d_{c2}+\ldots+d_{cq}$, given in partitioned form as follows:

3 $$\underline{A} \triangleq \begin{bmatrix} \underline{a}'_{11} & \underline{a}'_{12} & \cdots & \underline{a}'_{1q} \\ \underline{a}'_{21} & \underline{a}'_{22} & \cdots & \underline{a}'_{2q} \\ \vdots & \vdots & & \vdots \\ \underline{a}'_{s1} & \underline{a}'_{s2} & \cdots & \underline{a}'_{sq} \end{bmatrix}.$$

Next choose an $x^- \in \mathscr{D}_t^{-q}$ and form the corresponding initial condition list $x^0 \in \mathbf{C}^{d_2}$ given in partitioned form by the following column matrix:

29 $$x^0 \triangleq \begin{bmatrix} x_1^0 \\ x_2^0 \\ \vdots \\ x_q^0 \end{bmatrix},$$

where $x_j^0 (j=1, 2, \ldots, q)$ represents the initial condition list $(x_j^-(t), (px_j^-)(t), \ldots, p^{d_{cj}-1}x_j^-)(t)) \in \mathbf{C}^{d_{cj}}$ corresponding to x_j^-. Compute the corresponding coefficient list $c \in \mathbf{C}^{d_1}$ from the matrix equation

30 $$c = \underline{A}x^0,$$

or more explicitly from

31 $$\begin{bmatrix} c_1 \\ c_2 \\ \vdots \\ c_s \end{bmatrix} = \begin{bmatrix} \underline{a}'_{11} & \underline{a}'_{12} & \dots & \underline{a}'_{1q} \\ \underline{a}'_{21} & \underline{a}'_{22} & \dots & \underline{a}'_{2q} \\ \vdots & \vdots & & \vdots \\ \underline{a}'_{s1} & \underline{a}'_{s2} & \dots & \underline{a}'_{sq} \end{bmatrix} \begin{bmatrix} x_1^0 \\ x_2^0 \\ \vdots \\ x_q^0 \end{bmatrix}.$$

Form the corresponding list $C(p) \triangleq (c_1(p), c_2(p), \dots, c_s(p)) \in \mathbf{C}_{d_{r1}-1}[p] \times \mathbf{C}_{d_{r2}-1}[p] \times \dots \times \mathbf{C}_{d_{rs}-1}[p]$ of polynomial operators so that $c_i(p) \in \mathbf{C}_{d_{ri}-1}[p]$ $(i = 1, 2, \dots, s)$ is obtained from $c_i = (c_{i0}, c_{i1}, \dots, c_{i(d_{ri}-1)})$ according to

32 $$c_i(p) = c_{i0} + c_{i1}p + \dots + c_{i(d_{ri}-1)}p^{d_{ri}-1}.$$

Then A^0x^- as given by (19) can be expressed as

33 $$A^0x^- = -C(p)\,\delta_t = -\begin{bmatrix} c_1(p) \\ c_2(p) \\ \vdots \\ c_s(p) \end{bmatrix} \delta_t.$$

Now when x^- is varied over the whole of $\mathcal{D}_t^{-q}$, the corresponding initial condition list x^0 in (29), ..., (31) varies over the whole of $\mathbf{C}^{d_2}$. The corresponding variations of $c \in \mathbf{C}^{d_1}$ and $C(p) \in \mathbf{C}[p]^s$ in (30), (31) and (33) depend on the properties of the coefficient matrix $\underline{A}$ as given by (28).

Using the results obtained in item (iii) above it is thus seen that the range of A^0 in this case is a finite dimensional subspace of $\mathcal{D}_t^{+s}$ spanned by some finite set of linearly independent s-lists of the form $(\delta_t, 0, \dots, 0)$, $(\delta_t^{(1)}, 0, \dots, 0), \dots, (0, \delta_t, \dots, 0), (0, \delta_t^{(1)}, \dots, 0), \dots, (0, 0, \dots, \delta_t), (0, 0, \dots, \delta_t^{(1)}), \dots$. The dimension of this subspace is at most equal to the sum of the row degrees (*cf.* appendix A2) of $A(p)$ $(\leqslant d_1 = d_{r1} + d_{r2} + \dots + d_{rs})$.

The next item is devoted to some additional comments relating to the above results.

(v) Consider the problem discussed in the previous item, and let the notations introduced there be valid. Suppose that the polynomial matrix operator $A(p): \mathcal{D}^q \to \mathcal{D}^s$ considered in such that $q \geqslant s$ and that $A(p)$ is of full rank s. Suppose furthermore that $A(p)$ is row proper and that all the row degrees of $A(p)$ are $\geqslant 1$, i.e. $A(p)$ does not contain any constant rows. It is then readily concluded that the corresponding coefficient matrix $\underline{A} \in \mathbf{C}^{d_1 \times d_2}$ as given by (28) has the following properties:

$d_{r1}, d_{r2}, \dots, d_{rs}$ are equal to the corresponding row degrees of $A(p)$,

34 denoted by $r_1, r_2, \ldots, r_s$ respectively ($r_i \triangleq$ the degree of row i),

$d_2 \geqslant d_1$,

$\underline{A}$ is of full rank d_1.

In this case it thus follows that c and $C(p)$ appearing in (30), . . . , (33) vary over the whole of $\mathbf{C}^{d_1}$ and $\mathbf{C}_{d_{r1}-1}[p] \times \mathbf{C}_{d_{r2}-1}[p] \times \ldots \times \mathbf{C}_{d_{rs}-1}[p]$, respectively when the initial condition list x^0 appearing in (29), . . . , (31) varies over the whole of $\mathbf{C}^{d_2}$. The range of A^0 is now a $d_1 = d_{r1} + d_{r2} + \ldots + d_{rs}$-dimensional subspace of $\mathcal{D}_t^{+s}$ spanned by a set of linearly independent s-lists of the form mentioned in the previous item. It is left as an exercise for the reader to pick a basis for the subspace in question.

A nonzero constant row in $A(p)$ influences the above results only to the extent that c and $C(p)$ in (30), . . . , (33) exhibit a corresponding zero component.

If $A(p)$ happens to be square, then the dimension of the range of A^0 is just $\partial(\det A(p))$—this property holding even if $A(p)$ contains nonzero constant rows.

(vi) In the course of time a number of authors have discussed the meaning and interpretation of the initial values at $t = 0(0+, 0-?)$ appearing in connection with the Laplace transform of derivatives of suitable time functions (*cf.* Fischer, 1961, Ramar, Ramaswami, and Murti, 1972). There has been considerable confusion concerning this matter. We shall briefly comment on this problem in view of the above results.

Let $x \in \mathcal{D}$, $t = 0$, and consider the integral (we may here assume that $T = \mathbf{R}$)

35 $$\mathcal{L}(x)(s) \triangleq \int_{-\infty}^{\infty} (\lceil_0 x)(\tau)\, e^{-s\tau}\, d\tau$$

for some $s \in \mathbf{C}$.

Now if $\lceil_0 x$ is regular, then the integral (35) may exist for some $s \in \mathbf{C}$ as an ordinary integral (note that there are $x \in \mathcal{D}$ with $\lceil_0 x$ regular such that the integral (35) does not exist for any $s \in \mathbf{C}$, for instance $x: \mathbf{R} \to \mathbf{C}$, $x(t) = e^{t^2}$, as well as functions $x: \mathbf{R} \to \mathbf{C}$, $x \notin \mathcal{D}$, for which (35) exists for $s \in C$ with C some nonempty region of $\mathbf{C}$). But (35) can be given a sensible interpretation even if $\lceil_0 x$ contains (say a finite number of) delta functions of various orders. In this case it turns out that the exponential function appearing in (35) qualifies as an "improper" test function, and the integral can be interpreted according to (A4.6). For instance, let $x = \lceil_0 x = \delta_0$. Then on the basis of (A4.6) (with $k = 0$) (35) yields

36 $$\mathcal{L}(\delta_0)(s) = \int_{-\infty}^{\infty} \delta_0(\tau)\, e^{-s\tau}\, d\tau = e^{-s0} = 1$$

for every $s \in \mathbf{C}$.

With this generalized interpretation we shall call $\mathscr{L}(x)(s)$, if it exists, the (generalized) "Laplace transform" of $x \in \mathscr{D}$ at $s \in \mathbf{C}$. It is seen that (36) coincides with the usual engineering interpretation of the Laplace transform of δ_0. Exercise for the reader: Determine the Laplace transform of a delta function of order k at $t \geqslant 0$.

To illustrate the use of the (generalized) Laplace transform as defined above, let $x \in \mathscr{D}$ be such that the (generalized) Laplace transforms $\mathscr{L}(x)(s) \triangleq X(s)$ and $\mathscr{L}(px)(s)$ of x and px, respectively exist for every $s \in C$, where C is some nonempty region of $\mathbf{C}$. Consider now the equality (5) for $t = 0$ and $A(p) = p$, and note that $\lceil_0(p\rceil_0 x) = -x^-(0)\delta_0 = -x(0)\delta_0$ (*cf.* (14) and (15)). Then (5) reads:

37 $$\lceil_0(px) = -x(0)\,\delta_0 + p\lceil_0 x.$$

On taking the Laplace transform of both sides of this equality at $s \in C$ we obtain (note the linearity of the transform and the fact that $\mathscr{L}(\lceil_0(px))(s) = \mathscr{L}(px)(s)$)

38 $$\mathscr{L}(px)(s) = -x(0) + \mathscr{L}(p\lceil_0 x)(s),$$

where we have also used the result (36). There remains a crucial step, namely the expansion of the integral determining $\mathscr{L}(p\lceil_0 x)(s)$, i.e.

39 $$\int_{-\infty}^{\infty} (\lceil_0(p\lceil_0 x))(\tau)e^{-s\tau}\,d\tau = \int_{-\infty}^{\infty} (p\lceil_0 x)(\tau)e^{-s\tau}\,d\tau,\ s \in C.$$

The difficulty here arises from the fact that (39) means an "integration in the generalized sense" (*cf.* note (A4.1), (ii)). Without going into details we shall here rely on the fact that the method of integration by parts can formally be used also in this generalized case. Applying this method to (39) we arrive at

40 $$\mathscr{L}(p\lceil_0 x)(s) = s\mathscr{L}(x)(s) = sX(s),\ s \in C.$$

The final form of (38) is thus

41 $$\mathscr{L}(px)(s) = -x(0) + sX(s),\ s \in C.$$

This is now the usual form for the Laplace transform of the derivative of a time function.

Within the present framework there should be no difficulty in interpreting the meaning and origin of the initial value $x(0)$ appearing above.

The (generalized) Laplace transform introduced above corresponds, in fact, to the transform tacitly used in engineering applications. □

8.5 The projection method

Now let us return to the question of how to assign a suitable system to a given regular differential input-output relation. The problem was briefly discussed already in sections 4.2 and 8.1, but the final answer could not be given at that stage—a special machinery had first to be devised for the purpose. This machinery is now available—it is based on the choice of $\mathscr{D}$ as the basic signal space and on the use of the projections introduced in section 8.2.

Assigning a linear dynamic system to a regular differential input-output relation

To begin with recall the reasoning presented in sections 4.2 and 8.1 concerning the problem of how to assign a linear dynamic system to a given regular differential input-output relation. Here we shall repeat this reasoning in a slightly modified manner making use of the results obtained in the previous paragraphs.

Thus let $S \subset \mathscr{D}^r \times \mathscr{D}^q$ be a regular differential input-output relation generated by the regular generator $[A(p) \vdots -B(p)]$, and let $t_0 \in T$ be fixed. As before we choose a pair $(u_1, y_1) \in S$, and follow the time evolution of (u_1, y_1) on the time interval up to t_0. The initial segments of (u_1, y_1) on this time interval can now be conveniently represented by the projected pair $\rceil_{t_0}(u_1, y_1) \triangleq (u_1^-, y_1^-) \in \mathscr{D}_{t_0}^{-r} \times \mathscr{D}_{t_0}^{-q}$. Note that the set of all possible pairs (u_1^-, y_1^-) of this kind is just the set

1 $$\rceil_{t_0} S \triangleq S^-.$$

The set $S_{(u_1^-, y_1^-)}$ corresponding to (8.1.1) and consisting of all the input-output pairs (u, y) belonging to S and coinciding with $(u_1^-, y_1^-) \in S^-$ up to t_0 is then formed according to

2 $$S_{(u_1^-, y_1^-)} \triangleq \{(u, y) \,|\, (u, y) \in S \quad \text{and} \quad \rceil_{t_0}(u, y) = (u_1^-, y_1^-)\}.$$

The interesting part of $S_{(u_1^-, y_1^-)}$ is given by the final segments from t_0 onwards of the input-output pairs contained in this set. These segments can thus be represented by the following projection of $S_{(u_1^-, y_1^-)}$:

3 $$\lceil_{t_0} S_{(u_1^-, y_1^-)} \triangleq S^+_{(u_1^-, y_1^-)} \subset \mathscr{D}_{t_0}^{+r} \times \mathscr{D}_{t_0}^{+q}.$$

From (2) and (3) it follows that

4 $$S^+_{(u_1^-, y_1^-)} = \{(u^+, y^+) \,|\, (u^+, y^+) \in \mathscr{D}_{t_0}^{+r} \times \mathscr{D}_{t_0}^{+q} \quad \text{and} \quad (u_1^- + u^+, y_1^- + y^+) \in S\}.$$

Clearly, (3) corresponds to (8.1.3), and it will be shown below that $S^+_{(u_1^-, y_1^-)}$ really is a mapping $\mathscr{D}^{+r}_{t_0} \to \mathscr{D}^{+q}_{t_0}$ implying that also (2) is a mapping.

It remains to form the system consisting of the set of all input-output mappings of the form (3). We shall denote this system by $\mathscr{S}^+$, i.e.

5 $$\mathscr{S}^+ \triangleq \{S^+_{(u_1^-, y_1^-)} \mid (u_1^-, y_1^-) \in S^-\}.$$

Clearly,

6 $$\cup\, \mathscr{S}^+ = \lceil_{t_0} S \triangleq S^+.$$

It is our intention to show that (5) is a finite dimensional linear dynamic (causal) system on T and that it thus gives the solution to our present problem.

7 **Note. (i)** The above result suggests the following parametric input-output mapping for the system (5) (*cf.* (2.1.1)):

$$h^+ : S^- \times \mathscr{D}^{+r}_{t_0} \to \mathscr{D}^{+q}_{t_0},$$

8 $$((u_1^-, y_1^-), u^+) \mapsto S^+_{(u_1^-, y_1^-)}(u^+),$$

implying that the set S^- of initial segment pairs plays the role of a state set.

The parametric input-output mapping (8) and the system (5) determined by (8) are uniquely determined by S, $t_0 \in T$, and the construction employed. They do not depend on the particular regular generator used to generate S.

Of course, (8) is not very suitable for explicit computations because it does not tell how $S^+_{(u_1^-, y_1^-)}(u^+)$ can be found. We shall return to this question in a moment.

(ii) The main difference between the mappings $S^+_{(u_1^-, y_1^-)}$ as given by (8.1.3) and (3), (4) above is that the former mapping is defined with the aid of signal restrictions to half-closed time intervals, whereas the latter mapping is defined with the aid of the special projections introduced for this purpose. The old formulation could not have been used in the present context, where generalized functions are involved, because restrictions of such functions to half-closed intervals are generally not well-defined.

Equivalent expressions for S

The regular differential input-output relation $S \subset \mathscr{D}^r \times \mathscr{D}^q$ considered above was supposed to be generated by the regular generator $[A(p) \,\vdots\, -B(p)]$, i.e. S was the set of all $(u, y) \in \mathcal{D}^r \times \mathcal{D}^q$ such that

9 $$A(p)y = B(p)u,$$

or equivalently, such that

10 $$(\rceil_{t_0} + \lceil_{t_0})(A(p)y) = (\rceil_{t_0} + \lceil_{t_0})(B(p)u),$$

where $\rceil_{t_0}$ and $\lceil_{t_0}$ are appropriately extended projections.

Next note again the fact that $\mathcal{D} = \mathcal{D}_{t_0}^- \oplus \mathcal{D}_{t_0}^+$, which means that (10) can be split into two simultaneous equations:

11 $$\rceil_{t_0}(A(p)y) = \rceil_{t_0}(B(p)u)$$

and

12 $$\lceil_{t_0}(A(p)y) = \lceil_{t_0}(B(p)u).$$

Further expanding the expressions appearing in (11), (12) on the basis of the results presented in section 8.4 (*cf.* (8.4.10), . . . , (8.4.12)) we get the following equations, which are equivalent to (11), (12) and thus also to (9):

13 $$-A^0\rceil_{t_0}y + A(p)\rceil_{t_0}y = -B^0\rceil_{t_0}u + B(p)\rceil_{t_0}u$$

and

14 $$A^0\rceil_{t_0}y + A(p)\lceil_{t_0}y = B^0\rceil_{t_0}u + B(p)\lceil_{t_0}u,$$

where A^0 and B^0 are the initial condition mappings at t_0 associated with $A(p)$ and $B(p)$ respectively as defined in section 8.4.

Thus S is the set of all $(u, y) \in \mathcal{D}^r \times \mathcal{D}^q$ such that (13) and (14) are satisfied, moreover $\mathsf{D}S = \mathcal{D}^r = \mathcal{D}_{t_0}^{-r} \oplus \mathcal{D}_{t_0}^{+r}$. It also follows that we have the following equivalent expressions for $S^- = \rceil_{t_0}S$ and $S^+ = \lceil_{t_0}S$:

15 $$\begin{aligned} S^- = \{(u^-, y^-) \mid &(u^-, y^-) \in \mathcal{D}_{t_0}^{-r} \times \mathcal{D}_{t_0}^{-q} \\ &\text{and} \quad -A^0 y^- + A(p)y^- = -B^0 u^- + B(p)u^-\} \end{aligned}$$

and

16 $$\begin{aligned} S^+ = \{(u^+, y^+) \mid &(u^+, y^+) \in \mathcal{D}_{t_0}^{+r} \times \mathcal{D}_{t_0}^{+q} \\ &\text{and} \quad A^0 y^- + A(p)y^+ = B^0 u^- + B(p)u^+ \\ &\text{for some} \quad (u^-, y^-) \in S^-\} \end{aligned}$$

with $\mathsf{D}S^- = \mathcal{D}_{t_0}^{-r}$ and $\mathsf{D}S^+ = \mathcal{D}_{t_0}^{+r}$. S, S^-, and S^+ are vector spaces over $\mathbf{C}$ (with respect to the natural operations).

Equivalent expression for $S^+_{(u_1^-, y_1^-)}$

Now (16) immediately also leads to an equivalent expression for $S^+_{(u_1^-, y_1^-)}$ as given by (3) and (4) for some $(u_1^-, y_1^-) \in S^-$:

17 $$S^+_{(u_1^-, y_1^-)} = \{(u^+, y^+) \mid (u^+, y^+) \in \mathcal{D}_{t0}^{+r} \times \mathcal{D}_{t0}^{+q} \text{ and } A^0 y_1^- + A(p)y^+ = B^0 u_1^- + B(p)u^+\}$$

with D $S^+_{(u_1^-, y_1^-)} = \mathcal{D}_{t0}^{+r}$.

$S^+_{(u_1^-, y_1^-)}$ is a mapping

Next consider the equality

18 $$A^0 y_1^- + A(p)y^+ = B^0 u_1^- + B(p)u^+$$

appearing in (17). It is seen that all the terms in (18) are elements of $\mathcal{D}_{t0}^{+q}$ and that the polynomial matrix operators $A(p): \mathcal{D}^q \to \mathcal{D}^q$ and $B(p): \mathcal{D}^r \to \mathcal{D}^q$ operate only on elements of $\mathcal{D}_{t0}^{+q}$ and $\mathcal{D}_{t0}^{+r}$ respectively. It follows that $A(p)$ and $B(p)$ in (18) can be replaced by the corresponding restrictions to $\mathcal{D}_{t0}^{+q}$ and $\mathcal{D}_{t0}^{+r}$ respectively. From now on we shall regard $A(p)$ and $B(p)$ in (18) and in expressions derived therefrom as restricted in this way—the original notations $A(p)$ and $B(p)$ are, nevertheless, used also in this restricted case. But then we have the situation covered by corollary (8.3.6), i.e. $\mathcal{D}_{t0}^{+}$ can be regarded as a vector space over $\mathbf{C}(p)$, and (18) can be interpreted as a vector space equation. Moreover, $A(p): \mathcal{D}_{t0}^{+q} \to \mathcal{D}_{t0}^{+q}$ (recall that $\det A(p) \neq 0$) has an inverse $A(p)^{-1}: \mathcal{D}_{t0}^{+q} \to \mathcal{D}_{t0}^{+q}$, and $y^+ \in \mathcal{D}_{t0}^{+q}$ can be uniquely solved from (18):

19 $$y^+ = A(p)^{-1}(-A^0 y_1^- + B^0 u_1^-) + A(p)^{-1}B(p)u^+.$$

This means that for any $(u_1^-, y_1^-) \in S^-$ and any $u^+ \in \mathcal{D}_{t0}^{+r}$ there is a unique corresponding $y^+ \in \mathcal{D}_{t0}^{+q}$ obtainable from (19) such that $(u^+, y^+) \in S^+_{(u_1^-, y_1^-)}$. Consequently, $S^+_{(u_1^-, y_1^-)}$ is a mapping $\mathcal{D}_{t0}^{+r} \to \mathcal{D}_{t0}^{+q}$ for every $(u_1^-, y_1^-) \in S^-$.

20 **Remark.** In order to determine the y^+ given by (19) as a time function, we have to solve the set of linear differential equations represented by (18) for y^+, given the other necessary quantities. This can, in principle, be done using any suitable method for solving such equations. Of course, any such method must in this case also be able to cope with generalized functions.

If the time functions involved happen to be Laplace transformable in the generalized sense discussed in note (8.4.20), (vi), then y^+ above can in many cases be found with the aid of standard tables over Laplace transform pairs. For this purpose we have merely to note that the Laplace transform of y^+ as given by (19) is (for $t_0 = 0, T = \mathbf{R}$)

21
$$\mathcal{L}(y^+)(s) = A(s)^{-1}(\mathcal{L}(-A^0y_1^-)(s) + \mathcal{L}(B^0u_1^-)(s)) + A(s)^{-1}B(s)\mathcal{L}(u^+)(s) \qquad (s \in C \subset \mathbf{C}),$$

where $A(s)$ and $B(s)$ are obtained from $A(p)$ and $B(p)$ by replacing p by s (*cf.* (8.4.40)). Note also that $-A^0y_1^-$ and $B^0u_1^-$ consist of delta functions of various orders at $t_0 = 0$, whose Laplace transforms are determined as outlined in note (8.4.20), (vi). Of course, standard tables over Laplace transform pairs contain only ordinary time functions defined on $[0, \infty)$ or $(0, \infty)$. To get the proper element of $\mathcal{D}_0^+$ corresponding to a given Laplace transform function we have thus to extend the time function found in the tables by adding suitable zero segments, taking into account our agreement of left continuity (*cf.* section 8.2). □

The final system $\mathcal{S}^+$

We are now in a position to give an explicit expression for the parametric input-output mapping h^+ outlined in (8) for the system $\mathcal{S}^+$ as given by (5). On the basis of (19) we can thus write:

22
$$h^+ : S^- \times \mathcal{D}_{t_0}^{+r} \to \mathcal{D}_{t_0}^{+q},$$
$$((u^-, y^-), u^+) \mapsto y^+$$

with

23
$$y^+ = A(p)^{-1}(-A^0y^- + B^0u^-) + A(p)^{-1}B(p)u^+$$

and

24
$$\mathcal{S}^+ = \{h^+((u^-, y^-), .) | (u^-, y^-) \in S^-\}.$$

It follows from the construction that (24) really defines a family of input-output mappings, i.e. a system. Moreover, this system is a uniform time system on T (*cf.* section 2.2). Next we shall see that $\mathcal{S}^+$ as given by (24) is a linear system in the sense of definition (2.3.1), and also a dynamic (causal) system.

Linearity and causality

Let s be an arbitrary element of $\mathcal{S}^+$, i.e. $s = h^+((u^-, y^-), .)$ for some $(u^-, y^-) \in S^-$. From the definition of h^+ (*cf.* (22), (23)) it follows that

$\mathsf{D}s = \mathscr{D}_{t_0}^{+r}$, and $\mathsf{R}s \subset \mathscr{D}_{t_0}^{+q}$. Moreover, for every $u^+ \in \mathscr{D}_{t_0}^{+r}$ $s(u^+) = h^+((u^-, y^-), u^+)$ can be written as

25 $$s(u^+) = s(0) + \gamma(u^+)$$

with

26 $$s(0) \triangleq h^+((u^-, y^-), 0) = A(p)^{-1}(-A^0 y^- + B^0 u^-)$$

and

27 $$\gamma(u^+) = h^+(0, u^+) = A(p)^{-1} B(p) u^+.$$

This means that γ can be identified with the transfer matrix operator

28 $$\mathscr{G}(p) \triangleq A(p)^{-1} B(p) : \mathscr{D}_{t_0}^{+r} \to \mathscr{D}_{t_0}^{+q}.$$

Finally, the zero-input response space $S^+(0)$ of $\mathscr{S}^+$ given by (*cf.* section 2.3)

29 $$S^+(0) \triangleq \{y^+ | y^+ = s(0) \quad \text{for some } s \in \mathscr{S}^+\}$$

30 $$= \mathsf{R}h^+(.\,, 0)$$

31 $$= \{y^+ | y^+ = A(p)^{-1}(-A^0 y^- + B^0 u^-) \quad \text{for some } (u^-, y^-) \in S^-\}$$

is a subspace of $\mathscr{D}_{t_0}^{+q}$.

The above results together imply that $\mathscr{S}^+$ is a linear system. Moreover, the causality of $\mathscr{G}(p)$ (*cf.* note (8.3.7), (iii)) implies the causality of γ. The elements of $\mathscr{S}^+$ are thus causal input-output mappings (*cf.* note (2.3.3), (vi)) and $\mathscr{S}^+$ qualifies as a dynamic (causal) system.

Dimension of $\mathscr{S}^+$

Consider the zero-input response space $S^+(0)$ of $\mathscr{S}^+$ as given by (29), . . . , (31) above, and note that $A(p)$ in (31) is an automorphism of $\mathscr{D}_{t_0}^{+q}$. It follows that the dimension of $S^+(0)$ is equal to the dimension of the set

32 $$\{z | z = -A^0 y^- + B^0 u^- \quad \text{for some } (u^-, y^-) \in S^-\},$$

which is a subspace of $\mathscr{D}_{t_0}^{+q}$. But according to note (8.4.20), (iv), the ranges of the initial condition mappings A^0 and B^0 in (32) are finite dimensional, and consequently so is $S^+(0)$ also. It follows that $\mathscr{S}^+$ is finite dimensional according to definition (2.3.1). We shall comment on the actual dimension of $S^+(0)$ and $\mathscr{S}^+$ in note (35) below.

Now we have shown that $\mathscr{S}^+$ as given by (5) and (24) satisfies all the conditions required. We shall call $\mathscr{S}^+$ the *differential system* determined by S, or alternatively by $[A(p) \,\vdots\, -B(p)]$, at $t_0 \in T$. $\mathscr{S}^+$ is said to be *controllable* if S is controllable (*cf.* section 6.4) and (*strictly*) *proper* if S is (strictly) proper (*cf.* section 6.3).

The projection method

Let us give a brief summary of the above results.

We started with a given regular differential input-output relation S generated by a regular generator $[A(p) \vdots -B(p)]$, and with a fixed $t_0 \in T$. Then we constructed in a definite way a corresponding parametric input-output mapping h^+, given by (8), and a resulting system $\mathcal{S}^+$, given by (5). These results were given in purely set theoretical terms on the basis of S and t_0 without the use of any generator for S. It is a special feature of the construction that the h^+ thus obtained uses the space S^- of initial segment pairs as the state set. Equivalent expressions for h^+ and $\mathcal{S}^+$ were then derived on the basis of the generator $[A(p) \vdots -B(p)]$ for S. These results are given by (22), . . . , (24). The construction implies that (22), . . . , (24) do not depend on the particular regular generator chosen for S, i.e. if $[A_1(p) \vdots -B_1(p)]$ is input-output equivalent to $[A(p) \vdots -B(p)]$, then it holds that $A_1(p)^{-1}B_1(p) = A(p)^{-1}B(p)$ and $A_1(p)^{-1}(-A_1^0y^- + B_1^0u^-) = A(p)^{-1}(-A^0y^- + B^0u^-)$ for every $(u^-, y^-) \in S^-$. In this context it was essential that S was regular—otherwise (17) would not have been a mapping.

The construction employed above was based on the use of the projection mappings discussed in section 8.2. The method is therefore called the *projection method.*

The mapping PM

To formalize a little further we shall interpret the assignment

33 $$[A(p) \vdots -B(p)] \overset{\text{PM}}{\mapsto} h^+$$

with h^+ given by (22), (23), as defining a mapping PM (alludes to the Projection Method) from the set of all regular generators—$\mathcal{M}$ in section 6.2—to the set of all parametric input-output mappings. $\text{PM}(M_1) = \text{PM}(M_2)$ then holds if and only if M_1 and M_2 are input-output equivalent generators.

In the sequel we shall make frequent use of the assignment chain

34 $$[A(p) \vdots -B(p)] \overset{\text{PM}}{\mapsto} h^+ \mapsto \mathcal{S}^+,$$

where the last assignment represents the forming of $\mathcal{S}^+$ from h^+ (*cf.* (24)).

35 **Note.** **(i)** Consider (22), (23) and let (u^-, y^-) be an element of S^-, i.e. a state. From (23) and from the definitions of the initial condition mappings A^0 and B^0 it is seen that the relevant parts of (u^-, y^-) are the values of u^-, y^- and of their derivatives of various orders at t_0. It is thus reasonably clear that generally there are also other elements of S^- giving the same resulting input-output mapping as a given (u^-, y^-), i.e. the state set S^- is generally not minimal (*cf.* section 2.1). There are different ways to replace the h^+ given by (22), (23) with an equivalent parametric input-output mapping where the state set is minimal. We shall discuss one such way in the next item.

(ii) Here and in items (iii) and (iv) below we shall assume that our regular differential input-output relation $S \subset \mathcal{D}^r \times \mathcal{D}^q$ generated by the regular generator $[A(p) \vdots -B(p)]$ is strictly proper (*cf.* section 6.3). We shall comment on more general cases in item (v). Without loss of generality we may further assume that $A(p)$ is of CRP-form (*cf.* appendix A2, in particular (A2.65)). Then the following statements hold:

36 The degree of any row of $B(p)$ is less than the degree of the corresponding row of $A(p)$ (*cf.* note (A2.87)).

37 $A(p)$ is also column proper, and the degree of row i of $A(p)$ is equal to the degree of column i of $A(p)$, and also equal to $\partial(a_{ii}(p))$, the degree of the diagonal entry $a_{ii}(p)$. Moreover, the coefficient matrix having as its rows the leading coefficients of the rows of $A(p)$ (corresponding to C_A in (A2.31)) is (say) upper triangular, whereas the corresponding coefficient matrix having as its columns the leading coefficients of the columns of $A(p)$ is diagonal $(= I)$.

Finally we shall assume that $A(p)$ does not contain any constant rows. If the original generator does not fulfil this condition, then it can (provided that the output components have been ordered in a suitable way; *cf.* (7.1.12)) be assumed to be of the form

38
$$\begin{array}{ccc} y_1 & y & u \\ \left[\begin{array}{c|c|c} I & A_2 & -B_1(p) \\ \hline 0 & A(p) & -B(p) \end{array}\right], \end{array}$$

where $A(p)$ does not contain any constant rows. In (38) A_2 is a constant matrix, and $B_1(p) = 0$ in the strictly proper case.

We thus assume that we consider only the lower part of (38) in this context. The upper part can then be added when required.

Let $r_1, r_2, \ldots, r_q$ denote the row and column degrees of $A(p)$ i.e. $r_i \triangleq \partial(a_{ii}(p)) \geqslant 1$ $(i = 1, 2, \ldots, q)$. Next form the coefficient matrix

$[\underline{A} : -\underline{B}]$ corresponding to $[A(p) : -B(p)]$ as described in note (8.4.20), (iii), . . . , (iv), noting that $r_i = d_{ri}$ for all i. From the properties mentioned above it then follows that $\underline{A}$ is a square constant d_1 by d_1-matrix with $d_1 \triangleq d_{r1} + d_{r2} + \ldots + d_{rq}$. Moreover, $\underline{A}$ is of full rank d_1 and thus invertible as a constant matrix (*cf.* (8.4.34)).

Consider the parametric input-output mapping h^+ constructed as PM($[A(p) : -B(p)]$) according to (22), (23), and let $(u^-, y^-) \in S^-$ and $u^+ \in \mathcal{D}_{t_0}^{+r}$ ($t_0 \in T$ fixed). The corresponding $y^+ \in \mathcal{D}_{t_0}^{+q}$ is then obtained from (23) as

39 $$y^+ = A(p)^{-1}\left(-[A^0 : -B^0]\begin{bmatrix} y^- \\ \hdashline u^- \end{bmatrix}\right) + A(p)^{-1}B(p)u^+,$$

where we have written the zero input term in a slightly modified form.

Relying on the results obtained in note (8.4.20), (iii), . . . , (v), we can here write

40 $$-[A^0 : -B^0]\begin{bmatrix} y^- \\ \hdashline u^- \end{bmatrix} = C(p)\delta_{t_0},$$

where $C(p) \triangleq (c_1(p), c_2(p), \ldots, c_q(p)) \in \mathbf{C}_{d_{r1}-1}[p] \times \mathbf{C}_{d_{r2}-1}[p] \times \ldots \times \mathbf{C}_{d_{rq}-1}[p]$ is constructed according to (8.4.30), . . . , (8.4.33) in the following way.

Form the initial condition lists y^0 and u^0 corresponding to y^- and u^- according to $y^0 = (y_1^0, y_2^0, \ldots, y_q^0) \in \mathbf{C}^{d_1}$ and $u^0 = (u_1^0, u_2^0, \ldots, u_r^0)$ with y_j^0 $(j = 1, 2, \ldots, q)$ given by

$$y_j^0 = (y_j^-(t_0), (py_j^-)(t_0), \ldots, (p^{d_{rj}-1}y_j^-)(t_0)) \in \mathbf{C}^{d_{rj}}$$

and with the components of u^0 determined correspondingly.

Then compute the coefficient list $c \in \mathbf{C}^{d_1}$ from the matrix equation

41 $$c = [\underline{A} : -\underline{B}]\begin{bmatrix} y^0 \\ \hdashline u^0 \end{bmatrix},$$

i.e

42 $$\begin{bmatrix} c_1 \\ \hdashline c_2 \\ \hdashline \vdots \\ \hdashline c_q \end{bmatrix} = [\underline{A} : -\underline{B}]\begin{bmatrix} y_1^0 \\ \hdashline y_2^0 \\ \hdashline \vdots \\ \hdashline y_q^0 \\ \hdashline u_1^0 \\ \hdashline u_2^0 \\ \hdashline \vdots \\ \hdashline u_r^0 \end{bmatrix},$$

where the coefficient matrix $[\underline{A} \vdots -\underline{B}]$ is partitioned correspondingly.

Finally $C(p) = (c_1(p), c_2(p), \ldots, c_q(p))$ is formed from c as shown in (8.4.32).

In view of the results obtained in note (8.4.20), (v), it is now clear that c as given by (41), (42) could be varied over the whole of $\mathbf{C}^{d_1}$, and $C(p)$ in (40) correspondingly over the whole of $\mathbf{C}_{d_{r1}-1}[p] \times \mathbf{C}_{d_{r2}-1}[p] \times \ldots \times \mathbf{C}_{d_{rq}-1}[p]$ if y^0 and u^0 could be varied freely, i.e. if (u^-, y^-) could be varied over the whole of $\mathscr{D}_{t0}^{-r} \times \mathscr{D}_{t0}^{-q}$. But we must have $(u^-, y^-) \in S^-$, i.e. the initial condition lists u^0 and y^0 are restricted to correspond to pairs (u, y) satisfying $A(p)y = B(p)u$. Now there is a result which ensures that c in (41), (42) can be varied over the whole of $\mathbf{C}^{d_1}$ even in this restricted case, in fact, full variation of c can be achieved by putting $u^0 = 0$ and forming various y^0 corresponding to the elements of the set of all $y \in \mathscr{D}^q$ satisfying $A(p)y = 0$. This follows from the fact that $\underline{A}$ is invertible, and that there is, for every $\eta \in \mathbf{C}^{d_1}$, a corresponding (unique) $y \in \mathscr{D}^q$, $A(p)y = 0$, such that the initial condition list $y^0 \in \mathbf{C}^{d_1}$ corresponding to this y coincides with η. We shall comment on the last point in item (iv) below.

As a result it is thus concluded that (39) can always be replaced by an expression of the form

43 $$y^+ = A(p)^{-1}C(p)\delta_{t0} + A(p)^{-1}B(p)u^+,$$

where $C(p)$ is determined as described above. Moreover, the parametric input-output mapping h^+ for the system $\mathscr{S}^+$ under consideration can be replaced by an equivalent parametric input-output mapping k^+ given by

44 $$k^+ : \Sigma \times \mathscr{D}_{t0}^{+r} \to \mathscr{D}_{t0}^{+q},$$
$$(C(p), u^+) \mapsto y^+$$

with

45 $$y^+ = A(p)^{-1}C(p)\delta_{t0} + A(p)^{-1}B(p)u^+,$$

where the state set Σ is given as

46 $$\Sigma \triangleq \mathbf{C}_{d_{r1}-1}[p] \times \mathbf{C}_{d_{r2}-1}[p] \times \ldots \times \mathbf{C}_{d_{rq}-1}[p].$$

It is readily concluded that the state set (46) thus obtained is minimal (*cf.* section 2.1). Moreover Σ is a $d_1 = d_{r1} + d_{r2} + \ldots + d_{rq}$-dimensional vector space over $\mathbf{C}$ and so are also the sets $\{z | z = C(p)\delta_{t0}$ for some $C(p) \in \Sigma\}$ (*cf.* (32)) and $S^+(0) \triangleq \{y^+ | y^+ = A(p)^{-1}C(p)\delta_{t0}$ for some $C(p) \in \Sigma\}$ (*cf.* (29), ..., (31)). The system $\mathscr{S}^+$ under consideration is thus d_1-dimensional, i.e. the dimension of $\mathcal{S}^+$ coincides in the present strictly proper case with $\partial(\det A(p))$ = the order of S and of $[A(p) \vdots -B(p)]$ (*cf.* section 6.1).

It is left as an exercise for the reader to discuss the more general strictly proper case where $A(p)$ also contains constant rows. In particular it should be shown—in view of (38)—that even in this case it holds that the dimension of $\mathcal{S}^+$ is equal to $\partial(\det A(p))$.

(iii) Let us continue the discussion of the system $\mathcal{S}^+$ above determined by the parametric input-output mapping k^+ as given by (44), . . . , (46). Consider the following control problem.

Let $C(p) \in \Sigma$ be arbitrary. The problem is: Find a $u^+ \in \mathcal{D}_{t_0}^{+r}$ composed solely of delta functions at t_0 such that the corresponding $y^+ \in \mathcal{D}_{t_0}^{+q}$ given by

47 $$y^+ = A(p)^{-1}C(p)\,\delta_{t_0} + A(p)^{-1}B(p)u^+$$

also is composed solely of delta functions at t_0. In other words, we are looking for a u^+ resulting in an input-output pair (u^+, y^+) such that $\Gamma_{t_1}(u^+, y^+) = (0, 0)$ for every $t_1 > t_0$.

If u^+ and y^+ are composed solely of delta functions at t_0 then there are polynomial column matrices $U(p) \in \mathbf{C}[p]^r$, $Y(p) \in \mathbf{C}[p]^q$ such that $u^+ = U(p)\,\delta_{t_0}$ and $y^+ = Y(p)\,\delta_{t_0}$. The problem thus amounts to finding $U(p)$ and $Y(p)$ such that

48 $$Y(p)\,\delta_{t_0} = A(p)^{-1}C(p)\,\delta_{t_0} + A(p)^{-1}B(p)U(p)\,\delta_{t_0},$$

or equivalently, such that

49 $$A(p)Y(p) - B(p)U(p) = C(p)$$

for an arbitrary $C(p) \in \Sigma$. But such $U(p)$, $Y(p)$ can clearly be found for any $C(p) \in \Sigma$ if and only if $A(p)$ and $B(p)$ are left coprime (*cf.* appendix A1), i.e. if and only if $\mathcal{S}^+$ is controllable. The reader is advised to give a formal proof of the above statement including the equivalence of (48) and (49), and to discuss the actual construction of the $U(p)$ and $Y(p)$ corresponding to a given $C(p)$ in (49). The formulated problem thus has a solution for every $C(p) \in \Sigma$ if and only if $\mathcal{S}^+$ is controllable.

The above control problem makes sense even if the original generator is of the more general form (38) with $B_1(p) = 0$, and the problem has then again always a solution if and only if the corresponding system $\mathcal{S}^+$ is controllable.

The control problem discussed above thus illuminates and motivates the controllability concept introduced in section 6.4. It is left as an exercise for the reader to discuss—in view of the above problem—the decomposition of the state space Σ into a controllable part and a part that is not controllable.

(iv) Consider again the strictly proper regular differential input-output relation $S \subset \mathfrak{D}^r \times \mathfrak{D}^q$ generated by the regular generator $[A(p) \vdots -B(p)]$ with $A(p)$ of CRP-form. To begin with suppose further that $A(p)$ does not contain any constant rows. Let the notations introduced in the previous items and concerning this particular case be valid. Further, let

50
$$\begin{array}{ccc} x & y & u \\ \left[\begin{array}{c|c|c} pI - F & 0 & -G \\ \hline -H & I & 0 \end{array}\right] \end{array}$$

be a state-space MR-system matrix as discussed in section 7.9 determining a state-space representation that is an observable decomposition of S in the sense explained in section 7.9. There is then also a row equivalent form of (50) having the structure:

51
$$\begin{array}{ccc} x & y & u \\ \left[\begin{array}{c|c|c} I & A_2(p) & -B_1(p) \\ \hline 0 & A(p) & -B(p) \end{array}\right] \end{array}.$$

The sizes of the matrices appearing in (50), (51) should be evident—in particular it is clear that $pI - F$ is a d_1 by d_1-matrix with $d_1 = \partial(\det A(p))$.

A much-discussed problem is that of how to find a suitable matrix (50) on the basis of the given generator $[A(p) \vdots -B(p)]$ (*cf.* note (7.10.59), (vi)). Of course, an equivalent problem would be to find suitable matrices $A_2(p), B_1(p)$ for (51). (50) could then be found in a simple way by premultiplication of (51) by a suitable unimodular matrix.

It is now interesting to note that the results obtained in this paragraph in fact suggest possible choices of $A_2(p)$ and $-B_1(p)$—note here that (51) implies a relation between x, y and u of the form

52
$$x = [-A_2(p) \vdots B_1(p)] \begin{bmatrix} y \\ u \end{bmatrix}.$$

To this end consider the expressions (41), (42) for c and let u, y be such that $u^- =]_{t_0} u$ and $y^- =]_{t_0} y$. The initial condition list y^0 in (41), (42) can then be written as

53
$$y^0 = (D_y(p)y)(t_0)$$

with

54 $$D_y(p)y$$

55 $$\triangleq \underbrace{\left[\begin{array}{c:c:c:c} \begin{matrix}1\\ p\\ \vdots\\ p^{d_{r1}-1}\end{matrix} & 0 & \cdots & 0 \\ \hdashline 0 & \begin{matrix}1\\ p\\ \vdots\\ p^{d_{r2}-1}\end{matrix} & \cdots & 0 \\ \hdashline \vdots & \vdots & & \vdots \\ \hdashline 0 & 0 & \cdots & \begin{matrix}1\\ p\\ \vdots\\ p^{d_{rq}-1}\end{matrix} \end{array}\right]}_{D_y(p)} \begin{bmatrix} y_1\\ y_2\\ \vdots\\ y_q\end{bmatrix}.$$

The initial condition list u^0 can be expressed in a corresponding way as

56 $$u^0 = (D_u(p)u)(t_0).$$

Thus c as given by (41) is just

57 $$c = \gamma(t_0)$$

with γ given by

58 $$\gamma \triangleq [\underline{A}D_y(p) \ \vdots\ -\underline{B}D_u(p)] \left[\begin{array}{c} y\\ \hdashline u\end{array}\right].$$

It is asserted that γ above qualifies as state in our state-space representation, i.e. that $\underline{A}D_y(p)$ can be taken as $-A_2(p)$ and $-\underline{B}D_u(p)$ as $B_1(p)$ in (51) and (52). Another natural choice would be (recall that $\underline{A}$ is invertible) to take

59 $$\underline{A}^{-1}\gamma = [D_y(p) \ \vdots\ -\underline{A}^{-1}\underline{B}D_u(p)] \left[\begin{array}{c} y\\ \hdashline u\end{array}\right]$$

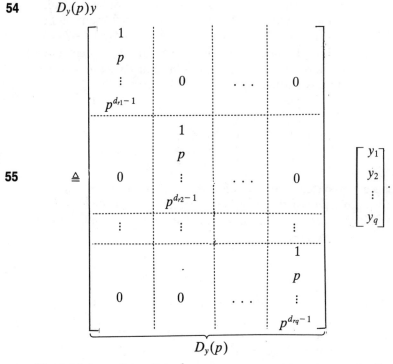

as state implying that $D_y(p)$ would qualify as $-A_2(p)$ and $-\underline{A}^{-1}\underline{B}D_u(p)$ as $B_1(p)$.

The "observability indices" (*cf.* Wolovich, 1974, section 3.6) of the state-space representations thus obtained coincide with the row degrees r_1, $r_2, \ldots, r_q$ of $A(p)$ above (suitably ordered).

It is left as a rather advanced exercise for the reader to prove the above assertions and to work out illustrative examples showing how the actual construction of (50) from (51) is carried out. The structures of the resulting matrices F and H in (50) are particularly interesting. The choice of (59) as state here leads to particularly simple structures.

The reader is also advised to consult a paper by Wolovich and Guidorzi, 1977, in which the choice of state variables actually coincides with (58) and (59) above.

As a further exercise the reader is advised to discuss $\ker A(p)$, i.e. the set $\{y | y \in \mathcal{D}^q \text{ and } A(p)y = 0\}$, in view of the state-space representation determined by (50) and (51), with (59) taken as state. In particular it should be shown that there is, for every $\eta \in \mathbf{C}^{d_1}$ a corresponding unique $y \in \ker A(p)$, such that $y^0 = (D_y(p)y)(t_0) = \eta$ (*cf.* item (ii) above). It also holds that

60 $$S^+(0) = \lceil_{t_0} \ker A(p),$$

i.e. $S^+(0)$ comprises, in the strictly proper case, functions of (projected) exponential type.

If we join to the output y above additional components y_1 of the form

61 $$y_1 = -A_2 y$$

corresponding to the upper part of (38) with $B_1(p) = 0$, then the corresponding state-space MR-system matrix (50) takes the form (follows from $y_1 = -A_2y = -A_2Hx$)

62 $$\begin{array}{c} \begin{array}{cccc} x & y_1 & y & u \end{array} \\ \left[\begin{array}{c:c:c::c} pI - F & 0 & 0 & -G \\ \hdashline A_2H & I & 0 & 0 \\ \hdashline -H & 0 & I & 0 \end{array}\right], \end{array}$$

i.e. the addition of algebraic relations of the form (61) influences the state-space MR-system matrix in a rather simple way.

It is also concluded that (60) still holds even if $A(p)$ contains constant rows.

(**v**) If the given regular differential input-output relation $S \subset \mathcal{D}^r \times \mathcal{D}^q$ is not strictly proper, then any regular generator $[A(p) \vdots -B(p)]$ for S can be written in the form (*cf.* (7.4.6))

63 $$[A(p) \vdots -B^0(p) - A(p)K(p)],$$

suggesting a decomposition of S of the form discussed in section 7.4 and shown in Fig. 7.4. Here $[A(p) \vdots -B^0(p)]$ thus represents a strictly proper generator. It is easily found that (50) and (51) in this case take the forms

64 $$\begin{array}{c} \begin{array}{ccc} x & y & u \end{array} \\ \left[\begin{array}{c|c|c} pI - F & 0 & -G \\ \hline -H & I & -K(p) \end{array}\right] \end{array}$$

and

65 $$\begin{array}{c} \begin{array}{ccc} x & y & u \end{array} \\ \left[\begin{array}{c|c|c} I & A_2(p) & -B_1(p) - A_2(p)K(p) \\ \hline 0 & A(p) & -B^0(p) - A(p)K(p) \end{array}\right] \end{array}$$

respectively. We may again assume that $A(p)$ is of CRP-form. In view of the results presented in the previous item, we may also allow $A(p)$ to contain constant rows. The reader is now advised to discuss this case in detail. In particular it should be shown that the control problem formulated in item (iii) above still makes sense and that the problem always has a solution if and only if the system $\mathscr{S}^+$ under consideration is controllable. Moreover, the relation (23) determining the parametric input-output mapping $h^+ = \mathrm{PM}([A(p) \vdots -B(p)])$ can be written in the equivalent form (with y_1 denoting $y - K(p)u$)

66 $$y^+ = A(p)^{-1}(-A^0 y_1^- + B^{00}u^-) + K^0 u^- + A(p)^{-1}B(p)u^+,$$

where $B^{00}: \mathscr{D}_{t_0}^{-r} \to \mathscr{D}_{t_0}^{+q}$ and $K^0: \mathscr{D}_{t_0}^{-r} \to \mathscr{D}_{t_0}^{+q}$ denote the initial condition mappings at t_0 associated with $B^0(p)$ and $K(p)$ in (63) respectively. It follows that the zero-input response space $S^+(0)$ is now of the form (*cf.* (60))

67 $$S^+(0) = \lceil_{t_0} \ker A(p) + \Delta,$$

where

68 $$\Delta \triangleq \mathrm{R}K^0.$$

Thus we have either $\Delta = \{0\}$, or else Δ is composed of delta functions of various orders at t_0, $\Delta = \{0\}$ holding if and only if $K(p)$ = a constant matrix (yielding $K^0 = 0$), i.e. if and only if S is proper. Further the dimension of $S^+(0)$ and of $\mathscr{S}^+$ is equal to or greater than the order of S ($= \partial(\det A(p))$), equality holding if and only if S is proper. It also follows that the results concerning $S^+(0)$ obtained for a strictly proper S in item (ii) above hold equally well even for a proper S.

(vi) Let us combine some of the various results presented above to get the following rather general result.

Let the given regular differential input-output relation $S \subset \mathcal{D}^r \times \mathcal{D}^q$ be generated by the regular generator $[A(p) \vdots -B(p)]$ with $A(p)$ of CRP-form. Let the row degrees of $A(p)$ be $r_1, r_2, \ldots, r_q$. We shall allow S to be proper or non-proper, and shall also allow $A(p)$ to contain (nonzero) constant rows, i.e. we shall allow some of the r_i to be zero. Further we shall write $[A(p) \vdots -B(p)]$ in the equivalent form (63) as

69 $$[A(p) \vdots -B^0(p) - A(p)\,K(p)].$$

In view of the above results there is then a parametric input-output mapping k^+ equivalent to $h^+ = \mathrm{PM}([A(p) \vdots -B(p)])$ (*cf.* (22), (23)) and given by

70 $$k^+ : (\Sigma_1 \times \Sigma_2) \times \mathcal{D}_{t0}^{+r} \to \mathcal{D}_{t0}^{+q},\ ((C_1(p), C_2(p)), u^+) \mapsto y^+$$

with

71 $$y^+ = A(p)^{-1} C_1(p)\,\delta_{t_0} - C_2(p)\,\delta_{t_0} + A(p)^{-1} B(p) u^+,$$

where the state set $\Sigma_1 \times \Sigma_2$ is given by

72 $$\Sigma_1 \triangleq \mathbf{C}_{r_1-1}[p] \times \mathbf{C}_{r_2-1}[p] \times \ldots \times \mathbf{C}_{r_q-1}[p]$$

and

73 $$\Sigma_2 \triangleq \{C_2(p) | C_2(p) \in \mathbf{C}[p]^q \quad \text{and} \quad C_2(p)\,\delta_{t_0} \in \mathrm{R}K^0\}.$$

Here $K^0 : \mathcal{D}_{t0}^{-r} \to \mathcal{D}_{t0}^{+q}$ denotes the initial condition mapping at t_0 associated with $K(p)$ in (69). If there is a row degree $r_i = 0$, then the corresponding $\mathbf{C}_{r_i-1}[p]$ in (72) is interpreted as $\{0\}$.

The reader is advised to derive the above result and to show that $\Sigma_1 \times \Sigma_2$ is a minimal state set. Consequently the dimension of $\mathcal{S}^+$ determined by k^+ above is equal to the dimension of $\Sigma_1 \times \Sigma_2$ regarded as a vector space over $\mathbf{C}$.

In the case of a proper S we have $\Sigma_2 = \{0\}$, and a simplified notation corresponding to (44), . . . , (46) can be used for k^+.

(vii) Finally, consider the assignment chain (34) $[A(p) \vdots -B(p)] \mapsto h^+ \mapsto \mathcal{S}^+$ according to which, using the projection method, we assigned a definite finite dimensional linear dynamic (causal) system $\mathcal{S}^+$ to a given regular differential input-output relation $S \subset \mathcal{D}^r \times \mathcal{D}^q$ generated by the regular generator $[A(p) \vdots -B(p)]$. The system $\mathcal{S}^+$ thus obtained is characterized by (*cf.* (25), . . . , (28))

74 $$\mathcal{S}^+ = \{s | s : \mathcal{D}_{t0}^{+r} \to \mathcal{D}_{t0}^{+q},\ u^+ \mapsto s(0) + \gamma(u^+),\ s(0) \in S^+(0)\},$$

where $\gamma = \mathscr{G}(p)$, the transfer matrix operator (28). Moreover

75 $$\cup \mathscr{S}^+ = \lceil_{t_0} S \triangleq S^+.$$

The causality of $\mathscr{S}^+$ follows from the causality of $\mathscr{G}(p)$.

Next suppose that there is another assignment chain $[A(p) \vdots -B(p)] \mapsto h_1^+ \mapsto \mathscr{S}_1^+$, $\cup \mathscr{S}_1^+ = \cup \mathscr{S}^+ = S^+$ with $\mathscr{S}_1^+$ again a linear dynamic (causal) system. The causality of $\mathscr{S}_1^+$ and note (2.3.3), (vii), then imply that $\mathscr{S}_1^+$ necessarily is of the form

76 $$\mathscr{S}_1^+ = \{s \mid s : \mathscr{D}_{t_0}^{+r} \to \mathscr{D}_{t_0}^{+q}, u^+ \mapsto s(0) + (\gamma + \Delta\gamma)(u^+), s(0) \in S^+(0)\},$$

where $\Delta\gamma : \mathscr{D}_{t_0}^{+r} \to S^+(0)$ is a linear causal mapping, and where γ is as in (74). The zero-state responses of $\mathscr{S}^+$ and $\mathscr{S}_1^+$ are thus determined by γ and $\gamma + \Delta\gamma$ respectively. It follows that $\mathscr{S}^+$ (for $t_0 \in T$ fixed) is the only linear dynamic (causal) system with $\cup \mathscr{S}^+ = S^+$ and with the zero-state response determined by the transfer matrix operator $\gamma = \mathscr{G}(p)$ that can be assigned to the given S. The $\mathscr{S}^+$ thus obtained is consequently very natural in a certain sense.

It is left as an exercise for the reader to show that there really are mappings $\Delta\gamma$ (other than the zero mapping) satisfying the above conditions. How does a nonzero $\Delta\gamma$ influence the role of S^- as a state set (*cf.* note (7), (i))?

(viii) The above considerations have shown the great applicability of the CRP-form in this context. In particular we have found that a regular generator $[A(p) \vdots -B(p)]$ with $A(p)$ of CRP-form is closely related to a corresponding state-space representation—in fact, such a generator could be said to be "almost in state-space form" (*cf.* (50), . . . , (59) above). It can also be expected that the CRP-form is very convenient from a physical point of view (*cf.* section 9.4). In what follows, therefore, we shall use the CRP-form on many occasions where a mere row proper form would probably suffice. □

9

Interconnections of differential systems

We shall devote this chapter to a study of some aspects of interconnections of families of differential systems of the kind introduced in section 8.5. For this purpose we shall apply the general theory for interconnections of families of systems as presented in chapter 3. It appears that here we generally have to distinguish between two main kinds of such interconnections—they differ from each other with respect to how the state constraint (*cf.* (3.1.7)) is formed. Under suitable conditions the two kinds of interconnections prove equivalent in a certain sense.

Next we shall explore the relationship between interconnections of differential systems and the corresponding compositions of differential input-output relations of the kind presented in section 7.2. The main result here is that —under certain conditions—an interconnection of differential systems can be studied on the basis of the corresponding composition of differential input-output relations. The usefulness and relevance of the polynomial systems theory as presented in part II of this book relies, in fact, on this fundamental result.

The projection method devised in the previous chapter plays a central role in the considerations mentioned above.

An illustrative example concludes the chapter.

Parts of the material presented in this chapter have previously been published in a laboratory report (Blomberg, 1975).

9.1 Two interconnections

Let us start by forming two interconnections in the following way.

Let $S \subset \mathcal{D}^r \times \mathcal{D}^q$ be a regular differential input-output relation generated by the regular generator

1 $$[A(p) \,\vdots\, -B(p)].$$

Furthermore, let $t_0 \in T$ be fixed and let $\mathcal{S}^+$ be the differential system determined by (1) according to the assignment chain (8.5.34), i.e.

2 $$[A(p) \,\vdots\, -B(p)] \overset{\text{PM}}{\longmapsto} h^+ \mapsto \mathcal{S}^+$$

with h^+ and $\mathcal{S}^+$ as in (8.5.22), . . . , (8.5.24).

Next we form the interconnections (*cf.* section 3.1)

3 $$(\mathcal{A}, C, S^-)$$

and

4 $$(\mathcal{A}, C, X),$$

where $\mathcal{A}$ is a singleton family of system descriptions containing a suitable description of $\mathcal{S}^+$ based on h^+, for instance according to (2.1.4)

5 $$\mathcal{A} = \{(\mathcal{D}_{t_0}^{+r}, \mathcal{D}_{t_0}^{+q}, S^-, S^+, \mathcal{S}^+, h^+)\}.$$

Recall that an interconnection comprising any number of systems can be reduced to this form (*cf.* note (3.1.8)).

Then we choose a suitable interconnection matrix (*cf.* (3.2.4))

6 $$C = \begin{array}{c} \\ y_0 \\ u \end{array} \begin{array}{c} \begin{array}{cc} u_0 & y \end{array} \\ \left[\begin{array}{c|c} 0 & C_2 \\ \hline C_3 & C_4 \end{array}\right] \end{array}$$

satisfying the conditions mentioned in section 3.1. C is a $(q_0 + r)$ by $(r_0 + q)$-matrix, where r_0 and q_0 denote the number of input and output components associated with the overall system.

State constraints

The interconnections (3) and (4) differ from each other with respect to the state constraints S^- and X.

Unrestricted state space S^-

Suppose that there are no interconnection constraints on the time interval up to t_0. Evidently any pair $(u^-, y^-) \in \rceil_{t_0} S \triangleq S^-$ of initial input-output segments can then appear. From t_0 on the interconnection constraints determined by C are valid, i.e. a change of mode, a "switching", occurs at t_0. In this case we say that we have an interconnection with unrestricted state space, i.e. with the state constraint represented by S^-, a vector space over **C** (*cf.* section 8.5). Figure 9.1 illustrates the situation.

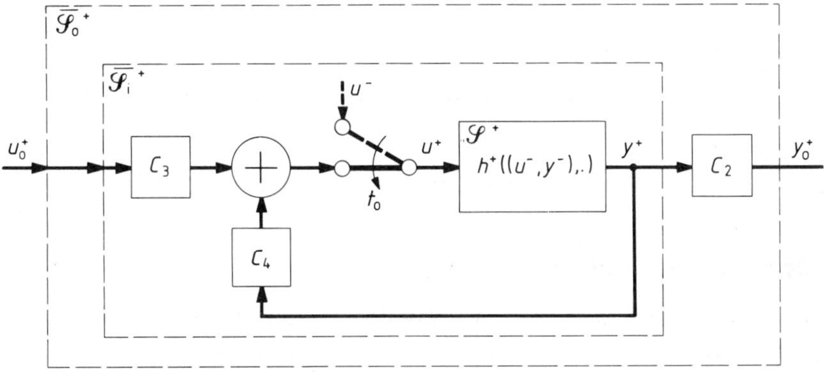

Fig. 9.1. Illustration to the interconnection ($\mathcal{A}$, C, S^-).

Restricted state space

Suppose now that the interconnection constraints determined by C have been valid "all the time". Then only those pairs $(u^-, y^-) \in S^-$ of initial input-output segments can appear which also satisfy the interconnection constraints (*cf.* (3.1.6))

7 $$\begin{bmatrix} y_0^- \\ u^- \end{bmatrix} = \begin{bmatrix} 0 & C_2 \\ C_3 & C_4 \end{bmatrix} \begin{bmatrix} u_0^- \\ y^- \end{bmatrix},$$

i.e. the set of possible (u^-, y^-) is a subspace X of S^- given by

8 $$X = \{(u^-, y^-) \mid (u^-, y^-) \in S^- \quad \text{and} \quad u^- = C_3 u_0^- + C_4 y^- \quad \text{for some} \quad u_0^- \in \mathcal{D}_{t_0}^{-r_0}\}.$$

In this case we say that we have an interconnection with restricted state space. Figure 9.2 illustrates the situation.

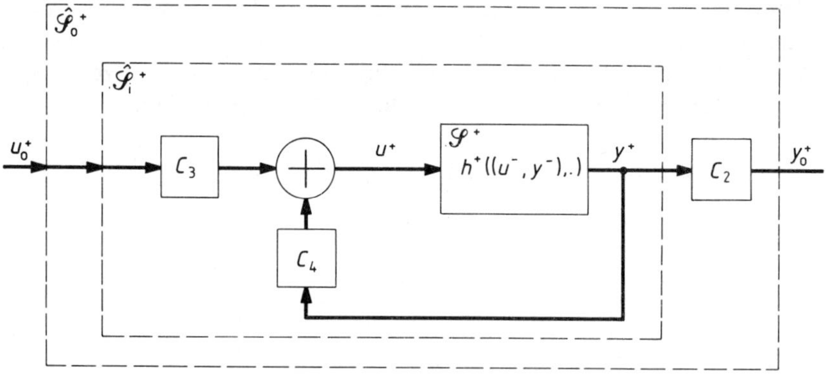

Fig. 9.2. Illustration to the interconnection $(\mathcal{A}, C, X)$.

Determinateness of the interconnections

The general results of chapter 3 can now be applied. Hence let $\alpha \triangleq (u^-, y^-)$ be an element of S^- or X. The set $S_{i\alpha}^+$ corresponding to $S_{i\alpha}$ in (3.2.7) is then obtained as

9
$$S_{i\alpha}^+ = \{(u_0^+, y^+) \,|\, u_0^+, y^+ \quad \text{are such that}$$

$$C_3 u_0^+ + C_4 y^+ \in Dh^+(\alpha,.) \quad \text{and} \quad y^+ = h^+(\alpha, C_3 u_0^+ + C_4 y^+)\}$$

with h^+ given by (8.5.22) and (8.5.23).

Noting that $Dh^+(\alpha,.) = \mathcal{D}_{t_0}^{+r}$ for all α it is concluded that (9) is equivalent to

10
$$S_{i\alpha}^+ = \{(u_0^+, y^+) \,|\, (u_0^+, y^+) \in \mathcal{D}_{t_0}^{+r_0} \times \mathcal{D}_{t_0}^{+q} \quad \text{and}$$

$$(A(p) - B(p)C_4)y^+ = -A^0 y^- + B^0 u^- + B(p)C_3 u_0^+\}.$$

Now causality was our only realizability condition. This means (*cf.* note (3.4.4)) that our interconnections (3) and (4) are determinate if and only if $S_{i\alpha}^+$ is a causal mapping for all α, i.e. if and only if $\det (A(p) - B(p)C_4) \neq 0$ (*cf.* sections 8.2 and 8.3).

Systems associated with the interconnections

Now suppose that our interconnections $(\mathcal{A}, C, S^-)$ and $(\mathcal{A}, C, X)$ are determinate, i.e. that $\det (A(p) - B(p)C_4) \neq 0$. Application of the general results of section 3.4 to the present case then leads to the following parametric input-output mappings for the internal and overall systems associated with our interconnections.

The interconnection $(\mathscr{A}, C, S^-)$.
The internal system $\bar{\mathscr{S}}_i^+$:

11 $$\bar{h}_i^+ : S^- \times \mathscr{D}_{t0}^{+r_0} \to \mathscr{D}_{t0}^{+q},$$
$$((u^-, y^-), u_0^+) \mapsto (A(p) - B(p)C_4)^{-1}(-A^0 y^- + B^0 u^-) + (A(p) - B(p)C_4)^{-1} B(p) C_3 u_0^+.$$

The overall system $\bar{\mathscr{S}}_o^+$:

12 $$\bar{h}_o^+ : S^- \times \mathscr{D}_{t0}^{+r_0} \to \mathscr{D}_{t0}^{+q_0},$$
$$\bar{h}_o^+ = C_2 \bar{h}_i^+.$$

The interconnection $(\mathscr{A}, C, X)$, $X \subset S^-$.

Parametric input-output mappings $\hat{h}_i^+$ and $\hat{h}_o^+$ determining the internal system $\hat{\mathscr{S}}_i^+$ and the overall system $\hat{\mathscr{S}}_o^+$ respectively are obtained as appropriate restrictions of $\bar{h}_i^+$ and $\bar{h}_o^+$:

13 $$\hat{h}_i^+ = \bar{h}_i^+ | X \times \mathscr{D}_{t0}^{+r_0},$$
$$\hat{h}_o^+ = \bar{h}_o^+ | X \times \mathscr{D}_{t0}^{+r_0}.$$

Of course, all these systems are finite dimensional linear dynamic (causal) systems.

From the results it is seen that $\hat{h}_i^+ \subset \bar{h}_i^+$, $\hat{\mathscr{S}}_i^+ \subset \bar{\mathscr{S}}_i^+$, $\hat{h}_o^+ \subset \bar{h}_o^+$, $\hat{\mathscr{S}}_o^+ \subset \bar{\mathscr{S}}_o^+$, $\hat{h}_i^+(0,.) = \bar{h}_i^+(0,.)$, and $\hat{h}_o^+(0,.) = \bar{h}_o^+(0,.)$, i.e. possible differences between the mappings and systems involved concern only the zero-input response spaces. It is easy to show by means of examples that such differences can occur (*cf.* section 9.4).

The transfer matrices determined by $\bar{h}_i^+$, $\bar{\mathscr{S}}_i^+$, $\hat{h}_i^+$, and $\hat{\mathscr{S}}_i^+$ are all equal and they are given by

14 $$\mathscr{G}_i(p) = (A(p) - B(p)C_4)^{-1} B(p) C_3.$$

The corresponding transfer matrices determined by $\bar{h}_o^+$, $\bar{\mathscr{S}}_o^+$, $\hat{h}_o^+$, and $\hat{\mathscr{S}}_o^+$ are also equal and they are given by

15 $$\mathscr{G}_o(p) = C_2 \mathscr{G}_i(p).$$

In view of the constructions employed it should be reasonably clear that the results obtained and given by (11), . . . , (15) do not depend on the particular regular generator (1) chosen for S—any regular generator input-output equivalent to (1) would yield the same results.

Summary

Let us summarize the above considerations with the aid of the assignment diagram shown in Fig. 9.3.

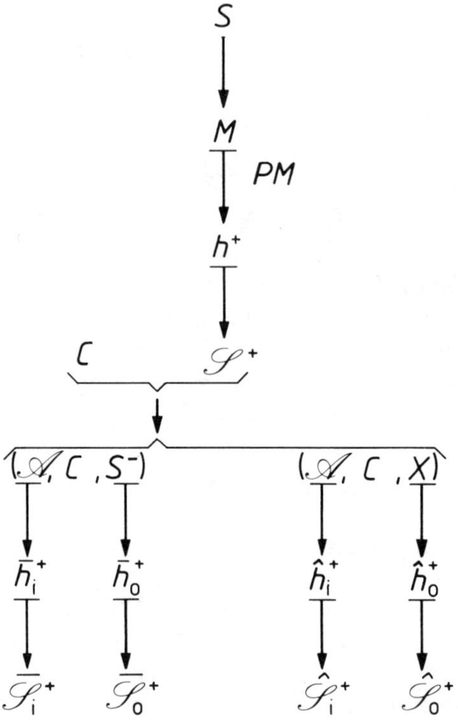

Fig. 9.3. Assignment diagram.

We started with the given regular differential input-output relation S (uppermost in Fig. 9.3) and chose a suitable regular generator M for S ($\to$ means that the assignment is not unique). With the aid of the projection method ($t_0 \in T$ fixed), we then assigned a definite parametric input-output mapping h^+ and a corresponding differential system $\mathcal{S}^+$ to M and S ($\mapsto$ means that we have a unique assignment). Recall that h^+ and $\mathcal{S}^+$ do not depend on the choice of M.

Next we chose an interconnection matrix C, a suitable description of $\mathcal{S}^+$, and two state constraints S^- and X, and formed the two interconnections $(\mathcal{A}, C, S^-)$ and $(\mathcal{A}, C, X)$. These interconnections were supposed to be determinate. S^- was said to be an unrestricted state space, X a restricted state space.

Finally we assigned parametric input-output mappings along with corresponding internal and overall systems to the two interconnections considered according to the general procedure explained in chapter 3. This completes the diagram in Fig. 9.3. By construction, the resulting $\bar{h}_i^+$, $\bar{h}_o^+$, $\hat{h}_i^+$, $\hat{h}_o^+$, $\bar{\mathcal{S}}_i^+$, $\bar{\mathcal{S}}_o^+$, $\hat{\mathcal{S}}_i^+$, and $\hat{\mathcal{S}}_o^+$ are uniquely determined by C and S.

It can now be seen that there is, in this case, very little use for the polynomial systems theory as developed in part II of this book, at least as far as the properties of the final systems $\bar{\mathcal{S}}_i^+$, $\bar{\mathcal{S}}_o^+$, $\hat{\mathcal{S}}_i^+$, and $\hat{\mathcal{S}}_o^+$ are concerned. This is a consequence of the fact that vector space expressions for $\bar{h}_i^+$, $\bar{h}_o^+$, $\hat{h}_i^+$, and $\hat{h}_o^+$ (*cf.* (11), . . . , (13), the actual vector space being the vector space $\mathcal{D}_{t_0}^+$ (or $\mathcal{D}^+$) over $\mathbf{C}(p)$), were obtained directly on the basis of the general theory for interconnections without the aid of the projection method applied to some corresponding regular differential input-output relations—the projection method was applied merely to form the original differential system $\mathcal{S}^+$ involved in the interconnection from the given differential input-output relation S. In fact, at this stage we do not even know if these final systems are differential systems at all in the sense of section 8.5.

In order to be able to utilize the polynomial systems theory to the full extent, we must thus replace the diagram in Fig. 9.3 by some other more suitable diagram. Such a diagram will be discussed in the next paragraph.

16 **Note.** Consider the assignment diagram in Fig. 9.3 and suppose that $M = [A(p) \vdots -B(p)]$ is such that $A(p)$ is of CRP-form. To generate the system $\mathcal{S}^+$ involved in the interconnection we can then also use—instead of $h^+ = \mathrm{PM}(M)$—an equivalent parametric input-output mapping k^+ of the form described in (8.5.69), . . . , (8.5.73). This would then lead to corresponding interconnections, say $(\tilde{\mathcal{A}}, C, \Sigma_1 \times \Sigma_2)$ and $(\tilde{\mathcal{A}}, C, \theta)$ with $\theta \subset \Sigma_1 \times \Sigma_2$ determining—provided that the interconnections are determinate—the same internal and overall systems as $(\mathcal{A}, C, S^-)$ and $(\mathcal{A}, C, X)$ respectively.

Here $\tilde{\mathcal{A}}$ denotes a singleton family of system descriptions containing a suitable description of $\mathcal{S}^+$ based on k^+ as given by (8.5.69), . . . , (8.5.73). Of course, the determination of the correct $\theta \subset \Sigma_1 \times \Sigma_2$ could require some effort. We shall therefore consider only $(\tilde{\mathcal{A}}, C, \Sigma_1 \times \Sigma_2)$.

Suppose that $(\tilde{\mathcal{A}}, C, \Sigma_1 \times \Sigma_2)$ is determinate. It is an easy matter to construct parametric input-output mappings $\bar{k}_i^+$ and $\bar{k}_o^+$ for the internal and overall systems respectively determined by $(\tilde{\mathcal{A}}, C, \Sigma_1 \times \Sigma_2)$. These systems are then clearly equal to the systems $\bar{\mathcal{S}}_i^+$ and $\bar{\mathcal{S}}_o^+$ appearing in Fig. 9.3.

Following the same lines as in the derivation of (11) and (12) above we find for $\bar{k}_i^+$:

17 $$\bar{k}_i^+ : (\Sigma_1 \times \Sigma_2) \times \mathcal{D}_{t_0}^{+r_0} \to \mathcal{D}_{t_0}^{+q},$$
$$((C_1(p), C_2(p)), u_0^+) \mapsto y^+,$$

with

18 $$y^+ = (A(p) - B(p)C_4)^{-1}(C_1(p)\,\delta_{t_0} - A(p)C_2(p)\,\delta_{t_0})$$
$$+ (A(p) - B(p)C_4)^{-1} B(p)C_3 u_0^+,$$

where we have used the notations and interpretations introduced in connection with (8.5.69), . . . , (8.5.73). A corresponding expression can be obtained for $\bar{k}_o^+$. The dimension of $\bar{\mathcal{S}}_i^+$ is seen to be equal to the dimension of $\mathcal{S}^+$. □

9.2 Systems associated with compositions of input-output relations

Let us consider the assignment diagram shown in Fig. 9.4.

We again start with the same regular differential input-output relation S—and a corresponding regular generator M for S—as in the previous paragraph (again → indicates non-unique and ↦ unique assignments).

At this stage we add the interconnection matrix C as given by (9.1.6) to the picture and form the composition $(\mathcal{B}, C)$ with $\mathcal{B}$ a singleton set comprising the generator M for S (*cf.* section 7.2).

Regularity of $(\mathcal{B}, C)$

Next suppose that $(\mathcal{B}, C)$ is regular, i.e. that $\det(A(p) - B(p)C_4) \neq 0$ (*cf.* section 7.2). The condition for the regularity of $(\mathcal{B}, C)$ thus coincides with the condition for the determinateness of the interconnections $(\mathcal{A}, C, S^-)$ and $(\mathcal{A}, C, X)$ considered in the previous paragraph. The order of $(\mathcal{B}, C)$ is $\partial(\det(A(p) - B(p)C_4))$.

Input-output relations determined by $(\mathcal{B}, C)$

The general theory for regular compositions of families of regular differential input-output relations as presented and developed in chapter 7 of this book is now applicable. In particular we can form the internal-overall input-output relation S_{io}, the internal input-output relation S_i, and the overall input-output relation S_o determined by $(\mathcal{B}, C)$ as described in

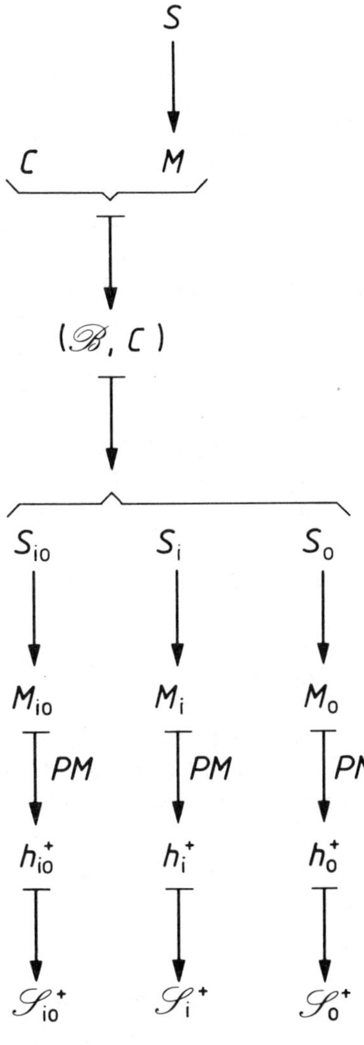

Fig. 9.4. Assignment diagram.

section 7.2. These relations are all regular differential input-output relations generated by corresponding regular generators M_{io}, M_i, and M_o respectively. M_{io} is given by (7.2.18) and M_i by (7.2.13). To obtain M_o, M_{io} is first transformed to a row equivalent upper triangular form $\tilde{M}_{io}$ as shown by (7.2.20), (7.2.21).

Systems associated with ($\mathcal{B}$, C)

In the diagram in Fig. 9.4 we finally apply the projection method to form the parametric input-output mappings h_{io}^{+}, h_{i}^{+}, and h_{o}^{+} as well as the corresponding differential systems $\mathcal{S}_{io}^{+}$, $\mathcal{S}_{i}^{+}$, and $\mathcal{S}_{o}^{+}$. The transfer matrix determined by h_{i}^{+} and $\mathcal{S}_{i}^{+}$ is (*cf.* (7.2.13)) equal to $\mathcal{G}_{i}(p)$ as given by (9.1.14), and the transfer matrix determined by h_{o}^{+} and $\mathcal{S}_{o}^{+}$ is (*cf.* (7.2.32)) equal to $\mathcal{G}_{o}(p)$ as given by (9.1.15). By construction, all these final quantities are again uniquely determined by C and S.

Now the assignment diagram in Fig. 9.4 clearly meets our requirements regarding the applicability of the polynomial systems theory as developed in part II of this book, because the final systems $\mathcal{S}_{io}^{+}$, $\mathcal{S}_{i}^{+}$, and $\mathcal{S}_{o}^{+}$ are associated directly with corresponding regular differential input-output relations and their generators. The final question remains: What is the relationship between $\bar{\mathcal{S}}_{i}^{+}$, $\bar{\mathcal{S}}_{o}^{+}$, $\hat{\mathcal{S}}_{i}^{+}$, $\hat{\mathcal{S}}_{o}^{+}$ in Fig. 9.3 and $\mathcal{S}_{io}^{+}$, $\mathcal{S}_{i}^{+}$, $\mathcal{S}_{o}^{+}$ in Fig. 9.4 above? One result was already noted: $\hat{\mathcal{S}}_{i}^{+}$, $\bar{\mathcal{S}}_{i}^{+}$, and $\mathcal{S}_{i}^{+}$ determine the same transfer matrix and therefore they also exhibit the same zero-state response; the corresponding result holds for $\hat{\mathcal{S}}_{o}^{+}$, $\bar{\mathcal{S}}_{o}^{+}$ and $\mathcal{S}_{o}^{+}$.

Additional results concerning this matter are presented in the next paragraph.

1 **Note.** Again it would be possible to replace the parametric input-output mappings $h_{io}^{+} = \mathrm{PM}(M_{io})$, $h_{i}^{+} = \mathrm{PM}(M_{i})$, and $h_{o}^{+} = \mathrm{PM}(M_{o})$ appearing in Fig. 9.4 with equivalent parametric input-output mappings of the form discussed in note (8.5.35), (vi) (*cf.* k^{+} as given by (8.5.69), . . . , (8.5.73)). These would then generate the same systems $\mathcal{S}_{io}^{+}$, $\mathcal{S}_{i}^{+}$, and $\mathcal{S}_{o}^{+}$ as h_{io}^{+}, h_{i}^{+}, and h_{o}^{+} respectively.

Consider, for instance, the internal input-output relation S_{i} and its regular generator (*cf.* (7.2.13))

2 $$M_{i} \triangleq [A_{i}(p) \vdots -B_{i}(p)] = [A(p) - B(p)C_{4} \vdots -B(p)C_{3}]$$

for which we have an equivalent form corresponding to (8.5.69):

3 $$M_{i} = [A_{i}(p) \vdots -B_{i}^{0}(p) - A_{i}(p)K_{i}(p)].$$

Finally suppose that $A_{i}(p)$ above is of CRP-form with the row degrees r_{i1}, r_{i2}, . . . , r_{iq}.

Application of (8.5.69), . . . , (8.5.73) to the present situation then yields a parametric input-output mapping k_{i}^{+} equivalent to h_{i}^{+} and given by

4 $$k_{i}^{+} : (\Sigma_{i1} \times \Sigma_{i2}) \times \mathcal{D}_{t0}^{+r_0} \to \mathcal{D}_{t0}^{+q},$$
$$((C_{i1}(p), C_{i2}(p)), u_{0}^{+}) \mapsto y^{+}$$

with

5 $$y^+ = A_i(p)^{-1}C_{i1}(p)\,\delta_{t_0} - C_{i2}(p)\,\delta_{t_0} + A_i(p)^{-1}B_i(p)u_0^+,$$

and

6 $$\Sigma_{i1} \triangleq \mathbf{C}_{r_{i1}-1}[p] \times \mathbf{C}_{r_{i2}-1}[p] \times \ldots \times \mathbf{C}_{r_{iq}-1}[p]$$

and

7 $$\Sigma_{i2} \triangleq \{C_{i2}(p) \mid C_{i2}(p) \in \mathbf{C}[p]^q \quad \text{and} \quad C_{i2}(p)\,\delta_{t_0} \in \mathsf{R}K_i^0\}.$$

Again, if there is a row degree $r_{ik} = 0$, then the corresponding $\mathbf{C}_{r_{ik}-1}[p]$ in (6) is interpreted as $\{0\}$. The dimension of $\mathcal{S}_i^+$ determined by k_i^+ above is equal to the dimension of $\Sigma_{i1} \times \Sigma_{i2}$ regarded as a vector space over $\mathbf{C}$.

If S_i happens to be proper, then $\Sigma_{i2} = \{0\}$, and a simplified notation can be used for k_i^+ (*cf.* note (8.5.35), (vi)).

9.3 The main results

We are now ready to present the first of the two main theorems concerning the relationship between the various systems introduced in the previous paragraphs.

The first main result

1 **Theorem.** *Let the quantities appearing in the two assignment diagrams in Fig. 9.3 and Fig. 9.4 be constructed as described above and under the assumptions mentioned starting from the same regular differential input-output relation S and the same interconnection matrix C. Then it holds that* $\hat{\mathcal{S}}_i^+ = \mathcal{S}_i^+$ *and* $\hat{\mathcal{S}}_o^+ = \mathcal{S}_o^+$.

2 **Proof.** To begin with we note the obvious fact that $\hat{\mathcal{S}}_i^+ = \mathcal{S}_i^+$ implies $\hat{\mathcal{S}}_o^+ = \mathcal{S}_o^+$, because of the simple relationship between the output components of the internal system and the output components of the overall system. In addition it was already noted above that $\hat{\mathcal{S}}_i^+$ and $\mathcal{S}_i^+$ exhibit the same zero-state response. In order to prove the theorem it therefore suffices to show that $\hat{\mathcal{S}}_i^+$ and $\mathcal{S}_i^+$ have the same zero-input response space (*cf.* note (2.3.3), (iii)).

Consider first the system $\mathcal{S}_i^+$ determined by $h_i^+ = \mathrm{PM}(M_i)$ with M_i given by (7.2.13). The zero-input response space $S_i^+(0)$ of $\mathcal{S}_i^+$ is then obtained as (*cf.* (8.5.31))

3 $$S_i^+(0) = \{y^+ | y^+ = A_i(p)^{-1}(-A_i^0 y^- + B_i^0 u_0^-)$$
$$\text{for some } (u_0^-, y^-) \in S_i^- \triangleq \rceil_{t_0} S_i\}$$

4 $$= \{y^+ | y^+ = (A(p) - B(p)C_4)^{-1}(-(A^0 - B^0 C_4)y^- + B^0 C_3 u_0^-)$$
$$\text{for some } (u_0^-, y^-) \in S_i^-\},$$

where the meaning of the various symbols should be clear from the notational agreements made previously. Note that (4) is obtained from (3) by expanding A_i^0 and B_i^0 using (8.4.21), . . . , (8.4.24).

The corresponding zero-input response space of $\hat{\mathcal{S}}_i^+$ is obtained from (9.1.11) and (9.1.13) as

5 $$\hat{S}_i^+(0) = \{y^+ | y^+ = (A(p) - B(p)C_4)^{-1}(-A^0 y + B^0 u^-)$$
$$\text{for some } (u^-, y^-) \in X\}$$

with X given by (9.1.8). But now it is clear that X can be described in an equivalent way as

6 $$X = \{(u^-, y^-) | (u^-, y^-) = (C_3 u_0^- + C_4 y^-, y^-)$$
$$\text{for some } (u_0^-, y^-) \in S_i^-\}.$$

Consequently (5) is equivalent to

7 $$\hat{S}_i^+(0) = \{y^+ | y^+ = (A(p) - B(p)C_4)^{-1}(-A^0 y^- + B^0(C_3 u_0^- + C_4 y^-))$$
$$\text{for some } (u_0^-, y^-) \in S_i^-\},$$

which is seen to coincide with (4).

The proof is complete. □

The second main result

Our second main result concerns the relationship between the interconnections $(\mathcal{A}, C, S^-)$ and $(\mathcal{A}, C, X)$ as described in section 9.1. A related result obtained in a more special setting can be found in Maeda, 1974.

8 **Theorem.** *In the assignment diagrams in Fig. 9.3 and Fig. 9.4 let* $S \subset \mathcal{D}^r \times \mathcal{D}^q$ *be a regular differential input-output relation generated by the regular generator* $M \triangleq [A(p) \vdots -B(p)]$. *Suppose that S is proper (strictly proper), i.e. that the transfer matrix* $\mathcal{G}(p)$ *determined by M is of the form*

(cf. (6.3.2))

9 $$\mathcal{G}(p) = A(p)^{-1}B(p) = \mathcal{G}^0(p) + K,$$

where $\mathcal{G}^0(p)$ *is the strictly proper part of* $\mathcal{G}(p)$ *and where* K *is a constant matrix (the zero matrix).* $B(p)$ *can thus be written as (cf.* (7.4.6) *and* (7.4.7))

10 $$B(p) = B^0(p) + A(p)K,$$

with

11 $$B^0(p) = A(p)\,\mathcal{G}^0(p).$$

Further let

12 $$C \triangleq \begin{array}{c} \\ y_0 \\ u \end{array}\begin{array}{c} \begin{array}{cc} u_0 & y \end{array} \\ \left[\begin{array}{c:c} 0 & C_2 \\ \hdashline C_3 & C_4 \end{array}\right] \end{array}$$

be a suitable interconnection matrix of appropriate size and let

13 $$\det\,(I - KC_4) \neq 0.$$

This condition is, of course, always fulfilled in the strictly proper case $(K = 0)$.

Next form the uppermost quantities appearing in the assignment diagrams in Figs. 9.3 *and* 9.4 *including the interconnections* $(\mathcal{A}, C, S^-)$ *and* $(\mathcal{A}, C, X)$ *as well as the composition* $(\mathcal{B}, C)$ *as described in sections* 9.1 *and* 9.2 *on the basis of* S *and* C *as specified above.*

Then the following statements are true:

(i) $(\mathcal{A}, C, S^-)$ *and* $(\mathcal{A}, C, X)$ *are determinate and* $(\mathcal{B}, C)$ *is regular, i.e.* $\det\,(A(p) - B(p)C_4) \neq 0$. *All the remaining quantities appearing in the assignment diagrams in Figs.* 9.3 *and* 9.4 *can thus be formed as described in sections* 9.1 *and* 9.2.
(ii) $\partial(\det\,(A(p) - B(p)C_4)) = \partial(\det A(p))$.
(iii) $(\mathcal{A}, C, S^-)$, $(\mathcal{A}, C, X)$ *and* $(\mathcal{B}, C)$ *are mutually equivalent in the sense that the systems associated with them according to the assignment diagrams in Fig.* 9.3 *and Fig.* 9.4 *are such that* $\bar{\mathcal{S}}_i^+ = \hat{\mathcal{S}}_i^+ = \mathcal{S}_i^+$ *and* $\bar{\mathcal{S}}_o^+ = \hat{\mathcal{S}}_o^+ = \mathcal{S}_o^+$.
(iv) *All the systems mentioned in the previous item are proper (strictly proper) differential systems in the sense of section* 8.5 *with the dimension of* $\bar{\mathcal{S}}_i^+ = \hat{\mathcal{S}}_i^+ = \mathcal{S}_i^+$ *equal to the dimension of* $\mathcal{S}^+$ $(= \partial(\det A(p))$ *= the order of* S), *and with the dimension of* $\bar{\mathcal{S}}_o^+ = \hat{\mathcal{S}}_o^+ = \mathcal{S}_o^+$ *equal to or smaller than the dimension of* $\bar{\mathcal{S}}_i^+ = \hat{\mathcal{S}}_i^+ = \mathcal{S}_i^+$, *equality holding if and only if* $(\mathcal{B}, C)$ *is observable.* □

14 **Proof.** To begin with note that some statements relating to the above theorem were already given (without proofs) in note (7.2.24), (v). We shall not, however, rely on those statements in this context; we prefer to discuss the present theorem from the beginning.

Let us start with a study of the regular generators M for S and M_i for S_i given by

15 $$M \triangleq [A(p) \vdots -B(p)]$$

and (*cf.* (7.2.13))

16 $$M_i \triangleq [A_i(p) \vdots -B_i(p)] \triangleq [A(p) - B(p)C_4 \vdots -B(p)C_3]$$

respectively. Without loss of generality we may assume that $A(p)$ is of CRP-form (*cf.* appendix A2). Here we shall not exclude (nonzero) constant rows in $A(p)$, i.e. there may be rows and columns in $A(p)$ of zero degree.

Next form the coefficient matrix C_M having as its rows the leading coefficients of the rows of M (*cf.* C_A in (A2.31)). It is asserted that C_M must be of the form

17 $$C_M = C_{A(p)}[I \vdots -K],$$

where $C_{A(p)}$ is the coefficient matrix having as its rows the leading coefficients of the rows of $A(p)$. $C_{A(p)}$ is upper triangular and invertible (*cf.* (8.5.37)). The above result is obtained as follows. Replace $B(p)$ by $B^0(p) + A(p)K$ according to (10) and note that the degree of any row of $B^0(p)$ is less than the degree of the corresponding row of $A(p)$ (*cf.* (8.5.36)). Hence C_M is equal to the corresponding coefficient matrix for $[A(p) \vdots -A(p)K] = A(p)[I \vdots -K]$ leading to (17). Note also that the row degrees of $A(p)K$ are equal to or less than the corresponding row degrees of $A(p)$, equality holding in particular if K happens to be square and invertible.

Consider then M_i as given by (16), and in particular the coefficient matrix C_{M_i} having as its rows the leading coefficients of the rows of M_i. Following the same reasoning as above it is found that C_{M_i} must be equal to the corresponding coefficient matrix for $A(p)[I - KC_4 \vdots -KC_3]$ provided that (13) holds. This yields

18 $$C_{M_i} = C_{A(p)}[I - KC_4 \vdots -KC_3].$$

Now with $\det(I - KC_4) \neq 0$ $C_{A(p)}(I - KC_4)$ is invertible and consequently $A(p) - B(p)C_4$ is row proper, but possibly not of CRP-form, having the same row degrees as $A(p)$. If $K = 0$ (strictly proper S) then (18) implies that $A(p) - B(p)C_4$ is, in fact, of CRP-form. Moreover, the row degrees of $B(p)C_3$ are (for any C_3) equal to or less than the corresponding row degrees of $A(p) - B(p)C_4$. This proves then that det $(A(p) - B(p)C_4) \neq 0$, that $\partial(\det(A(p) - B(p)C_4)) = \partial(\det A(p))$, and that M_i is regular and proper (strictly proper if $K = 0$). Further, a regular generator M_o for S_o can be constructed according to (7.2.20) and (7.2.21). The generator $M_o \triangleq [A_0(p) \vdots -B_0(p)]$ thus obtained is found to be proper (strictly proper if $K = 0$) with $\partial(\det A_0(p)) \leq \partial(\det(A(p) - B(p)C_4))$, equality holding if and only if $(\mathcal{B}, C)$ is observable ($\tilde{A}_{io1}(p)$ in (7.2.21) is unimodular).

Combining the above results with the statements of theorem (1) and the results presented in note (8.5.35) concerning the dimension of proper differential systems it is seen that we have herewith shown the validity of (i) and (ii) of the theorem. Likewise we have shown that (iii) and (iv) hold as far as $(\mathcal{A}, C, X)$, and $(\mathcal{B}, C)$ as well as the corresponding systems $\hat{\mathcal{S}}_i^+ = \mathcal{S}_i^+$, and $\hat{\mathcal{S}}_o^+ = \mathcal{S}_o^+$ are concerned.

It remains for us to show that $\bar{\mathcal{S}}_i^+ = \hat{\mathcal{S}}_i^+ = \mathcal{S}_i^+$, or rather that these systems have the same zero-input response space, implying that also $\bar{\mathcal{S}}_o^+ = \hat{\mathcal{S}}_o^+ = \mathcal{S}_o^+$. We shall do this in the following way.

To begin with take the row proper generator M_i constructed above and given by (16). Premultiply M_i if necessary by a unimodular matrix $Q(p)$ so that $Q(p)M_i = [Q(p)A_i(p) \vdots -Q(p)B_i(p)]$ is such that $Q(p)A_i(p)$ is of CRP-form. Let $Q(p)A(p)$, $Q(p)B(p)$, $Q(p)A_i(p)$, and $Q(p)B_i(p)$ be renamed $A(p)$, $B(p)$, $A_i(p)$, and $B_i(p)$ respectively.

Now it is seen that

19 $$[A(p) \vdots -B(p)] = [A_i(p) + B_i(p)C_4 \vdots -B(p)].$$

Using the same kind of reasoning as above (the condition (13) has to be rewritten in a suitable equivalent form) it is again concluded that the $[A(p) \vdots -B(p)]$ thus obtained is row proper but possibly not of CRP-form having the same row degrees as $[A_i(p) \vdots -B_i(p)]$.

Next consider the zero-input response spaces $\bar{S}_i^+(0)$, $\hat{S}_i^+(0) = S_i^+(0)$ of $\bar{\mathcal{S}}_i^+$, $\hat{\mathcal{S}}_i^+ = \mathcal{S}_i^+$ respectively using the above matrices.

$S_i^+(0)$ is obtained from (3) as

20 $$S_i^+(0) = \{y^+ | y^+ = A_i(p)^{-1}(-A_i^0 y^- + B_i^0 u_0^-) \text{ for some } (u_0^-, y^-) \in S_i^-\},$$

or equivalently, using the result (9.2.4), . . . , (9.2.7) and an obvious simplified notation,

21 $$S_i^+(0) = \{y^+ | y^+ = A_i(p)^{-1}C_i(p)\,\delta_{t_0} \text{ for some } C_i(p) \in \Sigma_i\},$$

where $\Sigma_i \triangleq \mathbf{C}_{r_{i1}-1}[p] \times \mathbf{C}_{r_{i2}-1}[p] \times \ldots \times \mathbf{C}_{r_{iq}-1}[p]$ with $r_{i1}, r_{i2}, \ldots, r_{iq}$ equal to the row (and column) degrees of $A_i(p)$. If there is an $r_{ik} = 0$, then $\mathbf{C}_{r_{ik}-1}[p]$ is interpreted simply as $\{0\}$.

$\hat{S}_i^+(0)$ is obtained from (5) as

22 $$\hat{S}_i^+(0) = \{y^+ | y^+ = A_i(p)^{-1}(-A^0y^- + B^0u^-) \text{ for some } (u^-, y^-) \in X\}$$

with $X \subset S^-$ given by (9.1.8). According to theorem (1), $\hat{S}_i^+(0) = S_i^+(0)$. $\bar{S}_i(0)$ is obtained from (9.1.11) as

23 $$\bar{S}_i^+(0) = \{y^+ | y^+ = A_i(p)^{-1}(-A^0y^- + B^0u^-) \text{ for some } (u^-, y^-) \in S^-\}.$$

Thus $\hat{S}_i^+(0) \subset \bar{S}_i^+(0)$.

Consider then the set

24 $$\{y^+ | y^+ = A_i(p)^{-1}(-A^0y^- + B^0u^-) \text{ for some } (u^-, y^-) \in \mathscr{D}_{t_0}^{-r} \times \mathscr{D}_{t_0}^{-q}\}.$$

Clearly then $\bar{S}_i^+(0)$ is included in the set (24).

On the other hand, the results of note (8.4.20), (v) imply, that the range of $[A^0 : -B^0]$ in (24) is just the set

25 $$\{f | f = -C_i(p)\delta_{t_0} \quad \text{for some } C_i(p) \in \Sigma_i\}$$

with Σ_i as given above, i.e. the set (24) coincides with the set (21). Hence we have the following relationship between the various sets appearing above:

26 $$\begin{aligned} &\text{the set (21)} = \text{the set (20)} = \text{the set (22)} \\ &\subset \text{the set (23)} \subset \text{the set (24)} = \text{the set (21)}. \end{aligned}$$

It follows that all the sets involved coincide implying that $\bar{\mathscr{S}}_i^+ = \hat{\mathscr{S}}_i^+ = \mathscr{S}_i^+$. □

27 **Note.** **(i)** The above theorem (8) thus gives sufficient conditions for the interconnections ($\mathscr{A}$, C, S^-) and ($\mathscr{A}$, C, X) to be equivalent. These conditions are very reasonable from a practical point of view—the properness of S, for instance, is for physical reasons often well motivated. Also the condition (13) is a very natural one. To see this note first that (13) holds for almost all K, i.e. if $\det(I - KC_4) = 0$, then $\det(I - KC_4) \neq 0$ can be obtained by an infinitely small variation of some suitable entries of K. Further it is found that if $\det(I - KC_4) = 0$, then $\partial(\det(A(p) - B(p)C_4)) < \partial(\det A(p))$. Small variations in K may in this case thus cause $\partial(\det A(p) - B(p)C_4))$ to vary—this makes the resulting system and its stability properties extremely sensitive to parameter variations.

Note though that the stability properties of the resulting system may be extremely sensitive to parameter variations even if $\det(I - KC_4) \neq 0$ (*cf.* sections 1.5, 7.12).

The reader is advised to discuss further consequences of $\det(I - KC_4) = 0$—what happens, for instance, if we in this case form our interconnections on the basis of a state-space representation of S?

(ii) If S in theorem (8) is strictly proper, then the equivalence between $(\mathscr{A}, C, S^-)$ and $(\mathscr{A}, C, X)$ follows directly from the fact that $\bar{k}_\mathrm{i}^+$ as given by (9.1.17) in this case coincides with k_i^+ as given by (9.2.4), . . . , (9.2.7).

(iii) If the interconnections $(\mathscr{A}, C, S^-)$ and $(\mathscr{A}, C, X)$ are determinate but not equivalent, then the corresponding zero-input response spaces $\bar{S}_\mathrm{i}^+(0)$ and $\hat{S}_\mathrm{i}^+(0)$ differ from each other to the extent that $\bar{S}_\mathrm{i}^+(0)$ exhibits additional delta functions of various orders at t_0 not present in $\hat{S}_\mathrm{i}^+(0)$.

(iv) When interconnections are studied in the literature it is generally not explicitly stated whether the study concerns an interconnection of the type $(\mathscr{A}, C, S^-)$ or of the type $(\mathscr{A}, C, X)$. The authors apparently tacitly assume that there is no need to make the distinction. Then everything works out nicely if the interconnections happen to be equivalent, but if they are not, then considerable confusion may arise (Rosenbrock, 1974a, c).

(v) Consider again our two interconnections $(\mathscr{A}, C, S^-)$ and $(\mathscr{A}, C, X)$ as described in section 9.1, and suppose that they are determinate. Further, let the systems determined by these interconnections and other relevant quantities be denoted as above.

Now depending on the situation at hand we may be faced either with the problem of finding resulting input-output mappings corresponding to some specified sets of initial condition lists, or with the task of studying the general structural properties of the resulting systems $\bar{\mathscr{S}}_\mathrm{i}^+, \bar{\mathscr{S}}_\mathrm{o}^+, \hat{\mathscr{S}}_\mathrm{i}^+ = \mathscr{S}_\mathrm{i}^+$, and $\hat{\mathscr{S}}_\mathrm{o}^+ = \mathscr{S}_\mathrm{o}^+$. In the former case it is natural to base the study on the parametric input-output mappings $\bar{h}_\mathrm{i}^+, \bar{h}_\mathrm{o}^+$, $\hat{h}_\mathrm{i}^+ = h_\mathrm{i}^+$, and $\hat{h}_\mathrm{o}^+ = h_\mathrm{o}^+$ derived above, because the initial condition lists appear explicitly in the expressions for these mappings. In the latter case it is most advantageous to base the study on parametric input-output mappings of the form represented by $\bar{k}_\mathrm{i}^+$ (determining $\bar{\mathscr{S}}_\mathrm{i}^+$) and k_i^+ (determining $\hat{\mathscr{S}}_\mathrm{i}^+ = \mathscr{S}_\mathrm{i}^+$) as given by (9.1.17), (9.1.18), and (9.2.4), . . . , (9.2.7) respectively. □

We shall illustrate some of the points mentioned above with the aid of an example.

9.4 Illustrative example

Consider the simple electrical network consisting of a connection of resistors and capacitors as depicted in Fig. 9.5. Rosenbrock has used this network in a number of papers (Rosenbrock, 1974a, c) to illustrate some aspects relating to the general structure of system descriptions.

The quantities appearing in Fig. 9.5 have to satisfy the following equations (we use obvious notations: R denotes resistance, C denotes capacitance, v_1, v_2 denote voltages, i_1, i_2 denote currents)

1
$$Cp(v_1 - R(i_1 - i_2)) = i_1 - i_2,$$
$$Cp(v_2 - Ri_3) = i_3.$$

If the switch SW is in position 2, then the following additional equations

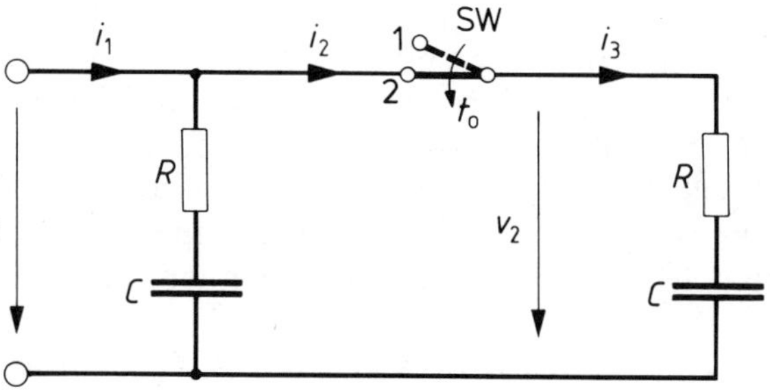

Fig. 9.5. Connection of two resistors and two capacitors.

must be satisfied:

2
$$i_2 = i_3.$$
$$v_1 = v_2.$$

The time instant $t_0 \in T$, with T some open interval of $\mathbf{R}$, is regarded as fixed.

The capacitance C is assumed positive (real) and fixed, the resistance R is permitted to take any nonnegative (real) value. Of course, the value $R = 0$—which plays a particular role in what follows—can never be exactly achieved in a real world network. Note also that we have—for the sake of simplicity—assumed that the two resistances involved are mutually exactly equal, as are the two capacitances. This can never be true in a real situation, and so it happens that this assumption leads to some unrealistic features.

v_1, v_2, i_1, i_2 are time functions (signals). They are assumed to be members of the signal space $\mathscr{D}$ (*cf.* section 8.2). The above equations imply that we have used some suitable consistent system of units for the physical quantities involved.

Now we can in different ways construct interconnections either of the kind $(\mathscr{A}, \underline{C}, S^-)$ or $(\mathscr{A}, \underline{C}, X)$ as described before corresponding to the network in Fig. 9.5. The former kind of interconnection is obtained if the switch SW is moved from position 1 to position 2 at the time instant t_0. The latter kind of interconnection is obtained when SW is held in position 2 "all the time". The structural details of the interconnections depend decisively on the choice of input and output variables. Here we shall discuss one particular choice of variables—the reader is advised to consider other possible choices. It turns out that the problem of how to choose suitable sets of input and output variables is far from trivial.

Let us make the following choice (we use the by now familiar notations).

The inputs u_0 and u are chosen according to

3 $$u_0 = i_1, \qquad u = (i_1, i_2, v_2),$$

and the outputs y_0 and y according to

4 $$y_0 = i_3, \qquad y = (v_1, i_3).$$

The above choice of inputs $u = (i_1, i_2, v_2)$ means that the basic network involved in the connection is such that it can be connected to suitable and mutually independent current and voltage sources from which the currents i_1 and i_2 and the voltage v_2 are fed into the network.

(2), (3), and (4) together imply an interconnection matrix $\underline{C}$ of the following form (we use an underlined $\underline{C}$ for the interconnection matrix to distinguish it from the capacitance C):

5 $$\underline{C} = \begin{array}{c} \\ y_0 \\ u \end{array}\!\!\begin{array}{c} \begin{array}{cc} u_0 & y \end{array} \\ \left[\begin{array}{c:c} 0 & C_2 \\ \hdashline C_3 & C_4 \end{array}\right] \end{array} = \begin{array}{c} \\ y_0 \\ i_1 \\ i_2 \\ v_2 \end{array}\!\!\begin{array}{c} \begin{array}{ccc} u_0 & v_1 & i_3 \end{array} \\ \left[\begin{array}{c:cc} 0 & 0 & 1 \\ \hdashline 1 & 0 & 0 \\ 0 & 0 & 1 \\ 0 & 1 & 0 \end{array}\right] \end{array}.$$

The set of all $(u, y) \in \mathscr{D}^3 \times \mathscr{D}^2$ satisfying the equations (1) then gives our basic differential input-output relation $S \subset \mathscr{D}^3 \times \mathscr{D}^2$. From (1) we obtain the following generator M for S:

6 $$M \triangleq [A(p) \; \vdots \; -B(p)]$$

$$= \begin{array}{c} \begin{array}{ccccc} v_1 & i_3 & i_1 & i_2 & v_2 \end{array} \\ \left[\begin{array}{cc:ccc} Cp & 0 & -RCp-1 & RCp+1 & 0 \\ 0 & RCp+1 & 0 & 0 & -Cp \end{array}\right] \end{array}.$$

With the relevant quantities thus chosen our interconnections $(\mathscr{A}, \underline{C}, S^-)$ and $(\mathscr{A}, \underline{C}, X)$ and related quantities can be depicted as shown in Fig. 9.6, where the system $\mathscr{S}^+$ involved in the interconnection is determined by $h^+ = \text{PM}([A(p) \,\vdots\, -B(p)])$, and where the other quantities are as described before.

We have $\det A(p) = Cp(RCp+1) \neq 0$ for every $R \geqslant 0$. M is consequently always regular and so is S. $\partial(\det A(p)) = 2$ for $R > 0$ and $=1$ for $R = 0$. The order of S thus depends on R. Further note that $A(p)$ above is of CUT-form as well as of CRP-form (up to a normalization of the diagonal entries) for every $R \geqslant 0$. For $R > 0$ M is seen to be proper, and the row degrees of $A(p)$ are $r_1 = r_2 = 1$. For $R = 0$ M is non-proper, and the row degrees of $A(p)$ are $r_1 = 1$ and $r_2 = 0$.

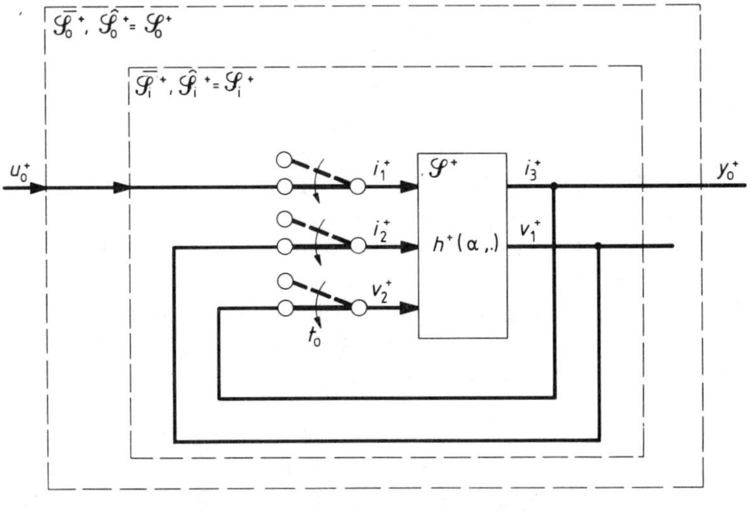

Fig. 9.6. Block diagram showing the interconnections $(\mathscr{A}, \underline{C}, S^-)$ and $(\mathscr{A}, \underline{C}, X)$ corresponding to the situation depicted in Fig. 9.5.

$A(p)$ and $B(p)$ in (6) are found to be left coprime for every $R \geqslant 0$. Our input-output relation S is consequently always controllable.

The transfer matrix $\mathscr{G}(p)$ determined by (6) is

7 $$\mathscr{G}(p) \triangleq A(p)^{-1}B(p) = \begin{bmatrix} \dfrac{RCp+1}{Cp} & -\dfrac{RCp+1}{Cp} & 0 \\ 0 & 0 & \dfrac{Cp}{RCp+1} \end{bmatrix}.$$

Consider the case $R > 0$.

In this case (6) and (7) are proper, and $\mathcal{G}(p)$ and $B(p)$ can be written in the familiar way as (*cf.* section 7.4).

8 $$\mathcal{G}(p) = \mathcal{G}^0(p) + K = \begin{bmatrix} \frac{1}{Cp} & -\frac{1}{Cp} & 0 \\ 0 & 0 & -\frac{1}{R(RCp+1)} \end{bmatrix} + \underbrace{\begin{bmatrix} R & -R & 0 \\ 0 & 0 & \frac{1}{R} \end{bmatrix}}_{K}$$

and

9 $$B(p) = B^0(p) + A(p)K = \underbrace{\begin{bmatrix} 1 & -1 & 0 \\ 0 & 0 & -\frac{1}{R} \end{bmatrix}}_{B^0(p)} \begin{bmatrix} Cp & 0 \\ 0 & RCp+1 \end{bmatrix} \underbrace{\begin{bmatrix} R & -R & 0 \\ 0 & 0 & \frac{1}{R} \end{bmatrix}}_{K}$$

respectively.

Hence

10 $$I - KC_4 = \begin{bmatrix} 1 & R \\ -\frac{1}{R} & 1 \end{bmatrix}$$

and det $(I - KC_4) = 2 \neq 0$.

It is seen that $R > 0$ thus gives a well-behaved case which fulfils the conditions of theorem (9.3.8). To determine the resulting systems $\bar{\mathcal{S}}_i^+ = \hat{\mathcal{S}}_i^+ = \mathcal{S}_i^+$ and $\bar{\mathcal{S}}_o^+ = \hat{\mathcal{S}}_o^+ = \mathcal{S}_o^+$ we may use any one of the parametric input-output mappings for these systems introduced above. The reader is advised to discuss the various possibilities in detail.

Next let $R = 0$.

In this case (6) and (7) are reduced to

11 $$M \triangleq [A(p) \,\vdots\, -B(p)] = \begin{array}{c} \begin{matrix} v_1 & i_3 & i_1 & i_2 & v_2 \end{matrix} \\ \left[\begin{array}{cc:ccc} Cp & 0 & -1 & 1 & 0 \\ 0 & 1 & 0 & 0 & -Cp \end{array}\right] \end{array}$$

and

12 $$\mathcal{G}(p) \triangleq A(p)^{-1}B(p) = \begin{bmatrix} \frac{1}{Cp} & -\frac{1}{Cp} & 0 \\ 0 & 0 & Cp \end{bmatrix}$$

respectively. Thus (11) and (12) are nonproper and (8) and (9) take the forms

13
$$\mathcal{G}(p) = \mathcal{G}^0(p) + K(p)$$
$$= \begin{bmatrix} \dfrac{1}{Cp} & -\dfrac{1}{Cp} & 0 \\ 0 & 0 & 0 \end{bmatrix} + \underbrace{\begin{bmatrix} 0 & 0 & 0 \\ 0 & 0 & Cp \end{bmatrix}}_{K(p)}$$

and

14
$$B(p) = B^0(p) + A(p)K(p)$$
$$= \underbrace{\begin{bmatrix} 1 & -1 & 0 \\ 0 & 0 & 0 \end{bmatrix}}_{B^0(p)} + \begin{bmatrix} Cp & 0 \\ 0 & 1 \end{bmatrix} \underbrace{\begin{bmatrix} 0 & 0 & 0 \\ 0 & 0 & Cp \end{bmatrix}}_{K(p)}$$

respectively.

The conditions of theorem (9.3.8) are in this case ($R = 0$) clearly not fulfilled, and it may well happen that $(\mathcal{A}, \underline{C}, S^-)$ and $(\mathcal{A}, \underline{C}, X)$ are not mutually equivalent. Thus they must be treated separately. Nevertheless, in this case too the resulting systems can be determined with the aid of a number of different parametric input-output mappings. The reader is again advised to consider the details.

Below we shall show in some detail how the treatment can be carried out in the two cases $R > 0$ and $R = 0$ following a particular course of advance. The reader should fill in the missing details.

The case $R > 0$. S is proper and of order 2. It follows (*cf.* note (8.5.35), (v)) that the dimension of $\mathcal{S}^+$ is also 2.

Theorem (9.3.8) is applicable, $(\mathcal{A}, \underline{C}, S^-)$, $(\mathcal{A}, \underline{C}, X)$, and $(\mathcal{B}, \underline{C})$ as described in the theorem are such that $(\mathcal{A}, \underline{C}, S^-)$ and $(\mathcal{A}, \underline{C}, X)$ are determinate and $(\mathcal{B}, \underline{C})$ regular. Further they are all mutually equivalent in the sense that they all determine the same resulting systems. Here we shall base our treatment on the composition $(\mathcal{B}, \underline{C})$, which thus allows us to use the polynomial systems theory as developed in part II of this book.

The first step is to determine a regular generator $M_i = [A_i(p) \vdots -B_i(p)]$ for the internal input-output relation S_i. We obtain (*cf.* (7.2.13), and (5), (6) above)

15 $$M_i \triangleq [A_i(p) \vdots -B_i(p)] = [A(p) - B(p)C_4 \vdots -B(p)C_3]$$

$$= \begin{bmatrix} \overset{v_1}{Cp} & \overset{i_3}{RCp+1} & \vdots & \overset{i_1}{-RCp-1} \\ -Cp & RCp+1 & \vdots & 0 \end{bmatrix}.$$

Note that the output components v_1 and i_3 are ordered in the way mentioned in note (7.2.24), (ii). To determine a regular generator $M_o \triangleq [A_0(p) \vdots -B_0(p)]$ for the overall input-output relation S_o, we have thus only to transform M_i to an upper triangular form. The CUT-form of (15) is found to be (up to a normalization of the diagonal entries)

16 $$\tilde{M}_i \triangleq [\tilde{A}_i(p) \vdots -\tilde{B}_i(p)] = \begin{bmatrix} 2Cp & 0 & \vdots & -RCp-1 \\ 0 & 2(RCp+1) & \vdots & -RCp-1 \end{bmatrix}.$$

As it happens, $\tilde{A}_i(p)$ above is also of CRP-form (up to a suitable normalization) having the row degrees $r_{i1} = r_{i2} = 1$.

A regular generator M_o for S_o is obtained from (16) as

17 $$M_o \triangleq [A_0(p) \vdots -B_0(p)] = [2(RCp+1) \vdots -RCp-1].$$

It is concluded that $(\mathcal{B}, \underline{C})$ is of order 2, which is equal to the order of S. S_o is of order 1.

S_i and S_o are proper, and the dimensions of the corresponding systems $\mathcal{S}_i^+$ and $\mathcal{S}_o^+$ determined by $h_i^+ = \mathrm{PM}(M_i)$ and $h_o^+ = \mathrm{PM}(M_o)$ respectively thus coincide with the orders of the corresponding input-output relations. In particular it thus holds that the order of S_i = the dimension of $\mathcal{S}_i^+$ = the order of S = the dimension of $\mathcal{S}^+$.

$(\mathcal{B}, \underline{C})$ is not observable, because the entry in the upper left-hand corner of $\tilde{A}_i(p)$ in (16) is not a constant.

$\tilde{A}_i(p)$ and $\tilde{B}_i(p)$ in (16) have a GCLD of the form

18 $$\begin{bmatrix} 1 & 0 \\ 0 & RCp+1 \end{bmatrix}$$

and $A_0(p)$ and $B_0(p)$ in (17) a GCLD of the form

19 $$RCp + 1.$$

It follows that neither S_i nor S_o is controllable.

As the parametric input-output mapping determining the resulting internal system $\mathcal{S}_i^+$ we choose a mapping of the form k_i^+ as given by (9.2.4), . . . , (9.2.7) (we again use an obvious simplified notation because we would have $\Sigma_{i2} = \{0\}$ in this case). In the present situation this yields (recall that $\tilde{A}_i(p)$ in (16) is of CRP-form)

20 $$k_i^+ : \Sigma_i \times \mathcal{D}_{t_0}^+ \to \mathcal{D}_{t_0}^{+^2}$$

with

21 $$\Sigma_i \triangleq \mathbf{C}_0[p] \times \mathbf{C}_0[p] = \mathbf{C}^2$$

and

$$((c_{i1}, c_{i2}), i_1^+) \mapsto (v_1^+, i_3^+)$$

with

22 $$\begin{bmatrix} v_1^+ \\ i_3^+ \end{bmatrix} = \tilde{A}_i(p)^{-1} \begin{bmatrix} c_{i1} \\ c_{i2} \end{bmatrix} \delta_{t_0} + \tilde{A}_i(p)^{-1}\tilde{B}_i(p) i_1^+$$

$$= \begin{bmatrix} \dfrac{c_{i1}}{2Cp} \\ \dfrac{c_{i2}}{2(RCp+1)} \end{bmatrix} \delta_{t_0} + \begin{bmatrix} \dfrac{RCp+1}{2Cp} \\ \dfrac{1}{2} \end{bmatrix} i_1^+$$

$$= \begin{bmatrix} \dfrac{c_{i1}}{2C} U_{t_0} \\ \dfrac{c_{i2}}{2RC} U_{t_0} e^{-\frac{1}{RC}(\cdot - t_0)} \end{bmatrix} + \begin{bmatrix} \dfrac{RCp+1}{2Cp} \\ \dfrac{1}{2} \end{bmatrix} i_1^+ .$$

A corresponding mapping k_o^+ for $\mathcal{S}_o^+$ is easily obtained from k_i^+ above (*cf.* (9.1.11), (9.1.12)).

At this stage the reader is advised to give a physical interpretation of the results obtained. In particular the reader should consider the physical meaning of the "integration constants" c_{i1} and c_{i2} appearing in (22). It would also be highly illuminating to construct a state-space representation of S_i above along the lines presented in note (8.5.35). Further, note that even if we assume that the interconnection constraints (2) have been valid "all the time" (SW in Fig. 9.5 in position 2 "all the time"), this does not imply that the two capacitors C also have the same charge "all the time".

The case $R = 0$. S is nonproper and of order 1.

Let us construct a parametric input-output mapping for $\mathcal{S}^+$ of the form k^+ as given by (8.5.69), . . . , (8.5.73). Using the values valid in the present case we obtain (recall that $A(p)$ in (11) is of CRP-form)

23 $$k^+ : (\Sigma_1 \times \Sigma_2) \times \mathcal{D}_{t_0}^{+^3} \to \mathcal{D}_{t_0}^{+^2}$$

with

24 $$\Sigma_1 = \mathbf{C}_0[p] \times \{0\} = \mathbf{C} \times \{0\}$$

and

25 $$\Sigma_2 = \{C_2(p) \mid C_2(p) \in \mathbf{C}[p]^2 \quad \text{and } C_2(p)\,\delta_{t_0} \in \mathrm{R}K^0\},$$

where $K^0 : \mathscr{D}_{t_0}^{-3} \to \mathscr{D}_{t_0}^{+2}$ is the initial condition mapping at t_0 associated with $K(p)$ in (13), (14). The results of section 8.5 yield:

26 $$\mathrm{R}K^0 = \{(0, c_2\delta_{t_0}) \mid c_2 \in \mathbf{C}\},$$

i.e.

27 $$\Sigma_2 = \{0\} \times \mathbf{C}.$$

Further k^+ is determined by the assignment

$$(((c_1, 0), (0, c_2)), (i_1^+, i_2^+, v_2^+)) \mapsto (v_1^+, i_3^+)$$

with

28 $$\begin{bmatrix} v_1^+ \\ i_3^+ \end{bmatrix} = A(p)^{-1}\begin{bmatrix} c_1 \\ 0 \end{bmatrix}\delta_{t_0} - \begin{bmatrix} 0 \\ c_2 \end{bmatrix}\delta_{t_0} + A(p)^{-1}B(p)\begin{bmatrix} i_1^+ \\ i_2^+ \\ v_2^+ \end{bmatrix}$$

$$= \begin{bmatrix} \dfrac{c_1}{Cp} \\ 0 \end{bmatrix}\delta_{t_0} - \begin{bmatrix} 0 \\ c_2 \end{bmatrix}\delta_{t_0} + \begin{bmatrix} \dfrac{1}{Cp} & -\dfrac{1}{Cp} & 0 \\ 0 & 0 & Cp \end{bmatrix}\begin{bmatrix} i_1^+ \\ i_2^+ \\ v_2^+ \end{bmatrix}$$

$$= \begin{bmatrix} \dfrac{c_1}{C}U_{t_0} \\ -c_2\delta_{t_0} \end{bmatrix} + \begin{bmatrix} \dfrac{1}{Cp} & -\dfrac{1}{Cp} & 0 \\ 0 & 0 & Cp \end{bmatrix}\begin{bmatrix} i_1^+ \\ i_2^+ \\ v_2^+ \end{bmatrix}$$

It follows that the dimension of $\mathscr{S}^+$ is equal to 2, i.e. equal to the dimension of $\Sigma_1 \times \Sigma_2$. Thus the order of S is one less than the dimension of $\mathscr{S}^+$.

Next let us consider the resulting internal system $\bar{\mathscr{S}}_{\mathrm{i}}^+$ associated with $(\mathscr{A}, \underline{C}, S^-)$. A parametric input-output mapping $\bar{k}_{\mathrm{i}}^+$ for $\bar{\mathscr{S}}_{\mathrm{i}}^+$ based on k^+ above can be constructed as described by (9.1.17) and (9.1.18). We obtain

29 $$\bar{k}_{\mathrm{i}}^+ : (\Sigma_1 \times \Sigma_2) \times \mathscr{D}_{t_0}^+ \to \mathscr{D}_{t_0}^{+2}$$

with Σ_1 and Σ_2 as in (24), (25) above, and with $\bar{k}_{\mathrm{i}}^+$ determined by the assignment

$$(((c_1, 0), (0, c_2)), i_1^+) \mapsto (v_1^+, i_3^+)$$

with

30 $$\begin{bmatrix} v_1^+ \\ i_3^+ \end{bmatrix} = (A(p) - B(p)C_4)^{-1}\left(\begin{bmatrix} c_1 \\ 0 \end{bmatrix}\delta_{t_0} - A(p)\begin{bmatrix} 0 \\ c_2 \end{bmatrix}\delta_{t_0}\right)$$
$$+ (A(p) - B(p)C_4)^{-1}B(p)C_3 i_1^+$$
$$= \begin{bmatrix} \frac{c_1}{2Cp} \\ \frac{c_1}{2} \end{bmatrix}\delta_{t_0} - \begin{bmatrix} -\frac{c_2}{2Cp} \\ \frac{c_2}{2} \end{bmatrix}\delta_{t_0} + \begin{bmatrix} \frac{1}{2Cp} \\ \frac{1}{2} \end{bmatrix} i_1^+$$
$$= \begin{bmatrix} \frac{c_1 + c_2}{2C} U_{t_0} \\ \frac{c_1 - c_2}{2}\delta_{t_0} \end{bmatrix} + \begin{bmatrix} \frac{1}{2Cp} \\ \frac{1}{2} \end{bmatrix} i_1^+.$$

A corresponding mapping $\bar{k}_o^+$ for $\bar{\mathcal{S}}_o^+$ can be read directly from (29), (30) above.

The dimension of $\bar{\mathcal{S}}_i^+$ is equal to the dimension of $\mathcal{S}^+$ and $\Sigma_1 \times \Sigma_2$, i.e. equal to 2, the dimension of $\bar{\mathcal{S}}_o^+$ is 1.

Finally, let us consider the resulting internal system $\hat{\mathcal{S}}_i^+ = \mathcal{S}_i^+$ associated with $(\mathcal{A}, \underline{C}, X)$ and $(\mathcal{B}, \underline{C})$ with $R = 0$. We shall construct a parametric input-output mapping k_i^+ for $\mathcal{S}_i^+$ as described in note (9.2.1) based on $(\mathcal{B}, \underline{C})$. The construction follows the corresponding construction for $R > 0$ (*cf.* (16), . . . , (22) above).

A suitable $\tilde{M}_i \triangleq [\tilde{A}_i(p) \vdots -\tilde{B}_i(p)]$ corresponding to (9.2.2) with $\tilde{A}_i(p)$ of CRP-form can, incidentally, be obtained from (16) by putting $R = 0$. The result is (up to a suitable normalization)

31 $$\tilde{M}_i \triangleq [\tilde{A}_i(p) \vdots -\tilde{B}_i(p)] = \left[\begin{array}{cc:c} 2Cp & 0 & -1 \\ 0 & 2 & -1 \end{array}\right],$$

further giving, instead of (17),

32 $$M_o \triangleq [A_0(p) \vdots -B_0(p)] = [2 \vdots -1].$$

The row degrees of $\tilde{A}_i(p)$ in (31) are $r_{i1} = 1$, $r_{i2} = 0$.

It is concluded that $(\mathcal{B}, \underline{C})$ is of order 1, which is equal to the order of S. S_o is of order 0.

S_i and S_o are proper, and the dimensions of the corresponding systems $\mathscr{S}_i^+$ and $\mathscr{S}_o^+$ thus coincide with the orders of the corresponding input-output relations. The dimensions of $\mathscr{S}_i^+$ and $\mathscr{S}_o^+$ are thus one less than the dimensions of $\bar{\mathscr{S}}_i^+$ and $\bar{\mathscr{S}}_o^+$ respectively. Clearly then we have here a situation where $\bar{\mathscr{S}}_i^+ \neq \hat{\mathscr{S}}_i = \mathscr{S}_i^+$ and $\bar{\mathscr{S}}_o^+ \neq \hat{\mathscr{S}}_o^+ = \mathscr{S}_o^+$.

$(\mathscr{B}, \underline{C})$ is again not observable, because the entry in the upper left-hand corner of $\tilde{A}_i(p)$ in (31) is not a constant. $\tilde{A}_i(p)$ and $\tilde{B}_i(p)$ in (31) are left coprime. It follows that S_i and S_o are controllable.

Now in the present case (9.2.4), . . . , (9.2.7) yield for k_i^+ (we again use a simplified notation because S_i is proper, *cf.* (16), . . . , (22) above):

33 $$k_i^+ : \Sigma_i \times \mathscr{D}_{t_0}^+ \to \mathscr{D}_{t_0}^{+2}$$

with

34 $$\Sigma_i \triangleq \mathbf{C}_0[p] \times \{0\} = \mathbf{C} \times \{0\}$$

and

$$((c_{i1}, 0), i_1^+) \mapsto (v_1^+, i_3^+)$$

with

35 $$\begin{bmatrix} v_1^+ \\ i_3^+ \end{bmatrix} = \tilde{A}_i(p)^{-1} \begin{bmatrix} c_{i1} \\ 0 \end{bmatrix} \delta_{t_0} + \tilde{A}_i(p)^{-1}\tilde{B}_i(p) i_1^+$$

$$= \begin{bmatrix} \dfrac{c_{i1}}{2Cp} \\ 0 \end{bmatrix} \delta_{t_0} + \begin{bmatrix} \dfrac{1}{2Cp} \\ \dfrac{1}{2} \end{bmatrix} i_1^+$$

$$= \begin{bmatrix} \dfrac{c_{i1}}{2C} U_{t_0} \\ 0 \end{bmatrix} + \begin{bmatrix} \dfrac{1}{2Cp} \\ \dfrac{1}{2} \end{bmatrix} i_1^+.$$

A corresponding mapping k_o^+ for $\mathscr{S}_o^+$ can be read directly from (33), . . . , (35) above. Comparison of (30) and (35) shows that the zero-input response space of $\bar{\mathscr{S}}_i^+$ contains delta functions in the second output component i_3^+ at t_0 which are not present in the zero-input response space of $\mathscr{S}_i^+$.

The reader is advised to construct other examples leading to $\bar{\mathscr{S}}_i^+ \neq \hat{\mathscr{S}}_i^+ = \mathscr{S}_i^+$.

Part IV

Difference Systems

In this part of the book we shall consider systems governed by linear time-invariant difference equations. These systems are very similar to time-invariant differential systems so that we can use most of the results obtained in the preceding parts. Nevertheless, there are also some dissimilarities which make difference systems a little harder to analyse. The underlying phenomenon behind these dissimilarities is that difference systems are not usually as "exact" models of real systems as e.g. differential systems are.

First the module structure of difference systems is considered following the lines of part II. The module structure can be constructed in many different ways depending on what kind of signal space and operator we can and must use. In order to apply the analogy to differential systems the time set bounded below and the unit prediction operator would be the most useful structure. However, from the systems theory point of view it has some drawbacks, so we prefer to use the unbounded time set and the unit delay operator even though this makes the structure a little more complicated.

Next we form the vector space structure for the consideration of difference systems. Again, from the systems theory point of view this structure cannot be based on the time set bounded below and we have to use the unbounded time set. Further, for the same reason the unit delay operator is better than the unit prediction operator.

The reader is also advised to consult the recent book by Kučera (Kučera, 1979). A number of results concerning difference systems resemble our results as presented below.

10

Generation of difference systems

In this chapter we shall study on which conditions difference equations can be considered as operator equations, i.e. how the signal space and the operator should be chosen.

10.1 Signal spaces and shift operators

Let $\mathbf{C}^T$ denote a (nontrivial) vector space of all complex-valued time functions (signals) defined on some nonempty time set $T \subset \mathbf{Z}$ ($\triangleq$ the integers). In order to define the *unit prediction operator* $\mathscr{q}$ on $\mathbf{C}^T$ by the equation

1 $$(\mathscr{q}u)(k) = u(k+1) \quad \text{with}$$
$$u \in \mathbf{C}^T, \quad k \in T$$

the time set T must be either the whole $\mathbf{Z}$ or a subset of $\mathbf{Z}$ of the form $\{k_0, k_0+1, \ldots\}$. Correspondingly, the *unit delay operator* $\mathscr{z}$ is well-defined on $\mathbf{C}^T$ by

2 $$(\mathscr{z}u)(k) = u(k-1) \quad \text{with}$$
$$u \in \mathbf{C}^T, \quad k \in T$$

if T is either $\mathbf{Z}$ or of the form $\{\ldots, k_0-1, k_0\} \subset \mathbf{Z}$.

Let $\mathscr{X}$ be a nontrivial subspace of $\mathbf{C}^T$. If T is either $\mathbf{Z}$ or of the form $\{\ldots, k_0-1, k_0\}$, and $\mathscr{X}$ is invariant under $\mathscr{z}$ defined by (2), then we shall call it a (*discrete-time*) *signal space*. Similarly, if T is of the form $\{k_0, k_0+1, \ldots\}$, and $\mathscr{X}$ is invariant under $\mathscr{q}$ defined by (1), it is also called a signal space.

There are, of course, many function spaces that qualify as signal spaces. In particular, the whole $\mathbf{C}^T$ on a suitable T is a signal space. In what follows, we shall use the notations q and z to denote also their restrictions to signal spaces if the meaning is clear from the context. Furthermore, we shall consider only the cases $T = \mathbf{Z}$ or $T = \{k_0, k_0 + 1, \ldots\}$; the latter can without loss of generality be taken as $\{0, 1, \ldots\} \triangleq \mathbf{N}_0$. In the case $T = \mathbf{Z}$ both q and z should be possible operators but we prefer the use of z for the reasons explained more thoroughly in chapter 14. As a consequence, we can make the notational convention that the use of z implies the time set $\mathbf{Z}$ and the use of q implies the time set $\mathbf{N}_0$.

Finally, in the case $T = \mathbf{Z}$ the operator z restricted to a signal space is always an injection but not necessarily a surjection.

According to chapter 4 the polynomials in q and z with complex coefficients are called "polynomial operators" and matrices over polynomial operators are "polynomial matrix operators".

3 **Remark.** **(i)** The concept of a signal space can easily be generalized. In fact every vector space V over the field $\mathbf{C}$ of complex numbers could be the codomain of the signals.
(ii) There are also some other operators which are used for consideration of difference systems. For instance, the operators defined by

4 $$(\nabla u)(k) = u(k+1) - u(k)$$

5 $$(\Delta u)(k) = u(k) - u(k-1)$$

are common. Because $\nabla = q - 1$ and $\Delta = 1 - z$, i.e. they are polynomial operators in q or z, respectively, there is no reason to consider them separately. □

10.2 Matrix difference equations

Let $\mathscr{X}$ be a signal space and suppose first that the time set $T = \mathbf{Z}$. Using the polynomial matrix operators a set of s difference equations can be written as

1 $$A(z)y = B(z)u,$$

where $u \in \mathscr{X}^r$, $y \in \mathscr{X}^q$ and $A(z)$ and $B(z)$ are polynomial matrix operators of sizes s by q and s by r, respectively.

Interpreting u as an input signal and y as an output signal, (1) generates a difference input-output relation

$$S = \{(u, y) \mid (u, y) \in \mathscr{X}^r \times \mathscr{X}^q \text{ and } A(\mathfrak{z})y = B(\mathfrak{z})u\}.$$

Suppose then that (u_1, y_1) is an input-output pair contained in S and let u_1^- and y_1^- be the initial segments of u_1 and y_1, respectively, observed on $\{\ldots, k_0 - 2, k_0 - 1\} = (-\infty, k_0) \cap \mathbf{Z}$. Then the set

2
$$S_{(u_1^-, y_1^-)} \triangleq \{(u, y) \mid (u, y) \in S \text{ and } u \mid (-\infty, k_0) \cap \mathbf{Z} = u_1^- \text{ and } y \mid (-\infty, k_0) \cap \mathbf{Z} = y_1^-\}$$

is a mapping if $\det A(\mathfrak{z}) \neq 0$. Consequently we can define a *difference input-output system* $\mathscr{S}$ as a family of all mappings $S_{(u_1^-, y_1^-)}$

3
$$\mathscr{S} = \{S_{(u_1^-, y_1^-)} \mid u_1^- = u \mid (-\infty, k_0) \cap \mathbf{Z} \text{ and } y_1^- = y \mid (-\infty, k_0) \cap \mathbf{Z} \text{ for some } (u, y) \in S\}.$$

However, the property $\det A(\mathfrak{z}) \neq 0$ does not suffice to make the system $\mathscr{S}$ causal in contrast with the differential systems. A difference input-output relation such that the above construction gives a causal system is called causal. The test for whether a given difference input-output relation is causal will be considered later on in chapter 12.

Suppose next that $T = \mathbf{N}_0$. A matrix difference equation can now be written as

4
$$A(\mathscr{q})y = B(\mathscr{q})u,$$

where $A(\mathscr{q})$ and $B(\mathscr{q})$ are polynomial matrix operators in $\mathscr{q}$, and the difference input-output relation S generated by (4) is defined in the obvious way. We can also define the set $S_{(u_1^-, y_1^-)}$ according to initial segments u_1^- and y_1^- on $\{0, 1, \ldots, k_0 - 1\}$ of u_1 and y_1 respectively, such that $(u_1, y_1) \in S$. However, in this case the property $\det A(\mathscr{q}) \neq 0$ alone does not suffice to make $S_{(u_1^-, y_1^-)}$ a mapping, for k_0 must be large enough, too. This dependence on the length of initial segments makes consideration of this kind of difference system rather complicated and is thus one of the reasons why we always prefer the time set $\mathbf{Z}$ if possible.

As in the case $T = \mathbf{Z}$, the system $\mathscr{S}$ consisting of mappings $S_{(u_1^-, y_1^-)}$ is not necessarily causal; the input-output relation S must have an additional property, causality, in order to obtain causal systems in this way.

11

The module structure

In this chapter we shall consider the basic algebraic structures for treating difference systems. The presentation follows the lines of chapter 5 for differential systems but there are some differences which make the structure a little more complicated.

The main structure is a module of signals over a ring of polynomial operators which has been developed in, for instance, Sinervo and Blomberg, 1971. Because the shift operators can be invertible in the case $T = \mathbf{Z}$, the ring of polynomial operators can, however, be extended to a ring of quotients. If the signal space is "sufficiently rich", the extension to a vector space is no longer possible.

In what follows we shall consider only the case $T = \mathbf{Z}$, i.e. we suppose that $T = \mathbf{Z}$ if nothing else is mentioned. However, the case $T = \mathbf{N}_0$ has some interesting properties which we shall comment on in additional remarks.

11.1 Suitable signal spaces

The main condition for a discrete-time signal space was that it should be invariant under the prediction operator q or the delay operator z. In order to develop the theory further, we need some additional conditions.

Regular signal spaces

A discrete-time signal space $\mathscr{X}$ is said to be regular if it possesses property (5.1.1) as well as property (5.1.2) with p replaced by z.

This means that $\mathscr{X}$ is regular if z is an indeterminate over $\mathbf{C}$ and every nonzero polynomial operator is a surjection with respect to $\mathscr{X}$. Note that the latter property makes the operator $\mathfrak{z}$ invertible.

A regular discrete-time signal space is analogous to a regular continuous-time signal space. Thus we can easily modify note (5.1.4) to hold for difference systems, so we will not present its discrete-time version here.

There are many regular signal spaces. In particular, the space $\mathbf{C}^{\mathbf{Z}}$ is regular. The subspace

1 $$\mathbf{C}^{\mathbf{Z}+} \triangleq \{x \mid x \in \mathbf{C}^{\mathbf{Z}} \text{ and there exists a } k \in \mathbf{Z} \text{ such that } x(l) = 0 \text{ for all } l < k\}$$

is also a regular signal space. However the space

2 $$\mathbf{C}_k^{\mathbf{Z}+} \triangleq \{x \mid x \in \mathbf{C}^{\mathbf{Z}} \text{ and } x(l) = 0 \text{ for all } l < k\}$$

with an arbitrary $k \in \mathbf{Z}$ is not regular even though it is a signal space and $\mathfrak{z}$ restricted to it is an indeterminate. This follows from the fact that the operators $\mathfrak{z}^n$, $n \in \mathbf{N}$, are not surjections.

Sufficiently rich signal spaces

A discrete-time signal space $\mathscr{X}$ is sufficiently rich if it possesses property (5.1.5) with p replaced by $\mathfrak{z}$, i.e. $\mathscr{X}$ contains all possible complex-valued solutions y to all equations $a(\mathfrak{z})y \neq 0$.

3 **Note.** **(i)** Let $\mathscr{X}$ be a sufficiently rich signal space and let $a(\mathfrak{z}) = a_0 + a_1\mathfrak{z} + \ldots + a_n\mathfrak{z}^n$, $a_n \neq 0$, be a polynomial operator $\mathscr{X} \to \mathscr{X}$. It is easily seen that the set of all possible solutions y to $a(\mathfrak{z})y = 0$, i.e. $\ker a(\mathfrak{z})$, forms a finite dimensional subspace of $\mathscr{X}$ whose dimension does not exceed n. In fact, if $a(\mathfrak{z})$ is of the form $a(\mathfrak{z}) = a_m\mathfrak{z}^m + a_{m+1}\mathfrak{z}^{m+1} + \ldots + a_n\mathfrak{z}^n$, with $m \leqslant n$, $a_m \neq 0$, $a_n \neq 0$, then the dimension of $\ker a(\mathfrak{z})$ is equal to $n - m$. It can also be seen that a sufficiently rich signal space must contain all signals of the form $k \mapsto ck^n\alpha^k$, $k \in \mathbf{Z}, n \in \mathbf{N}_0$, $\alpha, c \in \mathbf{C}$ and all finite sums of such functions.

(ii) Every sufficiently rich signal space must be infinite dimensional.

(iii) If $\mathscr{X}$ is sufficiently rich then $\mathfrak{z}$ is an indeterminate over $\mathbf{C}$.

(iv) If $\mathscr{X}$ is regular and sufficiently rich, then a polynomial operator $a(\mathfrak{z}): \mathscr{X} \to \mathscr{X}$ has an inverse $\mathscr{X} \to \mathscr{X}$ if and only if $a(\mathfrak{z})$ is of the form $a(\mathfrak{z}) = c\mathfrak{z}^n$ with $c \in \mathbf{C} - \{0\}$, $n \in \mathbf{N}_0$. □

4 **Remark.** In the case $T = \mathbf{N}_0$ the concepts of regular and especially sufficiently rich discrete-time signal spaces are more analogous to the corre-

sponding concepts of continuous-time signal spaces than in the case $T = \mathbf{Z}$. This is a consequence of the fact that $\mathscr{q}$ cannot be an injection on a sufficiently rich signal space. A sufficiently rich signal space must, in addition to the signals $k \mapsto ck^n\alpha^k$, $k, n \in \mathbf{N}_0$, $c, \alpha \in \mathbf{C}$, also contain all signals $x \in \mathbf{C}^{\mathbf{N}_0}$ such that there exists $k \in \mathbf{N}_0$ satisfying $x(l) = 0$ for all $l \geqslant k$. The latter belong to kernels of polynomial operators of the form $a(\mathscr{q}) = \mathscr{q}^k, k \in \mathbf{N}_0$. □

11.2 The rings $\mathbf{C}[\mathscr{z}]$, $\mathbf{C}(\mathscr{z}]$, $\mathbf{C}[1/\mathscr{z}]$ and $\mathbf{C}(\mathscr{z})$ and modules over them

The set of all polynomial operators in $\mathscr{z}$ form a ring of polynomials denoted by $\mathbf{C}[\mathscr{z}]$. $\mathbf{C}[\mathscr{z}]$ is completely analogous to the ring $\mathbf{C}[p]$ of polynomials in p introduced in section 5.2, and everything we presented there also holds in the discrete-time case with obvious modifications. In particular, the signal space $\mathscr{X}$ can be made a module over $\mathbf{C}[\mathscr{z}]$.

In the discrete-time case there is, however, an additional property, which is a consequence of the invertibility of $\mathscr{z}$. As shown in appendix A2 the set

1 $$S_0 = \{\mathscr{z}^n | n \in \mathbf{N}_0\}$$

is a denominator set, which means that we can form the ring of quotients

2 $$\mathbf{C}[\mathscr{z}]_{S_0} = \{a(\mathscr{z})/\mathscr{z}^n | a(\mathscr{z}) \in \mathbf{C}[\mathscr{z}] \quad \text{and}\, n \in \mathbf{N}_0\}.$$

This is denoted briefly by $\mathbf{C}(\mathscr{z}]$.

Furthermore, from the signal space $\mathscr{X}$ we can construct a module of quotients over $\mathbf{C}(\mathscr{z}]$ (*cf.* appendix A3)

3 $$\mathscr{X}_{S_0} = \{x/\mathscr{z}^n | x \in \mathscr{X} \quad \text{and}\, n \in \mathbf{N}_0\}.$$

Now, because every $\mathscr{z}^n \in S_0$ is a monomorphism, $\mathscr{X}$ can be identified with a $\mathbf{C}[\mathscr{z}]$-submodule of $\mathscr{X}_{S_0}$ by the assignment $x \mapsto x/1$. If, in particular, $\mathscr{X}$ is regular, then every $\mathscr{z}^n \in S_0$ is an automorphism of $\mathscr{X}$, and $\mathscr{X}$ can be identified with a $\mathbf{C}(\mathscr{z}]$-submodule of $\mathscr{X}_{S_0}$, i.e. $\mathscr{X}$ itself can be considered as a $\mathbf{C}(\mathscr{z}]$-module.

Furthermore, a regular $\mathscr{X}$ can also be made a module over the ring $\mathbf{C}[1/\mathscr{z}]$ (*cf.* appendix A2). In fact the multiplication $x \mapsto (1/\mathscr{z})x$ can be identified with the operator $\mathscr{z}^{-1} =$, but for the afore mentioned notational reasons we do not use $\mathscr{q}$ in connection with the case $T = \mathbf{Z}$.

If $\mathscr{X}$ is regular, then $\mathbf{C}[\mathscr{z}]$ as well as $\mathbf{C}(\mathscr{z}]$ and $\mathbf{C}[1/\mathscr{z}]$ are integral domains and have the fields of quotients which can all be identified with the field $\mathbf{C}(\mathscr{z})$ of quotients of $\mathbf{C}[\mathscr{z}]$. However, if $\mathscr{X}$ is also sufficiently rich, then the only automorphisms in $\mathbf{C}[z]$ are the elements of the form $cz^n, c \in \mathbf{C}^*, n \in \mathbf{N}_0$. Hence $\mathscr{X}$ cannot be made a vector space over $\mathbf{C}(z)$.

A suitable signal space which is regular and sufficiently rich is $\mathbf{C}^{\mathbf{Z}}$. The subspace $\mathbf{C}^{\mathbf{Z}+}$ is regular but not sufficiently rich. Every $a(z) \in \mathbf{C}[z]^*$ is easily shown to be an automorphism so that $\mathbf{C}^{\mathbf{Z}+}$ can be made a vector space over $\mathbf{C}(z)$. Instead $\mathbf{C}_k^{\mathbf{Z}+}$ for some $k \in \mathbf{Z}$ is neither regular nor sufficiently rich. However, every $a(z)$ which is not divisible by z, i.e. has no root equal to 0, is an automorphism. Thus $\mathbf{C}_k^{\mathbf{Z}+}$ can be made a module over the ring of quotients of $\mathbf{C}[z]$ by the set

4 $$S_{\mathbf{C}^*} \triangleq \{a(z) | a(z) \in \mathbf{C}[z], a(z) \text{ is monic and has all roots in } \mathbf{C}^*\}$$

(*cf.* appendix A2). This ring of quotients can be considered as a subring of $\mathbf{C}(z)$ and we shall denote it by $\mathbf{C}(z)^+$. We shall return to these modules and vector spaces of quotients in chapter 14.

5 **Remark.** In the case $T = \mathbf{N}_0$ the module structure of a signal space is analogous to the corresponding continuous-time structure, and if the signal space is regular and sufficiently rich it is impossible to extend the ring $\mathbf{C}[q]$ to a ring of quotients. The whole $\mathbf{C}^{\mathbf{N}_0}$ is a regular and sufficiently rich signal space but it has no sensible, proper subspace of the type $\mathbf{C}^{\mathbf{Z}+}$ and $\mathbf{C}_k^{\mathbf{Z}+}$, which would be a signal space. □

11.3 Polynomial and rational matrix operators and polynomial and rational matrices

The discrete-time polynomial matrix operators have the same properties as the continuous-time polynomial matrix operators (*cf.* sections 5.3 and 5.4). Thus we can regard the polynomial matrix operators as polynomial matrices over $\mathbf{C}[z]$. Furthermore, if $\mathscr{X}$ is a regular signal space, a polynomial matrix $A(z)$ can be identified with a rational matrix $A(z)/1$.

On the other hand, if $\mathscr{X}$ is regular, we can consider an s by q-rational matrix $A(z)$ over $\mathbf{C}(z]$ as a linear mapping $\mathscr{X}^q \to \mathscr{X}^s$. However, note that, if $\mathscr{X}$ is also sufficiently rich, we cannot identify rational matrices over $\mathbf{C}(z)$ with linear mappings.

Note (5.4.1) can easily be modified to hold for discrete-time polynomial matrix operators. The items (i), (ii) and (v) hold with p replaced by z. Items (iii) and (iv) can be changed to the following forms.

1 **Note. (iii′)** If $A(z)$ is a polynomial matrix operator $\mathscr{X}^q \to \mathscr{X}^q$, then $\ker A(z)$ is an infinite dimensional subspace of $\mathscr{X}^q$, if $\det A(z) = 0$, and an

$(n-m)$-dimensional subspace of $\mathscr{X}^q$ if $\det A(z)$ is of the form $a_m z^m + a_{m-1} z^{m+1} + \ldots + a_n z^n$ with $m \leqslant n$, $a_m \neq 0$, $a_n \neq 0$.
(iv′) A polynomial matrix operator $A(z): \mathscr{X}^q \to \mathscr{X}^q$ has an inverse $A(z)^{-1}$: $\mathscr{X}^q \to \mathscr{X}^q$ if and only if $\det A(z): \mathscr{X} \to \mathscr{X}$ has an inverse $(\det A(z))^{-1}$: $\mathscr{X} \to \mathscr{X}$, i.e. if and only if $\det A(z)$ is of the form cz^n, with $c \neq 0$, $n \in \mathbf{N}_0$. $A(z)$ thus has an inverse if and only if the polynomial matrix representing $A(z)$ is $\mathbf{C}(z]$-unimodular, i.e. invertible as a rational matrix over $\mathbf{C}(z]$. □

12

Difference input-output relations. Generators

In this chapter we shall briefly present fundamental properties of the difference input-output relations introduced in chapter 10. Because the theory is rather similar to the theory of differential input-output relations, we can pass over many details, only referring to chapter 6. However, there are some important dissimilarities which we shall study more thoroughly. The most important is the concept of causality, which was not explicitly needed in connection with differential systems. However, causality is closely related to the concept of properness of differential systems.

However, we shall also consider noncausal difference input-output relations, as we considered nonproper differential input-output relations. Here we are interested only in regular input-output relations and regular generators which determine unique transfer matrices. As with differential input-output relations, we shall define input-output equivalence and transfer equivalence on the set of regular generators and the latter also on the set of regular input-output relations. The canonical forms for these equivalences differ slightly from the corresponding canonical forms in the continuous-time case.

Choice of signal space $\mathscr{X}$

In this chapter the signal space $\mathscr{X}$ is assumed to be regular and sufficiently rich and we consider only the case $T = \mathbf{Z}$. Without loss of generality $\mathscr{X}$ can be taken as $\mathbf{C}^{\mathbf{Z}}$. The case $T = \mathbf{N}_0$ is almost completely analogous to the theory of differential input-output relations presented in chapter 6. Only the properness and controllability concepts have a slightly different meaning from the systems theory point of view.

12.1 Regular difference input-output relations and regular generators

In section 10.2 we defined the difference input-output relation generated by a pair $(B(z), A(z))$ of polynomial matrix operators as a set S of pairs $(u, y) \in \mathscr{X}^r \times \mathscr{X}^q$ satisfying the matrix difference equation

1 $$A(z)y = B(z)u.$$

This equation can be regarded as a $\mathbf{C}[z]$-module equation but also as a $\mathbf{C}(z)$-module equation. Because the latter structure has some advantages, we shall use it here. In particular, this implies that we can cancel powers of z if they exists on both sides of equation (1). This property is considered more precisely in section 12.2.

Now, the partitioned rational matrix operator $[A(z) \vdots -B(z)]$, $A(z) \in \mathbf{C}(z]^{s\times q}$, $B(z) \in \mathbf{C}(z]^{s\times r}$ is called a *generator* for the difference input-output relation S. Note that in what follows, we usually do not make any notational distinction between matrices over $\mathbf{C}[z]$ and $\mathbf{C}(z]$, neither their entries.

Regular generators for regular difference input-output relations. Order

A difference input-output relation S is said to be *regular* if it has a generator $[A(z) \vdots -B(z)]$ with $A(z)$ square and $\det A(z) \neq 0$. Consequently, such a generator is called regular. If $\det A(z) \in \mathbf{C}(z]$ is written in the form

$$a_m z^m + a_{m+1}z^{m+1} + \dots + a_n z^n = (a_m + a_{m+1}z + \dots + a_n z^{n-m})z^m$$

with $a_m \neq 0$, $a_n \neq 0$, m, $n \in \mathbf{Z}$, the number $n - m$ is called the *order* of the generator and of the corresponding regular difference input-output relation S.

A regular difference input-output relation has the property that it determines unique systems (*cf.* section 10.2), which, however, are not necessarily causal. Another important property is that a regular difference input-output relation $S \subset \mathscr{X}^r \times \mathscr{X}^q$ has the whole $\mathscr{X}^r$ as its domain.

2 **Note.** **(i)** A regular difference input-output relation $S \subset \mathscr{X}^r \times \mathscr{X}^q$ generated by the regular generator $[A(z) \vdots -B(z)]$ is a (linear) mapping $\mathscr{X}^r \to \mathscr{X}^q$ if and only if $A(z)$ is $\mathbf{C}(z]$-unimodular (*cf.* appendix A1), i.e. if and only if $\det A(z) = cz^n$ for some $c \neq 0$, $n \in \mathbf{Z}$.
(ii) Let S_1 and S_2 be two difference input-output relations generated by $[A_1(z) \vdots -B_1(z)]$ and $[A_2(z) \vdots -B_2(z)]$, respectively. If $[A_1(z) \vdots -B_1(z)]$

is regular, $A_2(z)$ is square and $S_2 \subset S_1$, then $[A_2(z) \vdots -B_2(z)]$ is also regular. □

12.2 Input-output equivalence. Canonical forms for input-output equivalence. Causality

Two fundamental theorems

The relationships between difference input-output relations and their generators are in principle similar to the corresponding properties of differential systems. There are, however, some dissimilarities, which make it necessary for us to present briefly the discrete-time counterparts of theorems (6.2.1) and (6.2.2). The theorems can be proved following the lines of section 6.5.

1 **Theorem.** *Let $S_1, S_2 \subset \mathscr{X}^r \times \mathscr{X}^q$ be the difference input-output relations generated by $[A_1(z) \vdots -B_1(z)]$ and $[A_2(z) \vdots -B_2(z)]$, respectively, and suppose that $[A_1(z) \vdots -B_1(z)] = L(z)[A_2(z) \vdots -B_2(z)]$ for some matrix $L(z)$ over $\mathbf{C}(z]$. Then*
(i) $S_2 \subset S_1$,
(ii) *if $L(z)$ is square and $\mathbf{C}(z]$-unimodular, then $S_1 = S_2$,*
(iii) *if $S_1 = S_2$, $L(z)$ is square, and $\det A_1(z) \neq 0$ (or $\det A_2(z) \neq 0$), then $\det A_2(z) \neq 0$ ($\det A_1(z) \neq 0$) and $L(z)$ is $\mathbf{C}(z]$-unimodular.* □

2 **Theorem.** *Let $S_1, S_2 \subset \mathscr{X}^r \times \mathscr{X}^q$ be the difference input-output relations generated by $[A_1(z) \vdots -B_1(z)]$ and $[A_2(z) \vdots -B_2(z)]$, respectively, with $[A_1(z) \vdots -B_1(z)]$ regular and $A_2(z)$ square. If $S_2 \subset S_1$, then $[A_2(z) \vdots - B_2(z)]$ is regular and there exists a unique matrix $L(z)$ over $\mathbf{C}(z]$, $\det L(z) \neq 0$, such that $[A_1(z) \vdots -B_1(z)] = L(z)[A_2(z) \vdots -B_2(z)]$, and if, in particular, $S_1 = S_2$, then $L(z)$ is $\mathbf{C}(z]$-unimodular.* □

Input-output equivalence = $\mathbf{C}(z]$-row equivalence

The above theorems lead to the following conclusion (*cf.* section 6.2).

Two regular generators $M_1 \triangleq [A_1(z) \vdots -B_1(z)]$ and $M_2 \triangleq [A_2(z) \vdots -B_2(z)]$ are said to be *input-output equivalent* if they generate the same difference input-output relation. As a consequence of theorems (1) and (2) we have the result that the regular generators M_1 and M_2 are input-output equivalent if and only if they are $\mathbf{C}(z]$-row equivalent, i.e. if and only if there exists a $\mathbf{C}(z]$-unimodular matrix $P(z)$ such that $[A_2(z) \vdots -B_2(z)] = P(z)[A_1(z) \vdots -B_1(z)]$.

Canonical forms for input-output equivalence

According to the previous result every set of canonical forms for $\mathbf{C}(z]$-row equivalence also constitutes a set of canonical forms for input-output equivalence. Thus, for instance, the set of CUT-forms ("canonical upper triangular"-forms) of regular generators is a set of canonical forms for input-output equivalence (see appendix A2).

Note that the CUT-form of a polynomial matrix considered as a matrix over $\mathbf{C}(z]$ can be different from its CUT-form when it is considered as a matrix over $\mathbf{C}[z]$. A regular generator $[A(z) \vdots -B(z)]$ of CUT-form over $\mathbf{C}(z]$ is such that $A(z)$ possesses the same properties with respect to the degrees of the entries as the CUT-form for polynomial matrices explained in appendix A2. The normalization is in this case most conveniently performed so that the constant terms of the diagonal entries of $A(z)$ are equal to 1. Then $A(z)$ and $B(z)$ can be written as

3 $$A(z) = A_0 + A_1 z + A_2 z^2 + \ldots + A_m z^m,$$

4 $$B(z) = (B_0 + B_1 z + \ldots + B_n z^n)/z^p,$$

where A_0, $A_1, \ldots, A_m$, B_0, $B_1, \ldots, B_n$ are matrices over $\mathbf{C}$ with $A_0, A_1, \ldots, A_m$ upper triangular and A_0 invertible with diagonal entries equal to 1.

Causality

The canonical form introduced above is closely related to causality (see Hirvonen *et al.*, 1975). If the generator $[A(\ ) \vdots -B(z)]$ for a difference input-output relation S is of CUT-form, and we write $A(z)$ and $B(z)$ in the forms (3) and (4), respectively, then S is generated also by a recursion equation

5 $$A_0 y(k) + A_1 y(k-1) + \ldots + A_m y(k-m)$$
$$= B_0 u(k+p) + B_1 u(k+p-1) + \ldots + B_n u(k+p-n), \quad k \in \mathbf{Z},$$

which can be uniquely solved for $y(k)$ because A_0 is invertible. Using this it is easily seen that S is causal if and only if $p \leqslant 0$, i.e. if and only if $[A(z) \vdots -B(z)]$ is a polynomial matrix over $\mathbf{C}[z]$. According to the foregoing we shall call a regular generator *causal* if its CUT-form is a polynomial matrix over $\mathbf{C}[z]$.

12.3 The transfer matrix. Properness and causality

Let $[A(z) \vdots -B(z)]$ be a regular generator for a difference input-output relation S. Considering $A(z)$ and $B(z)$ as matrices over the field $\mathbf{C}(z)$ of quotients of $\mathbf{C}[z]$ and using the fact that $A(z)$ is invertible, we can define the *transfer matrix*

1 $$\mathcal{G}(z) = A(z)^{-1}B(z)$$

determined by $[A(z) \vdots -B(z)]$. $\mathcal{G}(z)$ is also uniquely determined by S, which seems to be a straight consequence of theorem (12.2.2). However, the proof of theorem (12.2.2) is based on this property, so that it must be proven in a different way (*cf.* section 6.5).

Properness

The properness of a transfer matrix $\mathcal{G}(z) = A(z)^{-1}B(z)$ determined by a regular generator $[A(z) \vdots -B(z)]$ has no great importance from the systems theory point of view, in contrast with the continuous-time case. Instead, if we consider $\mathcal{G}(z)$ as a matrix $\tilde{\mathcal{G}}(1/z)$ over the field $\mathbf{C}(1/z)$ of quotients of $\mathbf{C}[1/z]$ (see appendix A2), then the properness of $\tilde{\mathcal{G}}(1/z)$ is closely related to the causality of $[A(z) \vdots -B(z)]$. In fact, it is quite easy to show that $[A(z) \vdots -B(z)]$ is causal if and only if $\tilde{\mathcal{G}}(1/z) = A(z)^{-1}B(z)$ is a proper matrix over $\mathbf{C}(1/z)$. As a consequence we shall say that $[A(z) \vdots -B(z)]$ is *strictly causal*, if $\tilde{\mathcal{G}}(1/z)$ is strictly proper, and *anticausal* if $\tilde{\mathcal{G}}(1/z)$ is a polynomial matrix over $\mathbf{C}(1/z)$.

2 **Remark.** In the case $T = \mathbf{N}_0$ we can also define a transfer matrix

3 $$\mathcal{G}(q) = A(q)^{-1}B(q)$$

determined by a regular generator $[A(q) \vdots -B(q)]$. Then $\mathcal{G}(q)$ is proper if and only if the input-output relation S generated by $[A(q) \vdots -B(q)]$ is causal (*cf.* section 10.2). This differs slightly from the continuous-time case, because there every regular input-output relation is also causal, irrespective of its properness. □

12.4 Transfer equivalence. Canonical forms for transfer equivalence. Controllability

Two regular generators $[A_1(z) \vdots -B_1(z)]$ and $[A_2(z) \vdots -B_2(z)]$ are called *transfer equivalent* if they determine the same transfer matrix, i.e.

1 $$A_1(z)^{-1}B_1(z) = A_2(z)^{-1}B_2(z).$$

This defines an equivalence relation on the set of all regular generators.

The set of regular generators $[A(z) \vdots -B(z)]$ of CUT-form such that $A(z)$ and $B(z)$ are left coprime matrices over $\mathbf{C}[z]$ constitutes a set of canonical forms for the transfer equivalence. Of course there are many other sets of canonical forms for the transfer equivalence, but coprimeness is a common property of them all.

Transfer equivalence on difference input-output relations

Because our signal space $\mathscr{X}$ is sufficiently rich it contains all signals $k \mapsto cz^k$, $k \in \mathbf{Z}$, c, $z \in \mathbf{C}$. Therefore the input-output relation $S \subset \mathscr{X}^r \times \mathscr{X}^q$ generated by a regular generator $[A(\imath) : -B(\imath)]$ contains all pairs $(u, y) \in \mathscr{X}^r \times \mathscr{X}^q$ such that

2 $$u = cz^{(\cdot)}, y = A(z^{-1})^{-1}B(z^{-1})cz^{(\cdot)} = \mathscr{G}(z^{-1})cz^{(\cdot)},$$

where $c \in \mathbf{C}^r$ and $z \in \mathbf{C} - \{0\}$ satisfies $\det A(z^{-1}) \neq 0$. Using this we can easily show that the transfer matrix $\mathscr{G}(\imath)$ is uniquely determined by the input-output relation S. More exactly, we have the following theorem.

3 **Theorem.** *Let S_1 and $S_2 \subset \mathscr{X}^r \times \mathscr{X}^q$ be the regular difference input-output relations generated by the regular generators $[A_1(\imath) : -B_1(\imath)]$ and $[A_2(\imath) : -B_2(\imath)]$, respectively. Let the corresponding transfer matrices be denoted by $\mathscr{G}_1(\imath) \triangleq A_1(\imath)^{-1}B_1(\imath)$ and $\mathscr{G}_2(\imath) \triangleq A_2(\imath)^{-1}B_2(\imath)$. Then the following statement is true*:

4 If $S_2 \subset S_1$, *or* $S_1 \subset S_2$, *then* $\mathscr{G}_1(\imath) = \mathscr{G}_2(\imath)$. □

As a consequence we can define the transfer equivalence on the set of regular difference input-output relations, too. The set of input-output relations S_m generated by the regular generators $[A(\imath) : -B(\imath)]$ with $A(\imath)$ and $B(\imath)$ left coprime constitute a set of canonical forms for this equivalence. Furthermore, every canonical form is the first (minimal) element of a transfer equivalence class, if this is partially ordered by set inclusion (*cf.* theorem (6.4.14)).

Controllability

Let S be a regular difference input-output relation generated by the regular generator $[A(\imath) : -B(\imath)]$. We shall say that S is *controllable* if $A(\imath)$ and $B(\imath)$ are left coprime matrices over $\mathbf{C}(\imath]$.

5 **Remark.** In the case $T = \mathbf{N}_0$ the transfer equivalence concepts are analogous to the case $T = \mathbf{Z}$. However, the controllability is a little more complicated. Let us say that the regular difference input-output relation S generated by $[A(\mathscr{q}) : -B(\mathscr{q})]$ is *reachable* if $A(\mathscr{q})$ and $B(\mathscr{q})$ are left coprime matrices over $\mathbf{C}[\mathscr{q}]$, and S is controllable if $A(\mathscr{q})$ and $B(\mathscr{q})$ are left coprime when

they are considered as matrices over $\mathbf{C}(\varphi]$. Thus S is controllable if all common left divisors $L(\varphi)$ of $A(\varphi)$ and $B(\varphi)$ have determinants of the form $\det L(\varphi) = c\varphi^n$, $c \in \mathbf{C} - \{0\}$, $n \in \mathbf{N}_0$. These definitions coincide with the ordinary definitions of controllability and reachability associated with the state-space representation. □

13

Analysis and synthesis problems

As we have shown in the previous chapter, difference and differential systems are very similar. There is thus no reason to transfer all the results concerning differential systems to the present situation, and we shall concentrate only on the slight differences that exist. The reader can easily fill in the remaining gaps.

The main tools in dealing with analysis and synthesis problems, i.e. the use of input-output equivalence for modifying generators and the decomposition of a given input-output relation into a suitable composition of a family of other input-output relations, are the same as in the differential system case, as are the related concepts. The most important differences lie in the fact that we use the delay operator $\mathscr{z}$ instead of the prediction operator $\mathscr{q}$, which implies that causality and properness are not equivalent properties. In order to use, for instance, synthesis methods for differential feedback and estimator systems, we therefore present the generators as polynomial matrices in $1/\mathscr{z}$.

We shall also make the assumption concerning the signal space as in the previous chapter, i.e. the signal space $\mathscr{X}$ is assumed to be regular and sufficiently rich. $\mathscr{X}$ may, in particular, be equal to $\mathbf{C}^{\mathbf{Z}}$.

13.1 Compositions and decompositions of regular difference input-output relations. Observability

A composition of a family of regular difference input-output relations and a decomposition of a regular difference input-output relation are defined precisely in the same way as in section 7.2 we defined the corresponding concepts of differential input-output relations. The different input-output relations determined by compositions are also analogous as well as the equivalences of compositions. However, it should be remembered that the generators of difference input-output relations are not necessarily polynomial matrices but rational matrices over $\mathbf{C}(z)$, even though this does not cause any remarkable difficulties or extra work.

Observability of a composition of difference input-output relations is also an analogous property in the case $T = \mathbf{Z}$.

1 **Remark.** In the case $T = \mathbf{N}_0$ the observability of a composition is a little more complicated property. Suppose that the internal-overall input-output relation S_{io} is generated by the regular generator

2
$$\begin{array}{c} \begin{matrix} y & & y_0 & u_0 \end{matrix} \\ \left[\begin{array}{cc|c|c} A(q) & -B(q)C_4 & 0 & -B(q)C_3 \\ \hline \multicolumn{2}{c|}{-C_2} & I & 0 \end{array}\right] \end{array} = \begin{array}{c} \begin{matrix} y & y_0 & u_0 \end{matrix} \\ \left[\begin{array}{c|c|c} A_i(q) & 0 & -B_i(q) \\ \hline -C_2 & I & 0 \end{array}\right] \end{array}$$

(*cf.* (7.2.18)). Bring this to a $\mathbf{C}[q]$-row equivalent form

3
$$\begin{array}{c} \begin{matrix} y & y_0 & u_0 \end{matrix} \\ \left[\begin{array}{c|c|c} \tilde{A}_{io1}(q) & \tilde{A}_{io2}(q) & -\tilde{B}_{io1}(q) \\ \hline 0 & A_0(q) & -B_0(q) \end{array}\right] \end{array}$$

with $\tilde{A}_{io1}(q)$ a GCRD of $A_i(q)$ and C_2, $\det \tilde{A}_{io1}(q) \neq 0$, and $\det A_0(q) \neq 0$.

Now we shall say that the composition is observable if $\tilde{A}_{io1}(q)$ is $\mathbf{C}[q]$-unimodular, and *reconstructible* if $\tilde{A}_{io1}(q)$ is $\mathbf{C}(q]$-unimodular, i.e. has $\det \tilde{A}_{io1}(q)$ of the form cq^n, $c \in \mathbf{C} - \{0\}$, $n \in \mathbf{N}_0$. These properties can be shown to be generalizations of the well-known concepts used in connection with the state-space representation. □

The concept of a causal composition

Because a regular difference input-output relation is not necessarily causal, we shall call a composition *causal* if the internal input-output relation determined by it is causal.

Suppose that the family of input-output relations consists of only one member S generated by $[A(z) \vdots -B(z)]$, which is causal. Then the transfer matrix $\mathcal{G}(z) \triangleq A(z)^{-1}B(z)$ determined by S can be written as

4
$$\mathcal{G}(z) = \tilde{\mathcal{G}}(1/z) = \tilde{\mathcal{G}}^0(1/z) + K = \mathcal{G}^0(z) + K,$$

where $\tilde{\mathcal{G}}^0(1/z)$ is strictly proper and K is a constant matrix. If $\det(I - KC_4) \neq 0$ (*cf.* section 7.2), then $\det(A(z) - B(z)C_4) \neq 0$, i.e. the composition is regular. Furthermore, its *order*, i.e. the order of the internal input-output relation S_i, is equal to the order of S, and S_i as well as the overall input-output relation S_o are causal. If S is strictly causal, i.e. $K = 0$, then so are S_i and S_o.

Parallel decompositions of noncausal difference input-output relations

Let S_o be a regular difference input-output relation generated by the regular generator

5 $$[A_0(z) \,\vdots\, -B_0(z)].$$

Suppose that S_o is not strictly causal, i.e. the transfer matrix

6 $$\tilde{\mathscr{G}}_o(1/z) = \mathscr{G}_o(z) = A_0(z)^{-1}B_0(z)$$

is not a strictly proper matrix over $\mathbf{C}(1/z)$. $\mathscr{G}_o(1/z)$ can then be written as

7 $$\begin{aligned}\tilde{\mathscr{G}}_o(1/z) &= \tilde{A}_0(1/z)^{-1}\tilde{B}_0(1/z)\\ &= \underbrace{\tilde{A}_0(1/z)^{-1}\tilde{B}_0^0(1/z)}_{\triangleq \tilde{\mathscr{G}}_o^0(1/z)} + \tilde{K}_o(1/z),\end{aligned}$$

where the matrix

8 $$[\tilde{A}_0(1/z) \,\vdots\, -\tilde{B}_0(1/z)]$$

is a polynomial matrix over $\mathbf{C}[1/z]$ and can be obtained from (5), for instance by multiplying it by sufficiently high power $(1/z)^n$ of $1/z$, $\tilde{B}_0^0(1/z)$ and $\tilde{K}_o(1/z)$ are polynomial matrices over $\mathbf{C}[1/z]$, and $\tilde{\mathscr{G}}_o^0(1/z)$ is a strictly proper matrix over $\mathbf{C}(1/z)$.

Denote

9 $$B_0^0(z) \triangleq A_0(z)\tilde{A}_0(1/z)^{-1}\tilde{B}_0^0(1/z) = A_0(z)\tilde{\mathscr{G}}_o^0(1/z).$$

Then

10 $$\tilde{\mathscr{G}}_o^0(1/z) = A_0(z)^{-1}B_0^0(z)$$

is strictly proper over $\mathbf{C}(1/z)$, which implies that the difference input-output relation S_1 generated by

11 $$[A_0(z) \,\vdots\, -B_0^0(z)]$$

is strictly causal.

On the other hand, the difference input-output relation S_2 generated by

12 $$[I \,\vdots\, -\tilde{K}_o(1/z)]$$

or by

13 $$[J(z) \vdots -K_o(z)] \triangleq A_0(z)\, \tilde{A}_0(1/z)^{-1}[I \vdots -\tilde{K}_o(1/z)]$$

is an anticausal mapping, which is also causal, if $\tilde{K}_o(1/z)$ is a constant matrix.

Now the input-output relation S_o can be decomposed into a parallel composition of the form of Fig. 7.4 comprising the strictly causal input-output relation S_1 generated by (11) and the anticausal input-output relation S_2 generated by (13). The composition is regular and observable.

13.2 The feedback composition

Consider the feedback composition depicted in Fig. 7.11 and suppose that S is now a causal, regular difference input-output relation generated by

1 $$\begin{array}{cccc} y_1 & y_2 & u_1 & u_2 \end{array}$$
$$\left[\begin{array}{c:c::c:c} A_1(z) & A_2(z) & -B_1(z) & -B_2(z) \\ \hdashline 0 & L(z)A_{41}(z) & -B_3(z) & -L(z)B_{41}(z) \end{array}\right].$$

The internal input-output relation determined by the composition is generated by

2 $$\begin{array}{ccc} y_1 & y_2 & u_0 \end{array}$$
$$\left[\begin{array}{c:c::c} A_1(z) & A_2(z) - B_2(z) & -B_1(z) \\ \hdashline 0 & L(z)\,(A_{41}(z) - B_{41}(z)) & -B_3(z) \end{array}\right].$$

Thus the composition is regular if $\det(A_{41}(z) - B_{41}(z)) \neq 0$. If S is strictly causal, this also guarantees the causality of the composition, but otherwise the composition can be noncausal.

As shown in section 12.2 the causality of a generator $[A(z) \vdots -B(z)]$ is equivalent to the properness of

3 $$[\tilde{A}(1/z) \vdots -\tilde{B}(1/z)],$$

where $\tilde{A}(1/z)$ and $\tilde{B}(1/z)$ are polynomial matrices over $\mathbf{C}[1/z]$ such that (3) is $\mathbf{C}(z]$-row equivalent to $[A(z) \vdots -B(z)]$.

Because we already have results concerning properness of differential input-output relations and their compositions, it is useful to consider the feedback compositions of difference input-output relations using generators of the form (3).

Feedback compensator synthesis

The feedback compensator synthesis problem can be defined in the same way as problem (7.12.26). Only the word "proper" has to be replaced everywhere by the word "causal" and the stability requirement reads: all the roots of the characteristic polynomial of S_i, i.e. the roots of $\det A_i(z)$ shall lay outside the closed unit circle.

In order to apply the results concerning the differential feedback compensator synthesis it is again useful to present the generators as polynomial matrices in $1/z$. After that the consideration is analogous to the consideration of section 7.12, so that we shall omit it here. The only slight difference is that now it is possible to get lower orders using the possibility to cancel common left divisors, which are $\mathbf{C}(1/z]$-unimodular.

14

The vector space structure. The projection method

In chapter 10 we already discussed the problem of how to choose a systems and their interconnections, passing from a module structure to a vector space structure using the projection method presented in chapter 8. The discrete-time case is a little simpler than the continuous-time case in the sense that we do not need generalized functions. On the other hand, the algebraic structures become a little more complicated.

14.1 Signal space

In chapter 10 we already discussed the problem of how to choose a suitable signal space $\mathcal{X}$ and how to assign a system to a given regular difference input-output relation $S \subset \mathcal{X}^r \times \mathcal{X}^q$ generated by the regular generator $[A(\imath) \ \vdots \ -B(\imath)]$. We chose the method which was analogous to the corresponding differential system construction, i.e. we formed the system $\mathcal{S}$ as a family of all mappings

1 $$S_{(u_1^-, y_1^-)} \triangleq \{(u, y | (u, y) \in S \quad \text{and}\, u|(-\infty, k_0) = u_1^- \quad \text{and} \quad y|(-\infty, k_0) = y_1^-\},$$

where

2 $$u_1^- \triangleq u_1|(-\infty, k_0) = u_1|(-\infty, k_0) \cap \mathbf{Z}, y_1^- \triangleq y_1|(-\infty, k_0) = y_1|(-\infty, k_0) \cap \mathbf{Z}$$

for some $(u_1, y_1) \in S$ and $k_0 \in \mathbf{Z}$ is an arbitrary, fixed time instant.

If our signal space $\mathscr{X}$ is sufficiently rich, the above construction does not lead to a linear system. Instead, if we replace the mappings $S_{(u_{\bar{1}},y_{\bar{1}})}$ by mappings $S^{+}_{(u_{\bar{1}},y_{\bar{1}})}$ of the form

3 $$S^{+}_{(u_{\bar{1}},y_{\bar{1}})}|(-\infty, k_0) \triangleq 0,$$
$$S^{+}_{(u_{\bar{1}},y_{\bar{1}})}|[k_0, \infty) \triangleq S_{(u_{\bar{1}},y_{\bar{1}})}|[k_0, \infty),$$

it is possible to obtain a linear system. This, however, requires that our signal space $\mathscr{X}$ allows arbitrary jumps at $k_0 \in \mathbf{Z}$. Because $\mathscr{X}$ has to be closed under $\imath$, it must allow arbitrary jumps at every $k \geqslant k_0$. If, in addition, we wish to choose k_0 arbitrarily, the only possible signal space is the whole $\mathbf{C}^{\mathbf{Z}}$.

Projection mappings

Let $k \in \mathbf{Z}$ be arbitrary and define the projection mappings $\lceil_k$ and $\rceil_k : \mathbf{C}^{\mathbf{Z}} \to \mathbf{C}^{\mathbf{Z}}$ as follows

4 $$(\lceil_k x)(l) = \begin{cases} 0 & \text{for } l < k, \\ x(l) & \text{for } l \geqslant k, \end{cases}$$

5 $$(\rceil_k x)(l) = \begin{cases} x(l) & \text{for } l < k, \\ 0 & \text{for } l \geqslant k. \end{cases}$$

The ranges of $\lceil_k$ and $\rceil_k$ are (*cf.* section 11.1)

6 $$\mathbf{C}_k^{\mathbf{Z}+} \triangleq \lceil_k \mathbf{C}^{\mathbf{Z}},$$
$$\mathbf{C}_k^{\mathbf{Z}-} \triangleq \rceil_k \mathbf{C}^{\mathbf{Z}},$$

and furthermore

7 $$\mathbf{C}^{\mathbf{Z}+} \triangleq \bigcup_{k \in \mathbf{Z}} \mathbf{C}_k^{\mathbf{Z}+},$$
$$\mathbf{C}^{\mathbf{Z}-} \triangleq \bigcup_{k \in \mathbf{Z}} \mathbf{C}_k^{\mathbf{Z}-}.$$

All the sets given above are subspaces of $\mathbf{C}^{\mathbf{Z}}$, and $\mathbf{C}^{\mathbf{Z}}$ can be written as the direct sum

8 $$\mathbf{C}^{\mathbf{Z}} = \mathbf{C}_k^{\mathbf{Z}-} \oplus \mathbf{C}_k^{\mathbf{Z}+}$$

for all $k \in \mathbf{Z}$.

Recall that $\mathbf{C}^{\mathbf{Z}+}$ is a regular signal space, but $\mathbf{C}_k^{\mathbf{Z}+}$ is not regular even though it is a signal space and $\imath$ restricted to it is an indeterminate. The space $\mathbf{C}^{\mathbf{Z}-}$ is also a regular signal space but $\mathbf{C}_k^{\mathbf{Z}-}$ is not even a signal space. None of the spaces (6), (7) above is sufficiently rich.

14.2 The modules and vector spaces of quotients

In section 11.2 we already discussed the possibilities of extending the scalar ring $\mathbf{C}[z]$ of an $\mathbf{C}[z]$-module $\mathscr{X}$ into a ring of quotients. As we have shown in appendix A3, this can be done if and only if every element of the denominator set is an automorphism of $\mathscr{X}$. Using this we extended the scalar ring of a regular signal space into the ring $\mathbf{C}(z]$. This was, in particular, applied to the space $\mathbf{C}^{\mathbf{Z}}$.

Furthermore, we showed that the space $\mathbf{C}^{\mathbf{Z}+}$ could be made a vector space over the field $\mathbf{C}(z)$ of quotients. Finally, we noted that the space $\mathbf{C}_k^{\mathbf{Z}+}$ for some $k \in \mathbf{Z}$ cannot be made a vector space over $\mathbf{C}(z)$ but it can be made a module over the ring $\mathbf{C}(z)^+$ of quotients of $\mathbf{C}[z]$ by the set

1 $$S_{\mathbf{C}^*} = \{a(z) \,|\, a(z) \in \mathbf{C}[z], a(z) \text{ is monic and has all roots in } \mathbf{C}^*\}.$$

Note that $\mathbf{C}_k^{\mathbf{Z}+}$ cannot be made a module over $\mathbf{C}(z]$ but $\mathbf{C}^{\mathbf{Z}+}$ can.

The space $\mathbf{C}^{\mathbf{Z}-}$ is analogous to the space $\mathbf{C}^{\mathbf{Z}+}$ and can be made a vector space over $\mathbf{C}(z)$. This is most easily seen using the identity $\mathbf{C}(z) = \mathbf{C}(1/z)$ and reversing the time set.

Rational matrix operators

Let $\mathscr{X}$ be one of the spaces $\mathbf{C}^{\mathbf{Z}+}$, $\mathbf{C}^{\mathbf{Z}-}$, or $\mathbf{C}_k^{\mathbf{Z}+}$ for some $k \in \mathbf{Z}$ so that $\mathscr{X}$ can be considered either as a vector space over $\mathbf{C}(z)$ or as a module over the ring $\mathbf{C}(z)^+$. In both cases a q by r-matrix $\mathscr{G}(z)$ with entries from the corresponding scalar ring clearly represents a linear mapping $\mathscr{X}^r \to \mathscr{X}^q$, $u \mapsto \mathscr{G}(z)u$ which is evaluated by interpreting u as a column matrix and using the ordinary matrix multiplication.

We shall use the same terminology as in connection with differential systems, i.e. we shall call the mappings represented by rational matrices "rational matrix operators".

2 **Note.** Suppose that $\mathscr{X} = \mathbf{C}^{\mathbf{Z}+}$. If a rational operator $g(z) = b(z)/a(z) \in \mathbf{C}(z)$ represents a causal mapping $\mathbf{C}^{\mathbf{Z}+} \to \mathbf{C}^{\mathbf{Z}+}$, then its restriction $g(z)|\mathbf{C}_k^{\mathbf{Z}+}$ is a causal mapping $\mathbf{C}_k^{\mathbf{Z}+} \to \mathbf{C}_k^{\mathbf{Z}+}$ for all $k \in \mathbf{Z}$. Furthermore, it is easily seen that the set of causal rational operators is a subring of the field $\mathbf{C}(z)$ of quotients. On the other hand, the elements of the ring $\mathbf{C}(z)^+$ of quotients also represent causal mappings $\mathbf{C}^{\mathbf{Z}+} \to \mathbf{C}^{\mathbf{Z}+}$. Because every causal $g(z) \in \mathbf{C}(z)$ can be written in the form $g(z) = b(z)/a(z)$ with $b(z) \in \mathbf{C}[z]$, $a(z) \in S_{\mathbf{C}^*}$, the subring of causal rational operators is equal to $\mathbf{C}(z)^+$. □

Transfer matrices as rational matrix operators

Let $\mathscr{X}$ be either $\mathbf{C}^{\mathbf{Z}+}$ or $\mathbf{C}_k^{\mathbf{Z}+}$ for some $k \in \mathbf{Z}$. Consider a difference input-output relation $S \in \mathscr{X}^r \times \mathscr{X}^q$ generated by a matrix difference equation

3 $$A(\mathfrak{z})y = B(\mathfrak{z})u,$$

where $u \in \mathscr{X}^r$, $y \in \mathscr{X}^q$, and where $A(\mathfrak{z})$ and $B(\mathfrak{z})$ are polynomial matrix operators of sizes q by q and q by r, respectively, with $\det A(\mathfrak{z}) \neq 0$.

If $\mathscr{X}$ is $\mathbf{C}^{\mathbf{Z}+}$, we can consider $A(\mathfrak{z})$ as an invertible matrix over $\mathbf{C}(\mathfrak{z})$ and multiply (3) by its inverse $A(\mathfrak{z})^{-1}$. Hence we obtain

4 $$y = A(\mathfrak{z})^{-1} B(\mathfrak{z})u = \mathscr{G}(\mathfrak{z})u,$$

where the transfer matrix $\mathscr{G}(\mathfrak{z})$ is considered as a rational matrix operator. Thus the relation S generated by (3) is a mapping $\mathscr{G}(\mathfrak{z}) : \mathscr{X}^r \rightarrow \mathscr{X}^q$.

If $\mathscr{X}$ is $\mathbf{C}_k^{\mathbf{Z}+}$ and S is causal, i.e. the transfer matrix

5 $$\mathscr{G}(\mathfrak{z}) = A(\mathfrak{z})^{-1} B(\mathfrak{z})$$

is a matrix over $\mathbf{C}(\mathfrak{z})^+$, then S is also a causal mapping $\mathscr{X}^r \rightarrow \mathscr{X}^q$. If S is not causal, then it is a mapping, too, but its domain is not the whole $\mathscr{X}^r$.

14.3 Compositions of projections and delay operators. Initial condition mappings

In this section we shall shortly adapt the main results of section 8.4 to difference systems.

A fundamental relationship

Let $A(\mathfrak{z}) : (\mathbf{C}^{\mathbf{Z}})^q \rightarrow (\mathbf{C}^{\mathbf{Z}})^s$ be a polynomial matrix operator and let $\lceil_k$ and $\rceil_k$ for some $k \in \mathbf{Z}$ be projections of the forms (14.1.4) and (14.1.5) respectively. Let $x \in (\mathbf{C}^{\mathbf{Z}})^q$ be arbitrary. Then we obtain

1 $$\lceil_k(A(\mathfrak{z})x) - A(\mathfrak{z}) \lceil_k x = \lceil_k(A(\mathfrak{z}) \rceil_k x),$$

2 $$\rceil_k(A(\mathfrak{z})x) - A(\mathfrak{z}) \rceil_k x = -\lceil_k(A(\mathfrak{z}) \rceil_k x).$$

If $A(\mathfrak{z})$ is not a constant matrix, then the signal $\lceil_k(A(\mathfrak{z}) \rceil_k x) \in (\mathbf{C}_k^{\mathbf{Z}+})^s$ is generally nonzero, which implies that the projection mappings $\lceil_k$ and $\rceil_k$ do not generally commute with polynomial matrix operators.

Initial condition mappings

The initial condition mapping $A_k^0 : (\mathbf{C}_k^{\mathbf{Z}-})^q \to (\mathbf{C}_k^{\mathbf{Z}+})^s$ at k associated with the polynomial matrix operator $A(\mathfrak{z}) : (\mathbf{C}^{\mathbf{Z}})^q \to (\mathbf{C}^{\mathbf{Z}})^s$ is defined by

3 $$A_k^0 \triangleq \lceil_k \circ A(\mathfrak{z}) | (\mathbf{C}_k^{\mathbf{Z}-})^q.$$

4 **Lemma.** *If $a(\mathfrak{z}) : \mathbf{C}^{\mathbf{Z}} \to \mathbf{C}^{\mathbf{Z}}$ is a polynomial operator of the form $a(\mathfrak{z}) = \mathfrak{z}^n$ for some $n \in \mathbf{N}_0$, the initial condition mapping $a_k^0 : \mathbf{C}_k^{\mathbf{Z}-} \to \mathbf{C}_k^{\mathbf{Z}+}$ at k associated with $a(\mathfrak{z})$ is given by*

5 $$\begin{aligned} a_k^0 x^- &= \lceil_k (\mathfrak{z}^n x^-) \\ &= (\mathfrak{z}^{n-1} x^-)(k-1)\,\delta_k + (\mathfrak{z}^{n-2} x^-)(k-1)\,\mathfrak{z}\delta_k \\ &\quad + \ldots + x^-(k-1)\,\mathfrak{z}^{n-1}\delta_k, \end{aligned}$$

where $x^- \in \mathbf{C}^{\mathbf{Z}-}$, and $\delta_k \in \mathbf{C}^{\mathbf{Z}}$ is the unit pulse at k, i.e.

6 $$\delta_k(l) = \begin{cases} 1 \text{ for } & l = k \\ 0 \text{ for } & l \neq k. \end{cases}$$

For $n = 0$ (5) is interpreted as $a_k^0 x^- = 0$. □

On the basis of lemma (4) and the linearity of the projection mapping $\lceil_k$ it is an easy matter to determine the initial condition mapping a_k^0 associated with an arbitrary polynomial operator $a(\mathfrak{z}) \triangleq a_0 + a_1 \mathfrak{z} + \ldots + a_n \mathfrak{z}^n$ and to obtain

7 $$\begin{aligned} a_k^0 x^- &= \sum_{i=1}^{n} a_i (\mathfrak{z}^{i-1} x^-)(k-1)\,\delta_k + \sum_{i=2}^{n} a_i (\mathfrak{z}^{i-2} x^-)(k-1)\mathfrak{z}\,\delta_k \\ &\quad + \ldots + a_n x^-(k-1)\mathfrak{z}^{n-1}\delta_k. \end{aligned}$$

Furthermore, the initial condition mapping A_k^0 associated with a polynomial matrix operator $A(\mathfrak{z})$ can be represented by the matrix of the initial condition mappings a_{ijk}^0 associated with the entries $a_{ij}(\mathfrak{z})$ of $A(\mathfrak{z})$ (*cf.* (8.4.19)).

As with the differential polynomial operators, the initial condition mapping a_k^0 associated with a polynomial operator $a(\mathfrak{z} \triangleq a_0 + a_1 \mathfrak{z} + \ldots + a_n \mathfrak{z}^n$ of degree $n \geqslant 1$ can also be written as

8 $$a_k^0 x^- = c(\mathfrak{z})\,\delta_k,$$

where $c(\mathfrak{z}) \triangleq c_0 + c_1 \mathfrak{z} + \ldots + c_{n-1} \mathfrak{z}^{n-1}$ is a polynomial, whose coefficient list $c \triangleq (c_0, c_1, \ldots, c_{n-1}) \in \mathbf{C}^n$ is obtained from the matrix equation

9 $$\underbrace{\begin{bmatrix} c_0 \\ c_1 \\ \vdots \\ c_{n-2} \\ c_{n-1} \end{bmatrix}}_{c} = \underbrace{\begin{bmatrix} a_1 & a_2 & \ldots & a_{n-1} & a_n \\ a_2 & a_3 & \ldots & a_n & 0 \\ \vdots & \vdots & & \vdots & \vdots \\ a_{n-1} & a_n & \ldots & 0 & 0 \\ a_n & 0 & \ldots & 0 & 0 \end{bmatrix}}_{\underline{a}} \underbrace{\begin{bmatrix} x^-(k-1) \\ x^-(k-2) \\ \vdots \\ x^-(k-n+1) \\ x^-(k-n) \end{bmatrix}}_{x^0} = \underline{a}x^0.$$

It is easily seen that the range of a_k^0 is an n-dimensional subspace of $\mathbf{C}_k^{\mathbf{Z}+}$ spanned by the set of independent functions $\{\delta_k, \delta_{k+1}, \ldots, \delta_{k+n-1}\}$.

The above presentation can be generalized to include polynomial matrix operators in exactly the same way as in section 8.4, so we shall omit the generalization here.

14.4 The projection method

Use of the projection mappings $\lceil_k$ and $\rceil_k$ and initial condition mappings in the same way as in section 8.5 gives the following expression for the set $S^+_{(u_1^-, y_1^-)}$ introduced in section 14.1 by (14.1.3)

1 $$S^+_{(u_1^-, y_1^-)} = \{(u^+, y^+) \mid (u^+, y^+) \in (\mathbf{C}_k^{\mathbf{Z}+})^r \times (\mathbf{C}_k^{\mathbf{Z}+})^q \text{ and } A^0 y_1^- + A(\imath)y^+ = B^0 u_1^- + B(\imath)u^+\}.$$

Because the input-output relation S was supposed to be regular, $S^+_{(u_1^-, y_1^-)}$ is clearly a mapping. For any $(u_1^-, y_1^-) \in S^- \triangleq \rceil_k S$ and for any

$$u^+ \in \mathrm{D}S^+_{(u_1^-, y_1^-)} \subset (\mathbf{C}_k^{\mathbf{Z}+})^r$$

the corresponding $y^+ \in (\mathbf{C}_k^{\mathbf{Z}+})^q$ can be uniquely solved from the equation

2 $$y^+ = A(\imath)^{-1}(-A^0 y_1^- + B^0 u_1^- + B(\imath)u^+),$$

where the rational matrix operator $A(\imath)^{-1}: (\mathbf{C}^{\mathbf{Z}+})^q \to (\mathbf{C}^{\mathbf{Z}+})^q$ is the inverse of the polynomial matrix operator $A(\imath): (\mathbf{C}^{\mathbf{Z}+})^q \to (\mathbf{C}^{\mathbf{Z}+})^q$. However, if S is not causal, the domain $\mathrm{D}S^+_{(u_1^-, y_1^-)}$ cannot be the whole $(\mathbf{C}_k^{\mathbf{Z}+})^r$.

On the other hand, if S is causal, then every $S^+_{(u_1^-, y_1^-)}$ is a causal mapping $(\mathbf{C}_k^{\mathbf{Z}+})^r \to (\mathbf{C}^{\mathbf{Z}+})^q$. Especially this holds for $S^+_{(0^-, 0^-)}$ which implies that (2) can also be written as

3 $$y^+ = A(\imath)^{-1}(-A^0 y_1^- + B^0 u_1^-) + A(\imath)^{-1}B(\imath)u^+.$$

Thus we have arrived at the same structure as was developed for differential systems in section 8.5. Because of the analogy we will leave the reader to complete the theory of difference systems along the lines of chapters 8 and 9.

APPENDICES

The following appendices contain some fundamental concepts of mathematics, especially of the abstract algebra used throughout this monograph. However, we have all the time assumed that the reader is already familiar with elementary set theory, real and complex matrix calculus and the elements of ordinary differential equations.

In particular, we assume that the reader knows at least the following set theory concepts: set, element, union, intersection and the (Cartesian) product of sets, relation, mapping and concepts related to these. If not, he would be advised to consult, for instance, Lipshutz, 1964, part I or some other basic textbook. For the sake of simplicity we identify relations and mappings with their graphs, i.e. we consider them as sets of ordered pairs. This implies that a mapping f has a (two-sided) inverse mapping f^{-1} if and only if f is injective. Then the composite mapping $f \circ g$ of two mappings g and f is always defined irrespective of their domains and ranges.

The theory of real and complex matrices is assumed to be familiar at, say, the level of Gantmacher, 1959, chapters I to V. This comprises concepts such as addition, multiplication, inversion, transposition, determinant, rank, minors, characteristic and minimal polynomials, eigenvalues, eigenvectors, etc. Furthermore, linear algebra on finite dimensional real and complex vector spaces closely related with matrices is also included there.

Finally, it is assumed that the reader has a knowledge of elementary differential and integral calculus and linear differential equations with constant (real or complex) coefficients. This is standard material in systems theory textbooks (e.g. Zadeh and Desoer, 1963; Kalman, Falb, and Arbib, 1969; Brockett, 1970; Padulo and Arbib, 1974).

A1

Fundamentals of abstract algebra

In this appendix we shall be giving a short review of the basic concepts of abstract algebra. Good reference books for a more thorough understanding are, for instance, MacLane and Birkhoff 1967; Lang 1965; Jacobson 1951, 1953, 1964.

Groups

A *semigroup* is a set G together with a *binary operation* $G \times G \to G$, $(a, b) \mapsto ab$, such that this operation is *associative*, i.e. for each $a, b, c \in G$

1 $$(ab)c = a(bc).$$

If there exists $u \in G$ such that for each $a \in G$

2 $$ua = au = a$$

the semigroup is called a *monoid.*

Furthermore, if for each $a \in G$ there exists $b \in G$ such that

3 $$ab = ba = u$$

the monoid is called a *group*. Finally, if the binary operation is *commutative*, i.e. for each $a, b \in G$

4 $$ab = ba$$

the group is said to be *commutative* or an *abelian group.*

If the binary operation of a group is clear from the context, the group can be notationally identified with the corresponding set of elements. Usually the binary operations are written as products $(a, b) \mapsto ab$ as before or as sums $(a, b) \mapsto a + b$; accordingly the corresponding groups are said to be *multiplicative* or *additive*. For a multiplicative group G the elements $u, b \in G$ satisfying (2), (3) are called the *unit element* or the *unity* of G and the *inverse* of $a \in G$ denoted by 1 and a^{-1}, respectively. For an additive group G they are the *zero element* of G and the *negative* of $a \in G$ denoted by 0 and $-a$, respectively. The element $a^n \in G$ with $n \in \mathsf{N}_0$ defined recursively by $a^n = a^{n-1}a$, $a^0 = 1$, is called the nth *power* of a. For an additive group this is usually written as na and called the nth *multiple* of a.

If G and G' are (e.g. multiplicative) groups, a mapping $f: G \to G'$ satisfying

5 $$f(ab) = f(a)f(b)$$

for each $a, b \in G$ is called a *morphism of groups.* If f is an injection, a surjection or a bijection, it is said to be a *monomorphism*, an *epimorphism* or an *isomorphism*, respectively. Furthermore, if $G' = G$, f is an *endomorphism* or an *automorphism*, if it is bijective.

If a subset H of a group G together with the binary operation of G restricted to H is a group, then it is called a *subgroup* of G.

Let $f: G \to G'$ be a morphism of groups. Then the range of f

6 $$\mathrm{R}f = \{b \mid b \in G', b = f(a) \quad \text{for some} \quad a \in G\}$$

and the *kernel* of f

7 $$\ker f = \{a \mid a \in G, \quad f(a) = 1\}$$

are subgroups of G' and G, respectively. Furthermore, f is a monomorphism if and only if $\ker f = \{1\}$.

Rings

A *ring* is a set R together with two binary operations, addition and multiplication such that

(i) R together with addition is an abelian group,
(ii) R together with multiplication is a semigroup,
(iii) multiplication is *distributive* on both sides over addition, i.e. for each a, b, $c \in R$

8 $$a(b + c) = ab + ac,$$

9 $$(a + b)c = ac + bc.$$

If a ring consists of just one element 0 it is called a *trivial ring*. If a ring contains a unit element $1 \neq 0$ it is said to be a *ring with unity*. A *commutative* ring is a ring in which the multiplication is commutative.

If two nonzero elements a, b of a ring satisfy $ab = 0$, they are called *zero divisors.* A commutative ring with unity and without zero divisors is an *integral domain.*

Let R be a ring with unity. An element $u \in R$ having a (multiplicative) inverse u^{-1} is called a *unit* of R. A commutative ring with unity in which every nonzero element is a unit is called a *field.*

A mapping $f: \mathrm{R} \to \mathrm{R}'$ of a ring R into a ring R' satisfying for each $a, b \in R$

10 $$f(a + b) = f(a) + f(b),$$

11 $$f(ab) = f(a)f(b)$$

is a *morphism of rings.* Hence a morphism of rings is also a morphism of additive groups. As for groups, monomorphisms, epimorphisms and isomorphisms are injective, surjective and bijective morphisms, respectively, and endomorphisms and automorphisms are morphisms and isomorphisms, respectively, of rings into themselves.

If a subset S of a ring R together with the binary operations of R restricted to S is a ring, then it is called a *subring* of R. Respectively, if R and S are integral domains or fields, then S is a *subdomain* or a *subfield* of R. For instance, $\mathsf{R}f$ and $\ker f$ for a morphism $f: R \rightarrow R'$ of rings are subrings of R' and R, respectively.

Note that some authors (e.g. MacLane and Birkhoff, 1967), define nontrivial rings as rings with unity and require that unity elements be preserved by morphisms and inherited by subrings. In this case $\ker f$ cannot be a subring unless the domain or the codomain of f is a trivial ring.

Let R be a commutative ring. If $a, b, c \in R$ satisfy $ab = c$, then we say that a (or b) *divides* c or a (or b) is a *divisor* of c denoted by $a|c$ (or $b|c$). Furthermore, c is a *multiple* of a (or b). If $a, b \in R$ satisfy $a|b$ and $b|a$, they are *associates* to each other. In an integral domain R $a, b \in R$ are associates if and only if there exists a unit $u \in R$ such that $a = ub$. If $a \neq 0$ has no other divisors apart from units and associates and it is not a unit, it is called *irreducible.* If a is irreducible and such that $a|bc$ implies either $a|b$ or $a|c$, then a is called a *prime.*

An element c of a commutative ring R is a *greatest common divisor* of $a_1, a_2, \ldots, a_n \in R$ if c is a common divisor of $a_1, a_2, \ldots, a_n$ and every other common divisor divides c. A nonzero element $d \in R$ is a *least common multiple* of $a_1, a_2, \ldots, a_n \in R$ if d is a common multiple of $a_1, a_2, \ldots, a_n$ and divides every other common multiple. In an integral domain a greatest common divisor and a least common multiple, if they exist, are unique up to associates. If $a, b \in R$ have no common divisors apart from units, they are said to be *coprime.*

A subset J of a commutative ring R is called a (*two-sided*) *ideal* if it is an additive subgroup of R and $ra \in J$ and $ar \in J$ hold for every $r \in R$, $a \in J$. Ideals are closely related to morphisms. In particular the kernels of morphisms of rings are ideals. An ideal of the form

12 $$Ra = \{b | b \in R, \quad b = ra \quad \text{for some} \quad r \in R\}$$

is said to be *principal.* An integral domain in which every ideal is principal is called a *principal ideal domain.*

Let R be a ring with unity. If $a, b \in R$ satisfy for some $c \in R$

13 $$Ra + Rb = Rc,$$

then c is a greatest common divisor of a, b. Furthermore, $d \in R$ is a least common multiple of $a, b \in R$ if and only if

14 $$Ra \cap Rb = Rd.$$

If $a, b \in R$ satisfy $Ra + Rb = R$ they are said to be *comaximal.* Comaximal elements are coprime. In a principal ideal domain a greatest common divisor and a least common multiple always exist.

Irreducible elements of a principal ideal domain are primes. Furthermore, a principal ideal domain is a *unique factorization domain*, which means that every nonzero element $a \in R$ is a unit or a finite product $a = p_1 p_2 \ldots p_n$ of primes $p_1, p_2, \ldots, p_n \in R$ unique up to changes of order and associates (see MacLane and Birkhoff, 1967, pp. 155–156).

Rings of quotients

Let R be a commutative ring with unity. An element of the form ba^{-1} with $b \in R$, $a^{-1} \in R$ a multiplicative inverse of $a \in R$ is called a *quotient* and denoted by b/a, where a is the *denominator* and b the *numerator* of b/a. The equality of two quotients b/a, $d/c \in R$ is equivalent to

15 $$ad = bc.$$

Moreover, the sum, product, negative and inverse—if it exists—are given by

16 $$b/a + d/c = (bc + ad)/(ac),$$

17 $$(b/a)(d/c) = (bd)/(ac),$$

18 $$-(b/a) = (-b)/a = b/(-a),$$

19 $$(b/a)^{-1} = a/b.$$

Obviously the set of all possible quotients is equal to the ring R and the set of all possible denominators is the set of units of R.

However, there is one way of extending the set of denominators by extending the whole ring R to a ring of quotients.

So let R be a commutative ring and S a subset of R closed under multiplication, i.e. for each a, $b \in S$ $ab \in S$. Define a relation $\sim$ on $R \times S$ by

20 $$(a, s) \sim (b, t) \Leftrightarrow \text{there exists } u \in S$$
$$\text{such that } u(ta - sb) = 0.$$

The relation $\sim$ is an equivalence relation, i.e. it is symmetric, reflexive and transitive. Denote the corresponding equivalence classes by $[a, s]$, $(a, s) \in R \times S$ and the factor set $R \times S/\sim$, i.e. the set of equivalence classes, shortly by R_S.

Next define two binary operations, addition $\oplus$ and multiplication $\odot$ on R_S as follows

21 $$[a, s] \oplus [b, t] = [ta + sb, st],$$

22 $$[a, s] \odot [b, t] = [ab, st].$$

These operations are well-defined, i.e. they are independent of the representatives of equivalence classes.

It is easily seen that R_S is a commutative ring with operations (21), (22), zero element $[0, s]$ and unit element $[s, s]$. The negative of $[a, s]$ is $[-a, s]$ and $[s, t]$, $s \in S$, has $[t, s]$ as its multiplicative inverse. R_S is a trivial ring if and only if S contains the zero 0 of R.

The assignment $a \mapsto [sa, s]$ defines a mapping $j: R \to R_S$ which is a morphism of rings. Every $[a, s] \in R_S$ can be written as a quotient

23 $$[a, s] = [ts, t]^{-1} \odot [ta, t] = j(a)/j(s).$$

If the set S does not contain the zero of R or any zero divisor, it is called a *denominator set*. The morphism j is a monomorphism if and only if S is a denominator set. In this case every $a \in R$ can be identified with its image $[sa, s] \in R_S$. Hence via this identification R is embedded in the ring R_S and the elements of R_S can be written as the quotients a/s, $a \in R$, $s \in S$. Accordingly, R_S is called the *ring of quotients of R by S*, irrespective of whether the morphism j is a monomorphism or not. The binary operations of R_S are also notationally identified with the binary operations of R.

It is easy to show that if R is an integral domain or a principal ideal domain, R_S is also an integral domain or a principal ideal domain, respectively. In particular, if R does not contain zero divisors, the set of all nonzero elements R^* qualifies as a denominator set. Then R_{R^*} is a field called the *field of quotients of R*.

24 **Note.** Every denominator set S included in the set of units of R satisfies $R_S = R$. If S and T are denominator sets in R, then T and the set T_S defined by

25 $$T_S = \{t/s \mid t/s \in R_S \quad \text{and} \quad t \in T\}$$

are denominator sets in R_S and, respectively, S and S_T are denominator sets in R_T. If $S \subset T$ then $R_S \subset R_T$ and

26 $$(R_S)_T = (R_S)_{T_S} = (R_T)_S = (R_T)_{S_T} = R_T. \qquad \square$$

Modules

Let X be an additive abelian group and R a ring with unity. X together with a mapping $R \times X \to X$, written $(a, x) \mapsto ax$, such that for each $a, b \in R$, $x, y \in X$

27 $$a(x + y) = ax + ay,$$

28 $$(a + b)x = ax + bx,$$

29 $$(ab)x = a(bx),$$

30 $$1x = x$$

is called a *(left) module over R* or a *(left) R-module*. The mapping $(a, x) \mapsto ax$ is called *scalar multiplication* and R is said to be the ring of *scalars*. Any ring with unity is obviously a module over itself. A module over a field is called a *vector space*. A module consisting of just one element 0 is called a *trivial module* or a *zero module*.

A mapping $f: X \to X'$ of an R-module X into an R-module X' is a *morphism of R-modules*, if for each $x, y \in X$, $a \in R$

31 $$f(x + y) = f(x) + f(y)$$

and

32 $$f(ax) = af(x).$$

A morphism of R-modules is also called an *R-linear mapping*. A morphism of R-modules is obviously a morphism of additive groups and monomorphisms, epimorphisms, etc., are defined analogously to morphisms of groups.

If a subset Y of an R-module X together with the addition and the scalar multiplication of X restricted to Y is an R-module, it is called a *submodule* of X.

Let B be a subset of an R-module X. X is said to be *generated* by B if every $x \in X$ can be written as a finite sum

33 $$x = \sum_{i=1}^{n} a_i x_i$$

with $n \in \mathbb{N}$, $a_1, a_2, \ldots, a_n \in R$, $x_1, x_2, \ldots, x_n \in B$.

X is *finitely generated* if it can be generated by a finite set and *cyclic* if it can be generated by a singleton set.

A subset B of an R-module X is *R-linearly independent* or *linearly independent over* R if for all $n \in \mathbf{N}$, $a_1, a_2, \ldots, a_n \in R$, $x_1, x_2, \ldots, x_n \in B$

34 $$\sum_{i=1}^{n} a_i x_i = 0 \Rightarrow a_1 = a_2 = \ldots = a_n = 0.$$

An R-module X generated by an R-linearly independent set B is said to be *free* (on B) and B is a *basis* for X. It is well-known that every vector space is free with unique cardinality of bases (see e.g. MacLane and Birkhoff, 1967, p. 231).

The set of *n-lists* $\mathbf{n} \to R$ over a ring R with unity denoted by R^n is a left module over R with pointwise operations. R^n is free on the set $\{e_1, e_2, \ldots, e_n\}$ defined by

35 $$e_i(j) = \begin{cases} 1 & \text{if } i = j, \\ 0 & \text{if } i \neq j. \end{cases}$$

If X is another R-module free on a finite set $\{x_1, x_2, \ldots, x_n\}$, the assignment $(a_1, a_2, \ldots, a_n) \mapsto a_1x_1 + a_2x_2 + \ldots + a_nx_n$ defines an isomorphism $R^n \cong X^n$ of R-modules. On the other hand an isomorphism $R^m \cong R^n$ of R-modules implies $m = n$ (see MacLane and Birkhoff, 1967, p. 307). Hence the number of basis elements is unique and is called the *rank* (or the *dimension*) of X.

If R is a principal ideal domain, the submodules of a finitely generated module X are also finitely generated. If, in addition, X is free, the submodules are also free and their rank at most the rank of X (see MacLane and Birkhoff, 1967, pp. 340, 358).

Modules of quotients

Let X be a module over a commutative ring R with unity, S a subset of R closed under multiplication, and R_S the ring of quotients of R by S. Suppose that R_S is nontrivial and the scalar ring R of X is extended to R_S so that for each $a \in R$, $x \in X$

36 $$(sa/s)x = ax.$$

Because every $(st/s) \in R_S$ with $t \in S$ is a unit, (36) implies that for each $t \in S$ the morphism $x \mapsto tx$ must be an automorphism. This requirement can be slightly weakened by extending X in the following way.

Define an equivalence relation $\sim$ on $X \times S$

37 $$(x, s) \sim (y, t) \Leftrightarrow \text{there exists } u \in S \text{ such that } u(tx - sy) = 0.$$

Denote the equivalence classes by $[x, s]$, $(x, s) \in X \times S$, and the factor set by X_S. Define the binary operations addition $\oplus$ on X_S and scalar multiplication $\odot: R_S \times X_S \to X_S$ by

38 $$[x, s] \oplus [y, t] = [tx + sy, st],$$

39 $$a/u \odot [x, s] = [ax, us].$$

These are well-defined and satisfy the axioms of an R_S-module. Hence X_S is a module over R_S.

X_S together with the scalar multiplication $R \times X_S \to X_S$,

40 $$a[x, s] = [ax, s]$$

is also a module over R and the mapping $j:X \to X_S$, $x \mapsto [sx, s]$, is a morphism of R-modules. The morphism j is a monomorphism if and only if for every $s \in S$ the morphism $x \mapsto sx$ is a monomorphism. In this case X can be identified with an R-submodule of X_S. The scalar ring of this module cannot, however, be extended to R_S unless every morphism $x \mapsto sx$ is an automorphism.

The module X_S over R_S is called the *module of quotients of X by S*, its elements are denoted like x/s and its addition and scalar multiplication are notationally identified with the operations of X.

Matrices over rings

Let R be a commutative ring with unity. The set of mappings $A: \mathbf{m} \times \mathbf{n} \to R$, $(i, j) \mapsto A(i, j) \triangleq a_{ij}$ is called the set of m by *n-matrices over R* and denoted by $R^{m \times n}$. The binary operations, in particular the pointwise addition and matrix multiplication, are defined in the usual way, i.e. for each $A, B \in R^{m \times n}$, $C \in R^{n \times p}$

41 $$(A + B)(i,j) = A(i,j) + B(i,j),$$

42 $$(AC)(i,j) = \sum_{k=1}^{n} A(i,k)\, C(k,j).$$

$R^{n \times n}$ with these operations is a ring with the *identity matrix* I $(\triangleq I_n)$, $I(i, j) = \delta_{ij} \triangleq$ Kronecker delta, as its unity. $R^{n \times 1}$ and sometimes also $R^{1 \times n}$ are identified with R^n, and $R^{1 \times 1}$ and R^1 are identified with R.

If R^m and R^n are regarded as free R-modules, every matrix $A \in R^{m \times n}$ represents a morphism $R^n \to R^m$ of R-modules defined by $x \mapsto Ax$. Conversely every morphism $R^n \to R^m$ of R-modules can be uniquely represented by a matrix. In particular, the identity mapping is represented by the identity matrix. The range of $A \in R^{m \times n}$ is usually denoted by AR^n.

Often we present a matrix in the *partitioned form*, for example

$$A = \left[\begin{array}{c:c} A_1 & A_2 \\ \hdashline A_3 & A_4 \end{array}\right],$$

where the matrix $A \in R^{m \times n}$ has been split into *blocks* $A_1 \in R^{p \times q}$, $A_2 \in R^{p \times (n-q)}$, $A_3 \in R^{(m-p) \times q}$, $A_4 \in R^{(m-p) \times (n-q)}$. In order to simplify many considerations it is useful to make the agreement that the partitioned form can also contain empty blocks. Thus in our example we can for instance have $p = m$ in which case we identify A with $[A_1 \vdots A_2]$.

As usual, a matrix $A \in R^{m \times n}$ is said to be *upper* (*right*) *triangular*, *lower* (*left*) *triangular* or *diagonal* if it satisfies, respectively,

43 $$i > j \Rightarrow a_{ij} = 0,$$

44 $$i < j \Rightarrow a_{ij} = 0,$$

45 $$i \neq j \Rightarrow a_{ij} = 0.$$

$A \in R^{m \times n}$ is called a *unit* or *invertible* or *R-unimodular* if there exists $B \in R^{n \times m}$ such that $AB = I_m$ and $BA = I_n$. It is found that if A is invertible, then it is necessarily square $(m = n)$.

The *determinant* $\det A$ of a square matrix $A \in R^{n \times n}$ is defined in the usual way, i.e. it satisfies

46 $$\det A = \sum_{\sigma \in S_n} (-1)^{\operatorname{sgn}\sigma} a_{\sigma_1 1} a_{\sigma_2 2} \ldots a_{\sigma_n n},$$

where S_n is the set of all permutations $\mathbf{n} \to \mathbf{n}$ and sgn σ the number of inversions for σ, i.e. the number of pairs $(i, j) \in \mathbf{n} \times \mathbf{n}$ such that $i < j$ and $\sigma_i > \sigma_j$. To be complete the determinant of a non-square matrix is defined as being equal to zero. As the product of two square matrices we have

47 $$\det(AB) = (\det A)(\det B).$$

The determinants of the p by p submatrices obtained by deleting $m - p$ rows and $n - p$ columns of $A \in R^{m \times n}$ are called p by *p-minors* of A. Using minors the determinant of $A \in R^{n \times n}$ can be written as

48 $$\det A = \sum_{i=1}^{n} (-1)^{i+j} a_{ij} \alpha_{ij}, \quad j \in \mathbf{n},$$

where α_{ij} for $i, j \in \mathbf{n}$ is the minor of A obtained by deleting the ith row and the jth column.

The classical *adjoint* adj A of $A \in R^{n \times n}$ is defined as an n by n-matrix with entries

49 $$(\operatorname{adj} A)(i, j) = (-1)^{i+j} \alpha_{ji}.$$

It holds that

50 $$(\operatorname{adj} A)A = A(\operatorname{adj} A) = (\det A)I.$$

This implies that A is invertible if and only if det A is invertible in R and the inverse A^{-1} can be written as

51 $$A^{-1} = (\det A)^{-1} \operatorname{adj} A.$$

The *determinantal rank* of a matrix $A \in R^{m \times n}$ is defined as being the largest $p \in \mathbf{N}$ such that A has a nonzero p by p-minor. Obviously the determinantal rank of an invertible n by n-matrix is n.

If $A \in R^{m \times n}$ can be written as $A = LM$ with $L \in R^{m \times p}$, $M \in R^{p \times n}$, then L and M are called a *left divisor* (in $R^{m \times p}$) and a *right divisor* (in $R^{p \times n}$) respectively of A, and A is called a *right multiple* (in $R^{m \times n}$) of L and *left multiple* (in $R^{m \times n}$) of M. A matrix $L \in R^{m \times q}$ is said to be a *greatest common left divisor* (GCLD) in $R^{m \times q}$ of $A \in R^{m \times n}$, $B \in R^{m \times p}$ if L is a common left divisor of A and B and every other common left divisor in $R^{m \times q}$ is a left divisor of L. A nonzero matrix $D \in R^{m \times q}$ is called a *least common right multiple* (LCRM) in $R^{m \times q}$ of $A \in R^{m \times n}$, $B \in R^{m \times p}$, if D is a common right multiple and every other common right multiple in $R^{m \times q}$ is a right multiple of D. Usually we are interested only in square greatest common left divisors, in which case the attribute "in $R^{m \times q}$" can be omitted. *Greatest common right divisors* (GCRD) and *least common left multiples* (LCLM) are defined correspondingly.

Two matrices $A \in R^{m \times n}$, $B \in R^{m \times p}$ are *left coprime* in $R^{m \times q}$ if they have no common left divisors in $R^{m \times q}$ apart from units. Obviously A, B cannot be left coprime in $R^{m \times q}$ if $q > m$ or $q > n + p$ and if they are left coprime in $R^{m \times m}$ they are left coprime in $R^{m \times q}$ with $q < m$. Hence the most important case is $q = m$, when we usually omit the attribute "in $R^{m \times q}$". The matrices A, B are *right comaximal* in $R^{m \times q}$ if

52 $$AR^{n \times q} + BR^{p \times q} = R^{m \times q}.$$

The matrices are right comaximal in $R^{m \times q}$ if and only if they are right comaximal in $R^{m\times 1} = R^m$. Hence the attribute "in $R^{m\times q}$" can again be omitted. *Right coprime* and *left comaximal* matrices are defined correspondingly.

53 **Note** Let $A \in R^{m\times n}$, $B \in R^{m\times p}$, $L \in R^{m\times q}$ be arbitrary.
(i) Suppose that

54 $$AR^n + BR^p = LR^q,$$

which implies $A = LA_1$, $B = LB_1$ with $A_1 \in R^{q\times n}$, $B_1 \in R^{q\times p}$. If $A = MA_2$, $B = MB_2$ with $M \in R^{m\times q}$, $A_2 \in R^{q\times n}$, $B_2 \in R^{q\times p}$, then

55 $$LR^q = MA_2R^n + MB_2R^p \subset MR^q,$$

which implies that $L = MN$ for some $N \in R^{q\times q}$. Hence L is a GCLD of A, B. In particular, if A, B are right comaximal they are left coprime in $R^{q\times q}$.
(ii) Let L be a GCLD of A, B and a monomorphism. If $M \in R^{m\times q}$ is another GCLD, then there exists a unimodular matrix $Q \in R^{q\times q}$ such that $L = MQ$. If A and B are written as $A = LA_1$, $B = LB_1$, the matrices A_1, B_1 are left coprime in $R^{q\times q}$.
(iii) L is a LCRM of A, B if and only if

56 $$AR^n \cap BR^p = LR^q.$$ □

Row and column equivalence

Two matrices A, $B \in R^{m\times n}$ are said to be (R-) *row equivalent* if there exists an R-unimodular matrix $P \in R^{m\times m}$ such that $A = PB$. Respectively, A and B are (R-) *column equivalent* if there exists an R-unimodular matrix $Q \in R^{n\times n}$ such that $A = BQ$. Finally A and B are (R-) *equivalent* if there exist R-unimodular matrices P and Q such that $A = PBQ$. All three equivalences define equivalence relations on the set of all matrices over R.

The following operations on matrices are called *elementary row operations*:
(i) interchange any two rows,
(ii) add to any row any other row multiplied by any element of R,
(iii) multiply any row by a unit of R.
The matrices obtained by applying the elementary row operations to identity matrices are called *elementary matrices* and the elementary row operations can be performed by premultiplying the matrix under consideration by elementary matrices. The elementary matrices corresponding to the row operations (i), (ii), and (iii) are

57 $$U_{ij} = \begin{array}{c} \\ 1 \\ 2 \\ \vdots \\ i \\ \vdots \\ j \\ \vdots \\ m \end{array} \begin{array}{c} \begin{array}{ccccccccc} 1 & 2 & \dots & i & \dots & j & \dots & m \end{array} \\ \left[\begin{array}{cccccccc} 1 & 0 & \dots & 0 & \dots & 0 & \dots & 0 \\ 0 & 1 & \dots & 0 & \dots & 0 & \dots & 0 \\ \vdots & \vdots & & \vdots & & \vdots & & \vdots \\ 0 & 0 & \dots & 0 & \dots & 1 & \dots & 0 \\ \vdots & \vdots & & \vdots & & \vdots & & \vdots \\ 0 & 0 & \dots & 1 & \dots & 0 & \dots & 0 \\ \vdots & \vdots & & \vdots & & \vdots & & \vdots \\ 0 & 0 & \dots & 0 & \dots & 0 & \dots & 1 \end{array}\right] \end{array}, \quad i \neq j,$$

58
$$T_{ij}(a) = \begin{array}{c} \\ 1 \\ 2 \\ \vdots \\ i \\ \vdots \\ j \\ \vdots \\ m \end{array} \begin{array}{c} \begin{array}{ccccccccc} 1 & 2 & \dots & i & \dots & j & \dots & m \end{array} \\ \left[\begin{array}{cccccccc} 1 & 0 & \dots & 0 & \dots & 0 & \dots & 0 \\ 0 & 1 & \dots & 0 & \dots & 0 & \dots & 0 \\ \vdots & \vdots & & \vdots & & \vdots & & \vdots \\ 0 & 0 & \dots & 1 & \dots & a & \dots & 0 \\ \vdots & \vdots & & \vdots & & \vdots & & \vdots \\ 0 & 0 & \dots & 0 & \dots & 1 & \dots & 0 \\ \vdots & \vdots & & \vdots & & \vdots & & \vdots \\ 0 & 0 & \dots & 0 & \dots & 0 & \dots & 1 \end{array}\right] \end{array}, i \neq j,$$

59
$$V_{ii}(u) = \begin{array}{c} \\ 1 \\ 2 \\ \vdots \\ i \\ \vdots \\ m \end{array} \begin{array}{c} \begin{array}{cccccc} 1 & 2 & \dots & i & \dots & m \end{array} \\ \left[\begin{array}{cccccc} 1 & 0 & \dots & 0 & \dots & 0 \\ 0 & 1 & \dots & 0 & \dots & 0 \\ \vdots & \vdots & & \vdots & & \vdots \\ 0 & 0 & \dots & u & \dots & 0 \\ \vdots & \vdots & & \vdots & & \vdots \\ 0 & 0 & \dots & 0 & \dots & 1 \end{array}\right] \end{array}.$$

The premultiplication of $A \in R^{m \times n}$ by U_{ij} interchanges the ith and the jth rows of A. Premultiplication by $T_{ij}(a)$ adds the jth row of A multiplied by a to the ith row of A and the premultiplication by $V_{ii}(u)$ multiplies the ith row by u.

The elementary matrices are obviously unimodular so that the matrices obtained by elementary row operations are row equivalent to the original ones. Furthermore, elementary row operation (i) permutes the minors and multiplies some of them by -1, row operation (ii) replaces some minors by sums of the original minor and another minor multiplied by an element $a \in R$, and row operation (iii) multiplies some minors by a unit of R. Hence the determinantal rank does not change in elementary row operations.

The *elementary column operations* are defined correspondingly in the obvious way and they can be performed by postmultiplying the matrix under consideration by elementary matrices.

Matrices over principal ideal domains

Let R be a principal ideal domain, and let $a = (a_1, a_2) \in R^2$ be arbitrary. Then there always exists an R-unimodular matrix $P \in R^{2 \times 2}$ such that the premultiplication of a by P replaces the entries of a by a greatest common divisor of a_1, a_2 and 0. This can be seen as follows. Let c and d be a greatest common divisor and a least common multiple, respectively, of a_1, a_2. Then there exist $s_1, s_2, t_1, t_2 \in R$ such that

60
$$s_1 a_1 + s_2 a_2 = c,$$

61
$$t_1 a_1 = t_2 a_2 = d.$$

Hence

62 $$\begin{bmatrix} s_1 & s_2 \\ t_1 & -t_2 \end{bmatrix} \begin{bmatrix} a_1 \\ a_2 \end{bmatrix} = \begin{bmatrix} c \\ 0 \end{bmatrix}.$$

The multiplication of (60) by t_1 and t_2 and the use of (61) gives

63 $$(s_1t_2 + s_2t_1)a_1 = ct_2,$$

64 $$(s_1t_2 + s_2t_1)a_2 = ct_1.$$

Because t_1 and t_2 must be coprime, c is a greatest common divisor of ct_1 and ct_2. On the other hand $(s_1t_2 + s_2t_1)c$ is a common divisor of them so that $s_1t_2 + s_2t_1$ is a unit. Hence the matrix

$$P = \begin{bmatrix} s_1 & s_2 \\ t_1 & -t_2 \end{bmatrix}$$

is invertible.

Let

$$P = \begin{bmatrix} p_{11} & p_{12} \\ p_{21} & p_{22} \end{bmatrix} \in R^{2\times 2}$$

be an arbitrary invertible matrix. Premultiplication of a matrix $A \in R^{m\times n}$ by a matrix $W_{ij}(P) \in R^{m\times m}$ defined by

65 $$W_{ij}(P) = \begin{array}{c} \\ 1 \\ 2 \\ \vdots \\ i \\ \vdots \\ j \\ \vdots \\ m \end{array} \begin{array}{c} \begin{array}{ccccccccc} 1 & 2 & \cdots & i & \cdots & j & \cdots & m \end{array} \\ \begin{bmatrix} 1 & 0 & \cdots & 0 & \cdots & 0 & \cdots & 0 \\ 0 & 1 & \cdots & 0 & \cdots & 0 & \cdots & 0 \\ \vdots & \vdots & & \vdots & & \vdots & & \vdots \\ 0 & 0 & \cdots & p_{11} & \cdots & p_{12} & \cdots & 0 \\ \vdots & \vdots & & \vdots & & \vdots & & \vdots \\ 0 & 0 & \cdots & p_{21} & \cdots & p_{22} & \cdots & 0 \\ \vdots & \vdots & & \vdots & & \vdots & & \vdots \\ 0 & 0 & \cdots & 0 & \cdots & 0 & \cdots & 1 \end{bmatrix} \end{array}, i \neq j,$$

is called a *secondary row operation* and $W_{ij}(P)$ is a *secondary matrix*. $W_{ij}(P)$ is obviously invertible and the premultiplication by it replaces some minors by sums of the original minor multiplied by p_{11} or p_{22} and another minor multiplied by p_{12} or p_{21}, respectively, and multiplies some minors by the unit det P. This implies that the determinantal rank remains unchanged in the secondary operations, too.

66 **Note.** Every matrix over a principal ideal domain is row equivalent to an upper triangular matrix, which can be found via the following procedure, called a *triangular reduction*. Let $A \in R^{m\times n}$ be arbitrary. Take the first, say the jth, of the nonzero columns, and bring one of the nonzero entries to the $(1, j)$-position by permuting the rows (operations U_{1k}). Then replace the new a_{1j} and a_{2j} by their greatest common divisor and 0, respectively, then the new a_{1j} and a_{3j} by their greatest

common divisor and 0, respectively, and so on (operations $W_{1k}(P)$). Thus a matrix is obtained such that a_{1j} is a greatest common divisor of the original entries and $a_{2j} = a_{3j} = \ldots = a_{mj} = 0$. Next apply the procedure to the submatrix formed by discarding the first row and carry the entries $a_{3k}, a_{4k}, \ldots, a_{mk}$ to 0, where the kth column is the first nonzero column of this submatrix ($k > j$). Continuing in this way we finally obtain an upper triangular matrix which is row equivalent to the original one.

As a consequence of the triangular reduction a matrix $A \in R^{n \times n}$ is invertible if and only if it is row equivalent to an upper triangular matrix $B \in R^{n \times n}$ with diagonal entries b_{ii}, $i \in \mathbf{n}$, units in R. In this case the entries above the diagonal in the last column can be brought to zero by the row operations $T_{in}(-b_{in}b_{nn}^{-1})$, $i < n$, then in the last but one column by the row operations $T_{i,n-1}(-b_{i,n-1}b_{n-1,n-1}^{-1})$ and so on until all the columns have been reduced. Finally the diagonal entries can be normalized equal to 1 by the operations $V_{ii}(b_{ii}^{-1})$, $i \in \mathbf{n}$. Hence the matrix $A \in R^{n \times n}$ is invertible if and only if it is row equivalent to the identity matrix $I \in R^{n \times n}$. Furthermore, this implies that an invertible matrix can be written as a finite product of elementary and secondary row operations. □

Because the determinant up to associates and the determinantal rank remain unchanged in the elementary and secondary row operations, they are *invariants* for the row equivalence (see MacLane and Birkhoff, 1967, pp. 287–288). There are some other important invariants. Let $A \in R^{m \times n}$ be of rank $r \leq \min\{m, n\}$ and det d_j denote a greatest common divisor of all j by j-minors of A for $j \in \mathbf{r}$. The elements d_j are called *determinantal divisors* of A. It is easily seen that they are invariants up to associates for the row equivalence. Furthermore $d_j | d_{j+1}$ for all $j \in \{1, 2, \ldots, r-1\}$, so that the quotients $i_{j+1} = d_{j+1}/d_j$, are elements of R. The elements $i_1 = d_1, i_2, \ldots, i_r$ are invariants up to associates for the row equivalence and they are called *invariant factors* of A.

There are of course, other triangular reductions depending on the corner chosen as the starting point. In addition, there are triangular reductions based on elementary and secondary column operations. If both the row and the column operations are used, we have a *diagonal reduction*.

67 **Note.** Every matrix $A \in R^{m \times n}$ over a principal ideal domain R is equivalent to a diagonal matrix $D \in R^{m \times n}$ such that the first r, $r \leq \min\{m, n\}$, diagonal entries of D are nonzero and satisfy

68 $$d_{ii} | d_{i+1,i+1}, i \in \{1, 2, \ldots, r-1\}$$

and the last ones are zero. Furthermore the diagonal entries are unique up to associates. The procedure for constructing a diagonal matrix D satisfying (68) is as follows. First bring some nonzero entry of A to the (1, 1)-position by permuting rows and columns. Then replace the first column by a column such that a_{11} is a greatest common divisor of the original entries and the others are zero. Thereafter the first row is also transformed to the form where a_{11} is a greatest common divisor of the entries and the others are zero. In this reduction some of the zero entries of the first column can become nonzero so that the operation must be repeated. This can happen only if the number of divisors of a_{11} is reduced, and because this number is finite the repetition stops after a finite number of times. Then apply the procedure to the submatrix formed by discarding the first row and the first column, and so on. Finally we obtain a diagonal matrix

69

$$\begin{array}{c} \\ 1 \\ 2 \\ \vdots \\ r \\ r+1 \\ \vdots \\ m \end{array}
\begin{array}{c} \begin{array}{cccccccc} 1 & 2 & \dots & r & r+1 & \dots & n \end{array} \\
\left[\begin{array}{cccccccc}
d_{11} & 0 & \dots & 0 & 0 & \dots & 0 \\
0 & d_{22} & \dots & 0 & 0 & \dots & 0 \\
\vdots & \vdots & & \vdots & \vdots & & \vdots \\
0 & 0 & \dots & d_{rr} & 0 & \dots & 0 \\
0 & 0 & \dots & 0 & 0 & \dots & 0 \\
\vdots & \vdots & & \vdots & \vdots & & \vdots \\
0 & 0 & \dots & 0 & 0 & \dots & 0
\end{array}\right] \end{array} .$$

This matrix does not necessarily satisfy the condition (68). If d_{11} fails to divide d_{22}, add the second row to the first row and start from the beginning. In this way the number of divisors in d_{11} is reduced, unless d_{11} divides d_{22}. Similarly then consider the entries d_{22} and d_{33}. Since the number of divisors in d_{22} can be reduced, it may be necessary to return to the entries d_{11} and d_{22}, but this can happen only a finite number of times. By continuing the procedure we finally obtain a matrix D which satisfies (68).

Because $d_{11}, d_{22}, \dots, d_{rr}$ are the invariant factors of D and they are invariants for both the row equivalence and the column equivalence, hence also for the equivalence, the matrix D is unique up to associates of the diagonal entries. □

A diagonal matrix D satisfying (68) is said to be in the *Smith* (*canonical*) *form*, and if D is equivalent to a matrix A it is called a *Smith* (*canonical*) *form* of A. The set of matrices of the Smith canonical form constitutes a set of *canonical forms* for the equivalence of matrices, i.e. two matrices are equivalent if and only if they have the same Smith form. Hence the Smith form is a complete invariant (*cf.* MacLane and Birkhoff, 1967) for the equivalence of matrices.

The *rank* of a matrix $A \in R^{m \times n}$ over a principal ideal domain R is defined as the rank of its range $\mathsf{R}A = AR^n$, which is a submodule of R^m generated by the columns of A. The Smith form shows that the rank of A is the maximum number of R-linearly independent columns and it is equal to the determinantal rank. Furthermore, the rank of the upper triangular matrix constructed in note (66) is equal to the number of nonzero rows.

70 **Note.** Let $A \in R^{m \times n}$ be arbitrary of rank r.
(i) A is an epimorphism if and only if $r = m \leqslant n$ and the invariant factors of A are units in R.
(ii) A is a monomorphism if and only if $m \geqslant n = r$.
(iii) A is an isomorphism if and only if $m = n = r$ and the invariant factors of A are units in R.
(iv) A has a right inverse, i.e. a matrix $B \in R^{n \times m}$ such that $AB = I_m$ if and only if $r = m \leqslant n$ and the invariant factors of A are units in R.
(v) A has a left inverse, i.e. a matrix $B \in R^{n \times m}$ such that $BA = I_n$ if and only if $m \geqslant n = r$ and the invariant factors of A are units in R.
(vi) A is invertible in $R^{m \times n}$ if and only if $m = n = r$ and the invariant factors of A are units in R. □

Greatest common divisors and least common multiples of matrices over principal ideal domains

Let R still be a principal ideal domain and let $A \in R^{m\times n}$, $B \in R^{m\times p}$ be arbitrary. The matrix $[A \vdots B]$ can by column operations be brought to the form $[L \vdots 0]$, where L is e.g. a lower triangular m by r-matrix of full rank r.

71 **Note.** Let $A \in R^{m\times n}$, $B \in R^{m\times p}$ be arbitrary and let the matrix $[A \vdots B]$ be of rank r.

(i) Suppose that $[A \vdots B]$ is column equivalent to a matrix $[L \vdots 0]$ with $L \in R^{m\times q}$ and $q \geqslant r$, i.e. there exists an R-unimodular matrix Q such that

72 $$[A \vdots B]\underbrace{\begin{bmatrix} Q_1 & Q_2 \\ Q_3 & Q_4 \end{bmatrix}}_{Q} = [L \vdots 0].$$

Then L is a GCLD in $R^{m\times q}$ of A, B. Furthermore, if $q = r$ then all GCLD's of A, B are column equivalent to L and $A = LA_1$, $B = LB_1$ with $A_1 \in R^{r\times n}$, $B_1 \in R^{r\times p}$ left coprime in $R^{r\times r}$. The proof follows.

Let P be the inverse of Q. Then

73 $$[A \vdots B] = [L \vdots 0]\underbrace{\begin{bmatrix} P_1 & P_2 \\ P_3 & P_4 \end{bmatrix}}_{P} = [LP_1 \vdots LP_2].$$

Thus L is a common left divisor. Let $M \in R^{m\times q}$ be another common left divisor of A, B, i.e.

74 $$[A \vdots B] = M[A_1 \vdots B_1]$$

with $A_1 \in R^{q\times n}$, $B_1 \in R^{q\times p}$. Equations (72), (74) together imply

75 $$[L \vdots 0] = M[A_1Q_1 + B_1Q_3 \vdots A_1Q_2 + B_1Q_4]$$

so that M is a left divisor of L. Hence L is a GCLD.

If $q = r$, then L is a monomorphism. Hence the second part is quite obvious (see note (53), (ii)). This further implies the next item.

(ii) If $L \in R^{m\times r}$ is a GCLD of A, B, then $[A \vdots B]$ and $[L \vdots 0]$ are column equivalent, i.e. there exists an R-unimodular matrix Q such that

76 $$[A \vdots B]\underbrace{\begin{bmatrix} Q_1 & Q_2 \\ Q_3 & Q_4 \end{bmatrix}}_{Q} = [L \vdots 0].$$

In particular, if A and B are left coprime in $R^{m\times m}$ and $m \leqslant n + p$, then $[A \vdots B]$ is column equivalent to $[I \vdots 0]$ and there exists an R-unimodular matrix $P \in R^{(n+p)\times(n+p)}$ of the form

77 $$P = \begin{bmatrix} A & B \\ P_3 & P_4 \end{bmatrix}$$

with $P_3 \in R^{(n+p-m)\times n}$, $P_4 \in R^{(n+p-m)\times p}$.

(iii) If $L \in R^{m \times r}$ is a GCLD of A, B, then by (76)

78
$$LR^r = (AQ_1 + BQ_3)R^r \subset AQ_1R^r + BQ_3R^r \subset AR^n + BR^p$$
$$\subset LR^r + LR^r = LR^r.$$

Hence

79
$$AR^n + BR^p = LR^q.$$

In particular, if A and B are left coprime in $R^{m \times m}$ and $m \leqslant n + p$, then A and B are right comaximal.

(iv) If $[A \vdots B]$ is column equivalent to a matrix $[L \vdots 0]$ with $L \in R^{m \times r}$ and $r < n + p$, i.e. there exists an R-unimodular matrix Q satisfying (76), then $AQ_2 = -BQ_4$ is a LCRM in $R^{m \times (n+p-r)}$ of A, B.

Suppose that $AE = -BF$ is another common right multiple. Let $M \in R^{s \times (n+p-r)}$ be a GCRD of E, F of full rank s and write $E = E_1M$, $F = F_1M$ with $E_1 \in R^{n \times s}$, $F_1 \in R^{p \times s}$ right coprime in $R^{s \times s}$. Then there exists an R-unimodular matrix $S \in R^{(n+p) \times (n+p)}$ of the form

80
$$S = \begin{bmatrix} S_1 & E_1 \\ S_3 & F_1 \end{bmatrix}$$

and satisfying

81
$$[A \vdots B] \begin{bmatrix} S_1 & E_1 \\ S_3 & F_1 \end{bmatrix} = [N \vdots 0],$$

where $N \in R^{m \times r}$ is a GCLD of A, B. Denote the inverses of Q and S by P and T, respectively. Then it holds that

82
$$[L \vdots 0] = [N \vdots 0] \underbrace{\begin{bmatrix} T_1 & T_2 \\ T_3 & T_4 \end{bmatrix} \begin{bmatrix} Q_1 & Q_2 \\ Q_3 & Q_4 \end{bmatrix}}_{\triangleq U}$$
$$= [N \vdots 0] \begin{bmatrix} U_1 & U_2 \\ U_3 & U_4 \end{bmatrix} = [NU_1 \vdots NU_2].$$

Now, N is a monomorphism, which implies that $U_2 = 0$, i.e. $U = TQ$ is lower triangular. Writing $Q = STQ = SU$ gives

83
$$\begin{bmatrix} Q_1 & Q_2 \\ Q_3 & Q_4 \end{bmatrix} = \begin{bmatrix} S_1 & E_1 \\ S_3 & F_1 \end{bmatrix} \begin{bmatrix} U_1 & 0 \\ U_3 & U_4 \end{bmatrix} = \begin{bmatrix} S_1U_1 + E_1U_3 & E_1U_4 \\ S_3U_1 + F_1U_3 & F_1U_4 \end{bmatrix}.$$

Because U is unimodular and lower triangular, U_4 is also unimodular. Hence

84
$$\begin{bmatrix} E \\ F \end{bmatrix} = \begin{bmatrix} E_1M \\ F_1M \end{bmatrix} = \begin{bmatrix} Q_2U_4^{-1}M \\ Q_4U_4^{-1}M \end{bmatrix}$$

and $AE = -BF$ is a common multiple of $AQ_2 = -BQ_4$, proving the statement. □

Matrices over rings of quotients

Let R be a principal ideal domain and R_{R^*} the field of quotients of R. Because R_{R^*} is also a principal ideal domain, all that has been presented above holds for matrices over R_{R^*}.

For each $\mathcal{G} \in R_{R^*}^{m\times n}$ there exist $A \in R^{m\times m}$, $B \in R^{m\times n}$, $T \in R^{n\times n}$, $V \in R^{m\times n}$ with A, T, R_{R^*}-unimodular such that

85 $$\mathcal{G} = A^{-1}B = VT^{-1}.$$

For instance, A and B can be constructed in the following way:

86 $$A = 1/dI,$$

87 $$B = A\mathcal{G},$$

where $d \in R^*$ is a common denominator of all entries of $\mathcal{G}$. Another pair satisfying (85) is

88 $$A = \begin{bmatrix} d_1 & 0 & \dots & 0 \\ 0 & d_2 & \dots & 0 \\ \vdots & \vdots & & \vdots \\ 0 & 0 & \dots & d_m \end{bmatrix},$$

89 $$B = A\mathcal{G},$$

where d_i, $i \in \mathbf{m}$, is a common denominator of all entries of the ith row of $\mathcal{G}$. (*cf.* Sinervo, 1972). The construction of T and V is quite obvious.

If A and B are given, T and V can also be found in the following way. Let $Q \in R^{(m+n)\times(m+n)}$ be an R-unimodular matrix such that

90 $$[A \vdots B]\underbrace{\left[\begin{array}{c:c} Q_1 & Q_2 \\ \hdashline Q_3 & Q_4 \end{array}\right]}_{Q} = [L \vdots 0],$$

where $L \in R^{m\times m}$ is a GCLD of A, B. Hence $AQ_2 = -BQ_4$. Because

91 $$\left[\begin{array}{c:c} A & B \\ \hdashline 0 & I \end{array}\right]\left[\begin{array}{c:c} Q_1 & Q_2 \\ \hdashline Q_3 & Q_4 \end{array}\right] = \left[\begin{array}{c:c} L & 0 \\ \hdashline Q_3 & Q_4 \end{array}\right]$$

is R_{R^*}-unimodular, L and Q_4 are R_{R^*}-unimodular. Then $A^{-1}B = -Q_2Q_4^{-1}$, i.e. we can take $T = Q_4$, $V = -Q_2$. These T and V are right coprime and $\det T | \det A$. Furthermore, $\det A$ and $\det T$ are associates if and only if A and B are left coprime.

Conversely, if T and V are given, A and B can be obtained from

92 $$\underbrace{\left[\begin{array}{c:c} P_1 & P_2 \\ \hdashline P_3 & P_4 \end{array}\right]}_{P}\left[\begin{array}{c} T \\ \hdashline V \end{array}\right] = \left[\begin{array}{c} M \\ \hdashline 0 \end{array}\right],$$

where $P \in R^{(n+m)\times(n+m)}$ is R-unimodular and $M \in R^{n\times n}$ is a GCRD of T and V by choosing $A = P_4$, $B = -P_2$. Then A and B are left coprime and $\det A | \det T$.

Let $A_1, A_2 \in R^{m\times m}$ be R_{R^*}-unimodular and $B_1, B_2 \in R^{m\times n}$ arbitrary. The pairs (B_1, A_1) and (B_2, A_2) as well as the partitioned matrices $[A_1 \vdots B_1]$ and $[A_2 \vdots B_2]$ are said to be *quotient equivalent* if

93 $$A_1^{-1}B_1 = A_2^{-1}B_2.$$

94 **Note.** Let $A_1, A_2 \in R^{m\times m}$, $B_1, B_2 \in R^{m\times n}$ be such that A_1, A_2 are R_{R^*}-unimodular. **(i)** $[A_1 \vdots B_1]$ and $[A_2 \vdots B_2]$ are quotient equivalent if and only if there exist R_{R^*}-unimodular matrices $C_1, C_2 \in R^{m\times m}$ such that

95 $$C_1[A_1 \vdots B_1] = C_2[A_2 \vdots B_2].$$

Suppose first that there exist C_1, C_2 satisfying (95). Then

96 $$A_1^{-1}B_1 = (C_1A_1)^{-1}C_1B_1 = (C_2A_2)^{-1}C_2B_2 = A_2^{-1}B_2.$$

Conversely, suppose that $[A_1 \vdots B_1]$ and $[A_2 \vdots B_2]$ are quotient equivalent. Let $d \in R^*$ be a common denominator of A_1^{-1} and A_2^{-1} (e.g. $d = \det A_1 \det A_2$) and write $C_1 = dA_1^{-1} \in R^{m\times m}$, $C_2 = dA_2^{-1} \in R^{m\times m}$. C_1 and C_2 are R_{R^*}-unimodular and satisfy (95).

(ii) Let A_1 and B_1 be left coprime. Then $[A_1 \vdots B_1]$ and $[A_2 \vdots B_2]$ are quotient equivalent if and only if there exists an R_{R^*}-unimodular matrix $L \in R^{m\times m}$ such that

97 $$[A_2 \vdots B_2] = L[A_1 \vdots B_1].$$

The part "if" is clear from the previous item. So suppose that $[A_1 \vdots B_1]$ and $[A_2 \vdots B_2]$ are quotient equivalent. By item (i) there exist R_{R^*}-unimodular matrices C_1, C_2 such that (95) is fulfilled. Because A_1 and B_1 are assumed left coprime, C_1 is a GCLD in $R^{m\times m}$ of C_1A_1 and C_1B_1. Hence C_2 is a left divisor of C_1, i.e. $C_1 = C_2L$ for some $L \in R^{m\times m}$. This further gives

98 $$C_2L[A_1 \vdots B_1] = C_2[A_2 \vdots B_2].$$

C_2 is R_{R^*}-unimodular so that it can be cancelled. L is obviously R_{R^*}-unimodular.

(iii) Let $L_1 \in R^{m\times m}$ be a GCLD of A_1, B_1 and $L_2 \in R^{m\times m}$ be a GCLD of A_2, B_2 and write

99 $$[A_1 \vdots B_1] = L_1[A_{11} \vdots B_{11}],$$

100 $$[A_2 \vdots B_2] = L_2[A_{21} \vdots B_{21}].$$

Then items (i), (ii) together imply that $[A_1 \vdots B_1]$ and $[A_2 \vdots B_2]$ are quotient equivalent if and only if $[A_{11} \vdots B_{11}]$ and $[A_{21} \vdots B_{21}]$ are R-row equivalent. □

Write an arbitrary $\mathcal{G} \in R_{R^*}^{m\times n}$ in the form $1/dC$, where $d \in R^*$ is a common denominator of all entries of $\mathcal{G}$ and $C \in R^{m\times n}$. Let $S = PCQ$ with P and Q R-unimodular be the Smith form of C. The matrix $1/dS = P\mathcal{G}Q$ can be presented in the form

101
$$\begin{bmatrix} n_{11}/d_{11} & 0 & \dots & 0 & 0 & \dots & 0 \\ 0 & n_{22}/d_{22} & \dots & 0 & 0 & \dots & 0 \\ \vdots & \vdots & & \vdots & \vdots & & \vdots \\ 0 & 0 & \dots & n_{rr}/d_{rr} & 0 & \dots & 0 \\ 0 & 0 & \dots & 0 & 0 & \dots & 0 \\ \vdots & \vdots & & \vdots & \vdots & & \vdots \\ 0 & 0 & \dots & 0 & 0 & \dots & 0 \end{bmatrix},$$

where r is the rank of C and $n_{ii}/d_{ii} = s_{ii}/d$ with n_{ii}, d_{ii} coprime for $i \in \mathbf{r}$, which is called the *Smith-McMillan* form of $\mathcal{G}$. Obviously it holds that

102
$$n_{ii} | n_{i+1,i+1},$$

103
$$d_{i+1,i+1} | d_{ii} \quad \text{for} \quad i \in \{1, 2, \dots, r-1\}.$$

Suppose that $1/d_1 C_1$ is another representation for $\mathcal{G}$ and $S_1 = P_1 C_1 Q_1$ the Smith form of C_1. Then it holds that

104
$$dS_1 = P_1 P^{-1} d_1 S Q^{-1} Q_1,$$

i.e. dS_1 and $d_1 S$ are equivalent, and because they are both of Smith form, they are equal up to associates of the diagonal entries. This implies that the Smith–McMillan form of $\mathcal{G}$ is unique up to associates of the numerators of the diagonal entries. Furthermore, the Smith–McMillan form is an invariant for the quotient equivalence. The matrices $[A \vdots B]$ constructed from matrices of Smith–McMillan form (101) in the following way:

105
$$A = \begin{bmatrix} n_{11} & 0 & \dots & 0 & 0 & \dots & 0 \\ 0 & n_{22} & \dots & 0 & 0 & \dots & 0 \\ \vdots & \vdots & & \vdots & \vdots & & \vdots \\ 0 & 0 & \dots & n_{rr} & 0 & \dots & 0 \\ 0 & 0 & \dots & 0 & 1 & \dots & 0 \\ \vdots & \vdots & & \vdots & \vdots & & \vdots \\ 0 & 0 & \dots & 0 & 0 & \dots & 1 \end{bmatrix} P^{-1},$$

106
$$B = \begin{bmatrix} d_{11} & 0 & \dots & 0 & 0 & \dots & 0 \\ 0 & d_{22} & \dots & 0 & 0 & \dots & 0 \\ \vdots & \vdots & & \vdots & \vdots & & \vdots \\ 0 & 0 & \dots & d_{rr} & 0 & \dots & 0 \\ 0 & 0 & \dots & 0 & 0 & \dots & 0 \\ \vdots & \vdots & & \vdots & \vdots & & \vdots \\ 0 & 0 & \dots & 0 & 0 & \dots & 0 \end{bmatrix} Q,$$

where $P \in R^{m \times m}$, $Q \in R^{n \times n}$ are R-unimodular, constitute a set of canonical forms for the quotient equivalence.

A2

Polynomials and polynomial matrices

This appendix contains some fundamental properties of polynomials and polynomial matrices, in particular of polynomials over complex numbers. We feel that the following material is very important in order to understand the central points of the present monograph. The material is to a large extent based on MacDuffee, 1933.

Polynomials

Let R be a subring and z an element of a ring R' with unity. Suppose that z satisfies $za = az$ for every $a \in R$. Then the set

1 $$R[z] = \{a(z) \,|\, a(z) = \sum_{i=0}^{n} a_i z^i,\ a_i \in R,\quad n \in \mathbf{N}_0\}$$

is a subring of R', called the ring of *polynomials in z with coefficients from R* or *polynomials in z over R*, and the binary operations of R' restricted to $R[z]$ satisfy

2 $$\sum_{i=0}^{n} a_i z^i + \sum_{i=0}^{m} b_i z^i = \sum_{i=0}^{\max\{m,n\}} (a_i + b_i) z^i,$$

3 $$\left(\sum_{i=0}^{n} a_i z^i\right)\left(\sum_{i=0}^{m} b_i z^i\right) = \sum_{i=0}^{m+n}\left(\sum_{j=0}^{i} a_j b_{i-j}\right) z^i,$$

where it is taken into account that each $a_0 + a_1 z + \ldots + a_n z^n + a_{n+1} z^{n+1} + a_{n+2} z^{n+2} + \ldots + a_m z^m$ with $a_{n+1} = a_{n+2} = \ldots = a_m = 0$ is equal to $a_0 + a_1 z + \ldots + a_n z^n$.

If it holds that

4 $$a_0 + a_1 z + \ldots + a_n z^n = 0 \Rightarrow a_0 = a_1 = \ldots = a_n = 0,$$

then z is an *indeterminate* (or *transcendental*) *over R*, otherwise z is *algebraic over R*. If z is an indeterminate over R, then the representation of an arbitrary

$a(z) \in R[z]$ in the form $a_0 + a_1z + \ldots + a_nz^n$ is unique up to the addition and subtraction of terms $0 \cdot z^k$. Note that if R does not contain a unity and z is an indeterminate over R, then z cannot belong to $R[z]$.

Let z be an indeterminate over R. Define the *degree* $\partial(a(z))$ of $a(z) \neq 0$ by

5 $$\partial(a(z)) = \max\{n \mid n \in \mathbf{N}_0, \quad a_n \neq 0\}$$

and make the agreement that $\partial(0) = -\infty$.

By defining $-\infty < n$ and $-\infty + n = -\infty$ for each $n \in \mathbf{N}_0$ and $-\infty + (-\infty) = -\infty$ we obtain

6 $$\partial(a(z) + b(z)) \leqslant \max\{\partial(a(z)), \partial(b(z))\},$$

7 $$\partial(a(z)b(z)) \leqslant \partial(a(z)) + \partial(b(z)).$$

The term a_nz^n of $a(z)$ of degree n is called the *leading term* of $a(z)$, and a_n is called the *leading coefficient* of $a(z)$, further $a(z)$ is *monic* if its leading coefficient is the unity of R (if it exists).

The following assertions are easy to prove.

8 **Note.** Let z be an indeterminate over R.
(i) $R[z]$ is commutative if and only if R is commutative.
(ii) $R[z]$ has a unity if and only if R has a unity.
(iii) $R[z]$ has zero divisors if and only if R has zero divisors.
(iv) $R[z]$ has no zero divisors if and only if for each $a(z), b(z) \in R[z]$

9 $$\partial(a(z)b(z)) = \partial(a(z)) + \partial(b(z)).$$

(v) $R[z]$ is an integral domain if and only if R is an integral domain.
(vi) If $R[z]$ is an integral domain, then the units of $R[z]$ are the units of R. □

Every ring with unity has an indeterminate. The set $R^{\mathbf{N}_0}$ of all sequences $\mathbf{N}_0 \to R$ with pointwise addition $(a+b)_i = a_i + b_i$ and convolution multiplication

$$(a * b)_i = \sum_{j=0}^{i} a_j b_{i-j}$$

is a ring with unity $(1, 0, 0, \ldots)$. By identifying each $a \in R$ with $(a, 0, 0, \ldots) \in R^{\mathbf{N}_0}$ R is made a subring of $R^{\mathbf{N}_0}$. Now, the element $z = (0, 1, 0, 0, \ldots) \in R^{\mathbf{N}_0}$ is easily shown to be an indeterminate over R.

Division algorithm

Suppose that z is an indeterminate over R and $a(z), b(z) \in R[z]$. If the leading coefficient of $b(z)$ is a unit of R, then there exist unique $q(z), r(z) \in R[z]$ such that

10 $$a(z) = q(z)b(z) + r(z), \; \partial(r(z)) < \partial(b(z)).$$

The existence of $q(z)$ and $r(z)$ can be shown by induction on $\partial(a(z))$. If $\partial(a(z)) < \partial(b(z))$, then we choose $q(z) = 0$, $r(z) = a(z)$. Suppose that the statement holds for polynomials of degree $n-1$ with $n \geqslant \partial(b(z)) \triangleq m$ and let $a(z)$ be of degree n. Because $\partial(a(z) - a_nb_m^{-1}z^{n-m}b(z)) \leqslant n-1$, there exist $q_1(z), r_1(z)$ such that

$$a(z) - a_nb_m^{-1}z^{n-m}b(z) = q_1(z)b(z) + r_1(z), \quad \partial(r_1(z)) < \partial(b(z)).$$

Hence $q(z) = a_n b_m^{-1} z^{n-m} + q_1(z)$ and $r(z) = r_1(z)$ fulfil the conditions (10).

Then let $q(z)$, $r(z)$ and $q_1(z)$, $r_1(z)$ be two pairs satisfying (10). Thus $(q(z) - q_1(z))b(z) + r(z) - r_1(z) = 0$, which implies

$$\partial(r(z) - r_1(z)) = \partial((q(z) - q_1(z))b(z)) = \partial(q(z) - q_1(z)) + \partial(b(z)).$$

On the other hand, $\partial(r(z) - r_1(z)) < \partial(b(z))$, so that the only possibility is $r(z) = r_1(z)$. Because $b(z)$ is not a zero divisor, this further yields $q(z) = q_1(z)$. Thus the uniqueness has been proved.

Finding $q(z)$, $r(z)$ corresponding to the given polynomials $a(z)$, $b(z)$ based on induction is called the *division algorithm*. The polynomial $q(z)$ is called the *quotient* of $a(z)$ by $b(z)$ and $r(z)$ is the *remainder*. If $b(z)$ is a divisor of $a(z)$, the remainder is zero.

If $a(z)$, $b(z) \in R[z]$ are such that there exist $q(z)$, $r(z) \in R[z]$ satisfying (10), they are said to satisfy the division algorithm. Moreover, if all $a(z)$, $b(z) \in R[z]^* \triangleq R[z]$-$\{0\}$ satisfy the division algorithm, $R[z]$ itself is said to satisfy the division algorithm. It is easily seen that $R[z]$ satisfies the division algorithm if and only if R is a field.

Then let R be a field and J an ideal in $R[z]$ and take a nonzero element $b(z) \in J$ of least degree. By the division algorithm for every $a(z) \in J$ there exist $q(z)$, $r(z) \in R[z]$ such that $\partial(r(z)) = \partial(a(z) - q(z)b(z)) < \partial(b(z))$. This is, however, possible only if $r(z) = 0$ because $r(z) \in J$. Hence $a(z) = q(z)b(z)$ and, furthermore, $J = R[z]b(z)$. Thus $R[z]$ is a principal ideal domain satisfying the division algorithm with unique quotients and remainders.

Furthermore, $R[z]$ then has the unique prime factorization property, i.e. every $a(z) \in R[z]$ can be written as a finite product of primes, unique up to permuting the factors and replacing them by associates. Obviously monic polynomials of degree 1 are primes.

Polynomial functions

Let $R[z]$ be a ring of polynomials in an indeterminate z. For each $a(z) = a_0 + a_1 z + \ldots + a_n z^n \in R[z]$ the assignment

11 $$c \mapsto a(c) = a_0 + a_1 c + \ldots + a_n c^n$$

defines a mapping $a: R \to R$ called a *polynomial function*. The set R^R of all mappings $R \to R$ together with the pointwise addition and pointwise multiplication is a ring and the assignment $a(z) \mapsto a$ defines a morphism of rings.

An element $c \in R$ with $a(c) = 0$ is called a *zero* or a *root* of $a(z)$. In particular every $c \in R$ is a zero of the zero polynomial $0 \in R[z]$.

12 **Note.** (i) Let $a(z) \in R[z]$ be arbitrary and $b(z) = z - c$ with $c \in R$. By the division algorithm $a(z) = q(z)(z - c) + r(z)$ with $\partial(r(z)) < \partial(z - c) = 1$. Hence $r(z) = r(c) = a(c)$. Consequently, $c \in R$ is a zero of $a(z) \in R[z]$ if and only if $z - c$ divides $a(z)$.

(ii) If R is an integral domain then a zero of $a(z)b(z)$ is either a zero of $a(z)$ or a zero of $b(z)$. Furthermore, then the item (i) implies that a polynomial $a(z) \in R[z]$ of degree $n \geqslant 0$ has at most n zeros. Thus, if $a(z)$, $b(z) \in R[z]$ are

of degree m, n, respectively, and there exist p distinct elements $c_1, c_2, \ldots, c_p \in R$ with $p > \max\{m, n\}$ such that $a(c_i) = b(c_i)$, $i \in \mathbf{p}$, then $a(z) = b(z)$.
(iii) If R is an infinite integral domain then the morphism $a(z) \mapsto a$ is a monomorphism, for if $a = 0$, i.e. a zero mapping, every $c \in R$ satisfies $a(c) = 0$, which implies that $a(z) = 0$. □

Polynomials over complex numbers

Let $\mathbf{C}$ be the field of complex numbers and $\mathbf{C}[z]$ a ring of polynomials in an indeterminate z. By the *fundamental theorem of algebra* (see e.g. MacLane and Birkhoff, 1967) every polynomial $a(z) \in \mathbf{C}[z]$ of degree $n > 0$ has at least one zero. By items (i), (ii) of note (12) $a(z) \in \mathbf{C}[z]$ of degree $n \geqslant 0$ can then be written uniquely up to permuting the factors in the form

13 $$a(z) = a_n(z - c_1)(z - c_2) \ldots (z - c_n),$$

where $a_n \in \mathbf{C}$ is the leading coefficient of $a(z)$ and $c_1, c_2, \ldots, c_n \in \mathbf{C}$ are zeros of $a(z)$. This further implies that $a(z) \in \mathbf{C}[z]$ is prime if and only if it is of degree 1. If a zero c_i occurs in (13) exactly m times, it is called a zero of *multiplicity m*.

Euclidean algorithm

Because $\mathbf{C}[z]$ is a principal ideal domain, a greatest common divisor $c(z) \in \mathbf{C}[z]$ of $a(z), b(z) \in \mathbf{C}[z]$ always exists and it can be written as

14 $$c(z) = d(z)a(z) + e(z)b(z)$$

for some $d(z), e(z) \in \mathbf{C}[z]$.

A greatest common divisor $c(z)$ as well as the corresponding $d(z), e(z)$ in the representation (14) can be found via the following procedure based on the division algorithm.

Write $r_0(z) = a(z)$, $r_1(z) = b(z)$ and use the division algorithm to produce a sequence of remainders $r_2(z), r_3(z), \ldots \in \mathbf{C}[z]$ such that

15 $$r_i(z) = q_{i+1}(z)r_{i+1}(z) + r_{i+2}(z),\ \partial(r_{i+2}(z)) < \partial(r_{i+1}(z)),$$

for $i \in \mathbf{N}_0$, until $r_{i+2} = 0$ for some i. Using the recursion formula (15) every combination $f(z)a(z) + g(z)b(z)$, $f(z), g(z) \in \mathbf{C}[z]$ can be presented in the form $h(z)r_{i+1}(z)$, $h(z) \in \mathbf{C}[z]$ and vice versa. Hence $r_{i+1}(z)$ is a greatest common divisor of $a(z), b(z)$ and using (15) it can be presented in the form (14). The procedure above is called the *Euclidean algorithm*.

16 **Note.** Suppose that $a(z), b(z) \in \mathbf{C}[z]$ are of degree m, n, respectively. If $a(z)$ and $b(z)$ are coprime, every $c(z) \in \mathbf{C}[z]$ of degree $p \leqslant m + n - 1$ can be uniquely presented in the form

17 $$c(z) = d(z)a(z) + e(z)b(z),$$

18 $$\partial(d(z)) < n,\ \partial(e(z)) < m.$$

Because $a(z)$ and $b(z)$ are coprime, every $c(z) \in \mathbf{C}[z]$ can be presented as a combination (17) for some $d(z), e(z)$. If $d(z), e(z)$ do not satisfy the conditions

(18) for degrees, we can use the division algorithm and obtain

19 $$d(z) = q_1(z)b(z) + r_1(z), \quad \partial(r_1(z)) < n,$$

20 $$e(z) = q_2(z)a(z) + r_2(z), \quad \partial(r_2(z)) < m.$$

Thus

21 $$c(z) = (q_1(z) + q_2(z))a(z)b(z) + r_1(z)a(z) + r_2(z)b(z).$$

If $q_1(z) + q_2(z) \neq 0$, the first term of the right hand side of (21) is of a degree greater than the other terms and greater than $m + n - 1$. Because $c(z)$ was assumed to be of degree $p \leqslant m + n - 1$, this leads to a contradiction. Hence $q_1(z) + q_2(z) = 0$ and $c(z) = r_1(z)a(z) + r_2(z)b(z)$, where $r_1(z), r_2(z)$ satisfy the conditions (18).

Suppose then that there are two pairs $d(z), e(z)$ and $d_1(z)$, $e_1(z)$ satisfying (17) and (18). This gives

22 $$(d(z) - d_1(z))a(z) = (e_1(z) - e(z))b(z),$$
$$\partial(d(z) - d_1(z)) < \partial(b(z)), \quad \partial(e_1(z) - e(z)) < \partial(a(z)).$$

If $d(z) - d_1(z) \neq 0$, this implies that $a(z)$ and $b(z)$ have a common divisor, which contradicts our assumption. Thus $d(z) = d_1(z)$ and consequently $e(z) = e_1(z)$. □

Rings of quotients of polynomials

Because $\mathbf{C}[z]$ is an integral domain, it has a field of quotients usually denoted by $\mathbf{C}(z)$. An element $g(z) = b(z)/a(z) \in \mathbf{C}(z)$, i.e. a quotient of polynomials is called a *rational form* (or a *rational*, in short) in z over $\mathbf{C}$. $\mathbf{C}[z]$ also has other considerable denominator sets, for instance the sets of powers of a single nonzero polynomial, in particular the set of powers of z

23 $$S_0 = \{z^i | i \in \mathbf{N}_0\}$$

is important, wherefore we denote the ring $\mathbf{C}[z]_{S_0}$ of quotients of $\mathbf{C}[z]$ by S_0 shortly by $\mathbf{C}(z]$. $\mathbf{C}(z]$ is an integral domain thus having the field of quotients which is equal to $\mathbf{C}(z)$.

Because $\mathbf{C}$ can be considered as a subring of $\mathbf{C}(z]$ and $1/z \in \mathbf{C}(z]$, we can construct the ring $\mathbf{C}[1/z]$ of polynomials in $1/z$ over $\mathbf{C}$. The element $1/z$ is obviously an indeterminate over $\mathbf{C}$. Furthermore, the ring $\mathbf{C}(1/z]$ of quotients is equal to $\mathbf{C}(z]$ via the equality

24 $$(a_0 + a_1(1/z) + \ldots + a_m(1/z)^m)/(1/z)^n = z^n(a_0 z^m + a_1 z^{m-1} + \ldots + a_m)/z^m.$$

The set S_1 of monic polynomials is also a denominator set in $\mathbf{C}[z]$ and $\mathbf{C}[z]_{S_1} = \mathbf{C}(z)$. Because every $b(z)/a(z) \in \mathbf{C}(z)$ is equal to exactly one $d(z)/c(z) \in \mathbf{C}[z]_{S_1}$ such that $c(z)$ and $d(z)$ are coprime, the set of pairs $(d(z), c(z)) \in \mathbf{C}[z] \times S_1$ with $d(z)$, $c(z)$ coprime constitutes a set of canonical forms for the "quotient equivalence" (A1.20).

Both S_0 and S_1 are examples of denominator sets S_C consisting of monic polynomials having all their zeros in a given domain $C \subset \mathbf{C}$.

A rational form $b(z)/a(z) \in \mathbf{C}(z)$ is said to be *proper* if $\partial(b(z)) \leqslant \partial(a(z))$ and *strictly proper* if $\partial(b(z)) < \partial(a(z))$. The concepts proper and strictly proper are obviously well-defined, i.e. they are independent of representatives of rational forms. Using the division algorithm every $b(z)/a(z) \in \mathbf{C}(z)$ can be written uniquely in the form

25 $$b(z)/a(z) = r(z)/a(z) + q(z)$$

with $r(z)/a(z) \in \mathbf{C}(z)$ strictly proper, $q(z) \in \mathbf{C}[z]$.

The sum and the product of two proper (strictly proper) rational forms are proper (strictly proper, respectively) again.

Let $b(z)/a(z) \in \mathbf{C}(z)$ be such that $b(z)$ and $a(z)$ are coprime and $\{c_1, c_2, \ldots, c_n\} \subset \mathbf{C}$ the set of zeros of $a(z)$. Then the assignment

26 $$c \mapsto b(c)/a(c)$$

defines a mapping $b/a : \mathbf{C} - \{c_1, c_2, \ldots, c_n\} \to \mathbf{C}$ called a *rational function*. The set of rational functions together with pointwise (almost everywhere) addition and multiplication is a field and the assignment $b(z)/a(z) \mapsto b/a$ defines a morphism of rings, in fact an isomorphism of rings.

Polynomial matrices

The matrices over a ring $R[z]$ of polynomials are called polynomial matrices and denoted as $A(z)$ with entries $a_{ij}(z)$. Sometimes it is useful to write $A(z) \in R[z]^{m \times n}$ in the form

27 $$A(z) = A_0 + A_1 z + \ldots + A_p z^p \quad \text{with } A_0, A_1, \ldots, A_p \in R^{m \times n}.$$

Let z be an indeterminate over R. If $A_p \neq 0$, p is said to be the *degree* of $A(z)$ denoted by $\partial(A(z))$. The degree of a zero matrix is defined as being $-\infty$. The usual polynomial terminology can be applied to polynomial matrices. In particular, the coefficient A_p of $A(z)$ of degree p is called the *leading coefficient*. The degree of polynomial matrices satisfies

28 $$\partial(A(z) + B(z)) \leqslant \max\{\partial(A(z)), \partial(B(z))\},$$

29 $$\partial(A(z)C(z)) \leqslant \partial(A(z)) + \partial(C(z)).$$

When $A(z)$ is an n by n-matrix of degree p, the determinant of $A(z)$ is of the form

30 $$\det A(z) = \det A_0 + d_1 z + \ldots + d_{np-1} z^{np-1} + \det A_p z^{np}.$$

Thus $\det A(z) \neq 0$ if $\det A_0 \neq 0$ or $\det A_p \neq 0$. Because the only units of $R[z]$ are the units of R, $A(z)$ is $R[z]$-unimodular if and only if $\det A(z) = \det A_0$ and a unit of R. In particular, if R is a field, say $\mathbf{C}$, $A(z)$ is $\mathbf{C}[z]$-unimodular if and only if $\det A(z) = \det A_0 \neq 0$.

Let $A_i(z) \in R[z]^n$ denote the ith row of $A(z) \in R[z]^{m \times n}$. The degree of $A_i(z)$ is called the ith *row degree* of $A(z)$ and denoted shortly by $\partial_{ri}(A(z))$. Respectively, the jth *column degree* is denoted by $\partial_{cj}(A(z))$. Now, let $\partial_{ri}(A(z)) = r_i \geqslant 0$, $i \in \mathbf{m}$, be the row degrees of $A(z)$. Then $A(z)$ can be written in the form

31
$$A(z) = \underbrace{\begin{bmatrix} c_{11}z^{r_1} & c_{12}z^{r_1} & \dots & c_{1n}z^{r_1} \\ c_{21}z^{r_2} & c_{22}z^{r_2} & \dots & c_{2n}z^{r_2} \\ \vdots & \vdots & & \vdots \\ c_{m1}z^{r_m} & c_{m2}z^{r_m} & \dots & c_{mn}z^{r_m} \end{bmatrix}}_{A^{\mathrm{r}}(z)} + \underbrace{\text{terms of lower degree}}_{A^{\mathrm{l}}(z)}$$

$$= \begin{bmatrix} z^{r_1} & 0 & \dots & 0 \\ 0 & z^{r_2} & \dots & 0 \\ \vdots & \vdots & & \vdots \\ 0 & 0 & \dots & z^{r_m} \end{bmatrix} \underbrace{\begin{bmatrix} c_{11} & c_{12} & \dots & c_{1n} \\ c_{21} & c_{22} & \dots & c_{2n} \\ \vdots & \vdots & & \vdots \\ c_{m1} & c_{m2} & \dots & c_{mn} \end{bmatrix}}_{C_A} + A^{\mathrm{l}}(z),$$

where the rows of $C_A \in R^{m\times n}$ are the leading coefficients of the corresponding rows of $A(z)$. More generally the matrices with zero rows can also be presented in the form (31) if the zero rows are, contrary to the usual practice, considered as polynomial matrices of degree 0 with zero matrices as leading coefficients.

An arbitrary p by p-minor $\alpha(z) \in R[z]$ of $A(z)$ obtained by deleting all but the rows $i_1, i_2, \dots, i_p$ and the corresponding minor $\gamma \in R$ of C_A satisfy

32
$$\alpha(z) = \gamma z^q + \text{terms of lower degree with}$$

33
$$q = \sum_{j=1}^{p} r_{i_j}.$$

Now the matrix $A(z)$ is said to be *row proper* if its determinantal rank is equal to the determinantal rank of C_A. In particular, a square matrix $A(z)$ with $\det A(z) \neq 0$ is row proper if and only if $\det C_A \neq 0$, i.e. if and only if the degree of $\det A(z)$ is equal to the sum of the row degrees (Wolovich, 1974). A *column proper* matrix is defined in the corresponding way.

Canonical upper triangular form

The ring $\mathbf{C}[z]$ with z an indeterminate is a principal ideal domain satisfying the division algorithm. Hence the triangular reduction procedure presented in appendix A1 can be applied to a matrix $A(z) \in \mathbf{C}[z]^{m\times n}$. Moreover, this can be carried out without secondary row operations because greatest common divisors can be found by the Euclidean algorithm, i.e. using the elementary row operations $T_{ij}(a(z))$. The degrees of the entries above an entry which is the last nonzero entry in its column and the first nonzero entry in its row can further be reduced by the division algorithm, i.e. using the elementary row operations $T_{ij}(a(z))$ starting with the first column. The leading coefficients of these "diagonal" entries can also be normalized equal to 1 by the operations $V_{ii}(u)$. The resulting matrix, i.e. an upper triangular matrix such that every entry which is the last nonzero entry in its column and the first nonzero entry in its row is monic and the entries above it are of lower degree, is said to be of *canonical upper triangular form* (CUT-*form*) and the *canonical upper triangular form* (CUT-*form*) of $A(z)$, because the set of matrices of this form constitutes a set of canonical forms for the $\mathbf{C}[z]$-row equivalence.

If the above triangular reduction algorithm is carried out correspondingly so that the resulting matrix is a lower triangular matrix, then this resulting matrix is said to be of *canonical lower triangular form* (CLT-*form*).

Canonical upper and lower triangular forms with respect to column equivalence are defined correspondingly in an obvious way. To avoid ambiguity, we shall *not* use the notations CUT and CLT to indicate these forms.

34 **Theorem.** *If $A(z), B(z) \in \mathbf{C}[z]^{m\times n}$ are of CUT-form and $\mathbf{C}[z]$-row equivalent then $A(z) = B(z)$.* □

35 **Proof.** (See also Ylinen, 1975, 1980; Hirvonen, Blomberg, Ylinen, 1975). Because the rank of a matrix of CUT-form is equal to the number of nonzero rows and an invariant for the row equivalence we can assume that $A(z)$ and $B(z)$ have the same number p of nonzero rows. Suppose first that $p < m$. By the row equivalence there exists a $\mathbf{C}[z]$-unimodular matrix $P(z) \in \mathbf{C}[z]^{m\times m}$ such that

36
$$\underbrace{\begin{bmatrix} A_1(z) \\ \hdashline 0 \end{bmatrix}}_{A(z)} = \underbrace{\begin{bmatrix} P_1(z) & P_2(z) \\ P_3(z) & P_4(z) \end{bmatrix}}_{P(z)} \underbrace{\begin{bmatrix} B_1(z) \\ \hdashline 0 \end{bmatrix}}_{B(z)},$$

where $A_1(z) \in \mathbf{C}[z]^{p\times n}$, $B_1(z) \in \mathbf{C}[z]^{p\times n}$. Because $B_1(z)$ is of full rank p, it is easily seen that the equation $0 = P_3(z)B_1(z)$ implies $P_3(z) = 0$. Hence $P(z)$ is of the upper block triangular form, which further implies that $P_1(z)$ and $P_4(z)$ are $\mathbf{C}[z]$-unimodular. Then $A_1(z)$ and $B_1(z)$ are of CUT-form and $\mathbf{C}[z]$-row equivalent, so that we have brought the proof to the case of matrices $A(z)$, $B(z)$ without zero rows.

Henceforth the proof is carried out by mathematical induction on m. Suppose first that $m = 1$. By the row equivalence there exists a $\mathbf{C}[z]$-unimodular $P(z)$ such that

37
$$\underbrace{\overset{1\,\ldots\,i-1\quad i\quad\ldots\quad n}{[0 \ldots\ 0\ \ a_{1i}(z) \ldots a_{1n}(z)]}}_{A(z)} = \underbrace{[p_{11}(z)]}_{P(z)} \underbrace{\overset{1\,\ldots\,j-1\quad j\quad\ldots\quad n}{[0 \ldots\ 0\ \ b_{1j}(z) \ldots b_{1n}(z)]}}_{B(z)},$$

where $i \leqslant n$, $j \leqslant n$ are the indices of the first nonzero entries of $A(z)$, $B(z)$, respectively. Because $p_{11}(z)$ is a unit of $\mathbf{C}[z]$, $i = j$ and because $a_{1i}(z)$ and $b_{1i}(z)$ are monic, $p_{11}(z)$ must be equal to 1. Consequently $A(z) = B(z)$.

Suppose next that the theorem holds for $m = k - 1 \geqslant 1$ and take $m = k$. By the row equivalence there again exists a $\mathbf{C}[z]$-unimodular $P(z)$ such that

38
$$\underbrace{\begin{bmatrix} A_1(z) & A_2(z) \\ 0 & A_4(z) \end{bmatrix}}_{A(z)} = \underbrace{\begin{bmatrix} P_1(z) & P_2(z) \\ P_3(z) & P_4(z) \end{bmatrix}}_{P(z)} \underbrace{\begin{bmatrix} B_1(z) & B_2(z) \\ 0 & B_4(z) \end{bmatrix}}_{B(z)},$$

where $A_1(z)$, $B_1(z)$ are $(k-1)$ by q-matrices of full rank $k-1$. The equation $0 = P_3(z)\, B_1(z)$ then implies that $P_3(z) = 0$, i.e. $P(z)$ is upper block triangular with $P_1(z)$, $P_4(z)$ unimodular. Then $A_1(z)$ and $B_1(z)$ are of CUT-form and

$\mathbf{C}[z]$-row equivalent, hence by the induction assumption $P_1(z) = I$ and $A_1(z) = B_1(z)$. Respectively, the 1 by $(n-q)$-matrices $A_4(z)$ and $B_4(z)$ are of CUT-form and $\mathbf{C}[z]$-row equivalent, so that $P_4(z) = I$ and $A_4(z) = B_4(z)$.

Finally, let i be the index of the first nonzero entry of the kth row of $A(z)$ and $B(z)$. Then the equation obtained from the ith column of (38) is

39
$$\begin{bmatrix} a_{1i}(z) \\ \vdots \\ a_{k-1,i}(z) \\ a_{ki}(z) \end{bmatrix} = \begin{bmatrix} b_{1i}(z) \quad + p_{1k}(z) b_{ki}(z) \\ \vdots \\ b_{k-1,i}(z) + p_{k-1,k}(z) b_{ki}(z) \\ b_{ki}(z) \end{bmatrix}.$$

Because the entries $a_{1i}(z), \ldots, a_{k-1,i}(z)$ are of lower degree than $a_{ki}(z)$ and the entries $b_{1i}(z), \ldots, b_{k-1,i}(z)$ of lower degree than $b_{ki}(z)$, it must be $p_{1k}(z) = \ldots = p_{k-1,k}(z) = 0$, i.e. $P_2(z) = 0$. Hence $P(z) = I$ and $A(z) = B(z)$. □

A matrix of CUT-form is obviously column proper. Hence as a byresult every matrix over $\mathbf{C}[z]$ is $\mathbf{C}[z]$-row equivalent to a column proper matrix.

Column permuted canonical upper triangular form

There is another upper triangular form whose structure lies somewhere between the CUT-form and the Smith form, as its construction based on elementary row operations and permutations of columns already suggests. Let S_n denote the set of all permutations $\sigma: \mathbf{n} \to \mathbf{n}$. A permutation σ of the columns of $A(z) \in \mathbf{C}[z]^{m \times n}$ giving the matrix $A(z)\sigma \in \mathbf{C}[z]^{m \times n}$, $(i, j) \mapsto a_{i\sigma(j)}$, can be performed by post-multiplying $A(z)$ by a permutation matrix with exactly one entry in each row and in each column equal to 1 and the others equal to 0.

Let $A(z) \in \mathbf{C}[z]^{m \times n}$ be of rank r. A list $(S_{n1}, S_{n2}, \ldots, S_{nr})$ of permutation sets is now constructed as follows.

40 $S_{ni} = \{\sigma | \sigma \in S_{n,i-1}$ and the (i, i) entry of the CUT-form of $A(z)\sigma$ is nonzero but attains the lowest possible degree$\}$, $i \in \mathbf{r}$, $S_{n0} = S_n$.

For an arbitrary $\sigma \in S_{nr}$ the CUT-form of $A(z)\sigma$ is called a *column permuted canonical upper triangular form* (CPCUT-*form*) of $A(z)$. It should be noted that, in general, a CPCUT-form of a matrix is not unique. Due to the construction the row degrees are nevertheless unique.

The CPCUT-form also has some other general features:

—going down the main diagonal, the degrees of the first r entries with r equal to the rank are nondecreasing,
—to the right of the diagonal entry, every entry in any of the first r rows is either a zero or of at least the same degree as the diagonal entry,
—in addition, the usual properties of the CUT-form hold so that a CPCUT-form has exactly r nonzero rows, and the entries above a nonzero diagonal entry are of lower degree than the diagonal entry.

Row proper form

Every matrix over $\mathbf{C}[z]$ is $\mathbf{C}[z]$-row equivalent to a matrix of row proper form. Let $A(z) \in \mathbf{C}[z]^{m \times n}$ be of rank p. If $p < m$, $A(z)$ is first brought by elementary row operations to the form

$$\begin{bmatrix} B(z) \\ \hdashline 0 \end{bmatrix},$$

where $B(z) \in \mathbf{C}[z]^{p\times n}$ is of full rank p, otherwise write $B(z) = A(z)$. Denote the degrees and the leading coefficients of the rows of $B(z)$ by $r_1, r_2, \ldots, r_p \in \mathbf{N}_0$ and $C_1, C_2, \ldots, C_p \in \mathbf{C}^n$, respectively. If $B(z)$ is not row proper, there exist $d_1, d_2, \ldots, d_p \in \mathbf{C}$, at least one of which is nonzero, satisfying

41 $$\sum_{i=1}^{p} d_i C_i = 0.$$

Take $k \in \mathbf{p}$ such that $d_k \neq 0$ and $r_k \geq r_j$ for every $j \in \mathbf{p}$ such that $d_j \neq 0$. Substitute the kth row of $B(z)$ by the row

42 $$\sum_{\substack{i=1 \\ d_i \neq 0}}^{p} (d_i/d_k)\, z^{r_k - r_i} B_i(z),$$

where $B_i(z) \in \mathbf{C}[z]^n$, $i \in \mathbf{p}$, denotes the ith row of $B(z)$. This can be made by using sequentially the row operations $T_{ki}((d_i/d_k)\, z^{r_k - r_i})$. The new row is of lower degree than the old one. If the new matrix is not row proper, proceed similarly until the desired form is obtained.

The row proper matrix constructed above has the least degree among the matrices $\mathbf{C}[z]$-row equivalent to it. Furthermore, its row degrees are unique up to permuting the rows.

43 **Theorem.** *Let $A(z)$, $B(z) \in \mathbf{C}[z]^{m\times n}$ be of full rank m and row proper, and denote their row degrees by $r_1, r_2, \ldots, r_m$ and $s_1, s_2, \ldots, s_m$, respectively. Suppose further that the row degrees are in increasing order, i.e. $r_1 \leq r_2 \leq \ldots \leq r_m$ and $s_1 \leq s_2 \leq \ldots \leq s_m$. If $A(z)$ and $B(z)$ are $\mathbf{C}[z]$-row equivalent, i.e. there exists a $\mathbf{C}[z]$-unimodular matrix $P(z)$, such that $A(z) = P(z)\, B(z)$, then $r_i = s_i$ for all $i \in \mathbf{m}$ and $P(z)$ is a lower block triangular matrix such that*

44 $$\partial(p_{ij}(z)) \leq r_i - r_j = s_i - s_j, \quad i, j \in \mathbf{m}. \qquad \square$$

45 **Proof.** Writing $A(z)$ and $B(z)$ in the form (31) gives

46 $$\begin{bmatrix} z^{r_1} & 0 & \ldots & 0 \\ 0 & z^{r_2} & \ldots & 0 \\ \vdots & \vdots & & \vdots \\ 0 & 0 & \ldots & z^{r_m} \end{bmatrix} C_A + A^{l}(z) = P(z) \begin{bmatrix} z^{s_1} & 0 & \ldots & 0 \\ 0 & z^{s_2} & \ldots & 0 \\ \vdots & \vdots & & \vdots \\ 0 & 0 & \ldots & z^{s_m} \end{bmatrix} C_B$$

$$+ (B^{l}(z)) = \begin{bmatrix} p_{11}(z) z^{s_1} & p_{12}(z) z^{s_2} & \ldots & p_{1m}(z) z^{s_m} \\ p_{21}(z) z^{s_1} & p_{22}(z) z^{s_2} & \ldots & p_{2m}(z) z^{s_m} \\ \vdots & \vdots & & \vdots \\ p_{m1}(z) z^{s_1} & p_{m2}(z) z^{s_2} & \ldots & p_{mm}(z) z^{s_m} \end{bmatrix} C_B + P(z) B^{l}(z).$$

Because

47 $$\partial\left(\sum_{k=1}^{m} p_{ik}(z)b^1_{kj}(z)\right) \leqslant \max\{\partial(p_{ik}(z)b^1_{kj}(z)) \mid k \in \mathbf{m}\}$$
$$< \max\{\partial(p_{ik}(z)) + s_k \mid k \in \mathbf{m}\}$$

for all $i \in \mathbf{m}$, $j \in \mathbf{n}$ and C_B is of full rank m we obtain

48 $$\partial(p_{ij}(z)) + s_j \leqslant r_i, \quad i, j \in \mathbf{m}.$$

First this gives $r_1 \geqslant s_1$, for otherwise $p_{11}(z) = p_{12}(z) = \ldots = p_{1m}(z) = 0$, which implies that $P(z)$ cannot be unimodular. By symmetry $s_1 \geqslant r_1$ so that $s_1 = r_1$. Suppose then that $s_i = r_i$ for each $i \in \mathbf{k}$ for some $k < m$. If $s_k = s_{k+1}$, then $r_{k+1} \geqslant s_{k+1}$. If $s_k < s_{k+1}$, suppose that $r_{k+1} < s_{k+1}$. Because $r_1 \leqslant r_2 \leqslant \ldots \leqslant r_k \leqslant r_{k+1}$, it holds that $p_{ij}(z) = 0$ for all $i \in \{1, 2, \ldots, k+1\}$, $j \in \{k+1, k+2, \ldots, m\}$. This implies that $P(z)$ cannot be unimodular. Hence $r_{k+1} \geqslant s_{k+1}$. By symmetry $s_{k+1} \geqslant r_{k+1}$ so that $s_{k+1} = r_{k+1}$. Hence by mathematical induction $r_i = s_i$ for each $i \in \mathbf{m}$. Finally, (48) then implies (44). □

49 **Note.** If the row degrees of $A(z)$ and $B(z)$ are not in increasing order, we can permute the rows without loosing the row properness. Let σ and τ be two permutations of rows such that

50 $$r_{\sigma^{-1}(1)} \leqslant r_{\sigma^{-1}(2)} \leqslant \ldots \leqslant r_{\sigma^{-1}(m)},$$

51 $$s_{\tau^{-1}(1)} \leqslant s_{\tau^{-1}(2)} \leqslant \ldots \leqslant s_{\tau^{-1}(m)}.$$

Then the row degrees of $\sigma A(z)$ and $\tau B(z)$, where σ and τ are identified with the corresponding permutation matrices, are in increasing order and

52 $$\sigma A(z) = \sigma P(z)\tau^{-1}\tau B(z) = Q(z)\tau B(z),$$

where $Q(z)$ is $\mathbf{C}[z]$-unimodular. Hence by theorem (43) $r_{\sigma^{-1}(i)} = s_{\tau^{-1}(i)}$ for each $i \in \mathbf{m}$ and $Q(z)$ is a lower block triangular matrix satisfying

53 $$\partial(q_{ij}(z)) \leqslant r_{\sigma^{-1}(j)} = s_{\tau^{-1}(i)} - s_{\tau^{-1}(j)}$$

for $i, j \in \mathbf{m}$. This implies that $P(z)$ satisfies

54 $$\partial(p_{ij}(z)) = \partial(q_{\sigma(i),\tau(j)}(z)) \leqslant r_i - s_j, i, j \in \mathbf{m}.$$ □

55 **Note.** Suppose that $P(z)$ is a $\mathbf{C}[z]$-unimodular matrix satisfying (54). Then the matrices $P'(z)$, $P''(z)$ and C defined by

56 $$p'_{ij}(z) = \begin{cases} p_{ij}(z) & \text{if } \partial(p_{ij}(z)) = r_i - s_j, \\ 0 & \text{else,} \end{cases}$$

57 $$p''_{ij}(z) = \begin{cases} p_{ij}(z) & \text{if } \partial(p_{ij}(z)) = r_i - s_j = 0, \\ 0 & \text{else,} \end{cases}$$

58 $$c_{ij} = \begin{cases} \text{the leading coefficient of } p_{ij}(z) \text{ if } \partial(p_{ij}(z)) = r_i - s_j, \\ 0 \quad \text{else} \end{cases}$$

for $i, j \in \mathbf{m}$ are also $\mathbf{C}[z]$-unimodular, in fact $P''(z)$ and C are $\mathbf{C}$-unimodular. Furthermore, if $A(z)$ and $B(z)$ in theorem (43) are written in the form (31), then

59 $$A^{\mathrm{r}}(z) = P'(z)\, B^{\mathrm{r}}(z),$$

60 $$C_A = C C_B. \qquad \square$$

61 **Note.** Suppose that $B(z) \in \mathbf{C}[z]^{m\times n}$ is row proper with row degrees $s_1, s_2, \ldots, s_m$ and $P(z) \in \mathbf{C}[z]^{m\times m}$ is a $\mathbf{C}[z]$-unimodular matrix satisfying

62 $$\partial(p_{ij}(z)) \leqslant s_i - s_j, \quad i, j \in \mathbf{m}.$$

It can quite easily be seen that the matrix $A(z) = P(z)B(z)$ is also row proper. Furthermore, its row degrees are equal to the row degrees of $B(z)$ and $C_A = CC_B$ where $C \in \mathbf{C}^{m\times m}$ is the invertible matrix defined by

63 $$c_{ij} = \begin{cases} \text{the leading coefficient of } p_{ij}(z) \text{ if } \partial(p_{ij}(z)) = s_i - s_j \\ 0 \text{ else} \end{cases}$$

for $i, j \in \mathbf{m}$. $\square$

Now let $B(z) \in \mathbf{C}[z]^{m\times n}$ be a row proper matrix with row degrees $s_1, s_2, \ldots, s_m$ and write $B(z)$ in the form (31)

64 $$B(z) = \underbrace{\begin{bmatrix} c_{11}z^{s_1} & c_{12}z^{s_1} & \ldots & c_{1n}z^{s_1} \\ c_{21}z^{s_2} & c_{22}z^{s_2} & \ldots & c_{2n}z^{s_2} \\ \vdots & \vdots & & \vdots \\ c_{m1}z^{s_m} & c_{m2}z^{s_m} & \ldots & c_{mn}z^{s_m} \end{bmatrix}}_{B^{\mathrm{r}}(z)} + B^{\mathrm{l}}(z).$$

Multiply $B(z)$ by a $\mathbf{C}[z]$-unimodular matrix $Q(z) \in \mathbf{C}[z]^{m\times m}$ such that $A^{\mathrm{r}}(z) = Q(z)B^{\mathrm{r}}(z)$ is of CUT-form. Because $Q(z)$ can be written as a product of elementary row operations $T_{ij}(cz^{s_i - s_j})$ with $c \in \mathbf{C}$, and U_{ij}, the resulting matrix $A(z) = Q(z)B(z)$ is also row proper and has the representation (31) $A(z) = A^{\mathrm{r}}(z) + A^{\mathrm{l}}(z)$ with $A^{\mathrm{r}}(z) = Q(z)B^{\mathrm{r}}(z)$, $A^{\mathrm{l}}(z) = Q(z)B^{\mathrm{l}}(z)$.

Next let $a_{kj}(z)$ be an entry of $A(z)$ such that the corresponding entry $a_{kj}^{\mathrm{r}}(z)$ of $A^{\mathrm{r}}(z)$ is the first nonzero entry in its row and the last nonzero entry in its column. Suppose that there is an entry $a_{ij}(z)$ below $a_{kj}(z)$ which is not of lower degree than $a_{kj}(z)$ (or $a_{kj}^{\mathrm{r}}(z)$). Because $a_{ij}^{\mathrm{r}}(z) = 0$ by construction, such a situation can arise only if $s_i > s_k$. The degree of $a_{ij}(z)$ can then be further lowered with the aid of $a_{kj}(z)$ by using row operations $T_{ik}(a(z))$. It is seen that the resulting new matrix $\tilde{A}(z)$ is still row proper with $\tilde{A}^{\mathrm{r}}(z)$ equal to $A^{\mathrm{r}}(z)$. Similarly, suppose that there is an entry $a_{lj}(z)$ above $a_{kj}(z)$ which is not of lower degree than $a_{kj}(z)$. Because $A^{\mathrm{r}}(z)$ was upper triangular by construction, such a situation can arise only if $s_l \geqslant s_k$. The degree of $a_{lj}(z)$ can then be further lowered with the aid of $a_{kj}(z)$ by using row operations $T_{lk}(a(z))$. The resulting new matrix $\tilde{A}(z)$ is still row proper, but $\tilde{A}^{\mathrm{r}}(z)$ may differ from the original $A^{\mathrm{r}}(z)$. If the above operations are appropriately performed column by column starting with the first column, a final matrix $A(z)$ having the corresponding matrix $A^{\mathrm{r}}(z)$ of CUT-form is obtained.

The matrix constructed above has an interesting form. In particular, if $B(z)$ is square and $\det B(z) \neq 0$, then the entries of the corresponding $A(z)$ satisfy

65
$$i<j \Rightarrow \partial(a_{ij}(z)) < \partial(a_{jj}(z)) \quad \text{and} \quad \partial(a_{ij}(z)) \leq \partial(a_{ii}(z)),$$
$$i>j \Rightarrow \partial(a_{ij}(z)) < \partial(a_{jj}(z)) \quad \text{and} \quad \partial(a_{ij}(z)) < \partial(a_{ii}(z)).$$

Put in another way, (65) means that the matrix having as its rows the leading coefficients of the rows of $A(z)$ is upper triangular, whereas the matrix having as its columns the leading coefficients of the columns of $A(z)$ is diagonal ($=I$). Thus $A(z)$ is both row and column proper, and the degree of any row i is equal to the degree of column i and equal to the degree of the diagonal entry $a_{ii}(z)$.

The form of polynomial matrices constructed above is called the *canonical row proper form* (CRP-*form*) because matrices of CRP-form constitute a set of canonical forms for the $\mathbf{C}[z]$-row equivalence.

66 **Theorem.** *If $A(z)$, $B(z) \in \mathbf{C}[z]^{m\times n}$ are of CRP-form and $\mathbf{C}[z]$-row equivalent, then $A(z) = B(z)$.* □

67 **Proof.** Obviously, the rank of a matrix of CRP-form is equal to the number of nonzero rows. Thus the proof can be brought to the case of matrices without zero rows in quite the same way as in proof (35). So suppose that the matrices $A(z)$ and $B(z)$ are of full rank m.

If $A(z) = P(z)B(z)$ with $P(z) \in \mathbf{C}[z]^{m\times m}$ $\mathbf{C}[z]$-unimodular, then $A^{\mathrm{r}}(z) = P'(z)B^{\mathrm{r}}(z)$ where $P'(z) \in \mathbf{C}[z]^{m\times m}$ is the $\mathbf{C}[z]$-unimodular matrix defined by (56). Because $A^{\mathrm{r}}(z)$ and $B^{\mathrm{r}}(z)$ are of CUT-form they must be equal, i.e. $P'(z) = I$. Hence the diagonal entries of $P(z)$ are equal to 1 and $A(z)$ and $B(z)$ have the same row degrees.

Suppose then that in the ith row of $P(z)$ there exist nonzero off-diagonal entries. Let n_i be their highest degree and suppose that $\partial(p_{ij}(z)) = n_i$ with $i \neq j$. Moreover, let $b_{jk}(z)$ be the first entry of degree $\partial_{\mathrm{r}j}(B(z))$ in the jth row of $B(z)$. Then

68
$$\begin{aligned}\partial(a_{ik}(z)) &= \partial(p_{ij}(z)b_{jk}(z)) = n_i + \partial_{\mathrm{r}j}(B(z)) \\ &= n_i + \partial_{\mathrm{r}j}(A(z)) \geq \partial_{\mathrm{r}j}(A(z)).\end{aligned}$$

Because $a_{jk}(z)$ is of degree $\partial_{\mathrm{r}j}(A(z))$ this cannot be true. Thus $P(z)$ must be diagonal. Consequently $A(z) = B(z)$. □

Every matrix $A(z) \in \mathbf{C}[z]^{m\times n}$ is $\mathbf{C}[z]$-row equivalent to a unique matrix of CRP-form called the CRP-form of $A(z)$ (see also Beghelli and Guidorzi, 1976; Forney, 1975).

A corresponding canonical row proper form exists where the matrix having as its rows the leading coefficients of the rows of the form is lower triangular. We shall occasionally use the notation CRP also for this form. This slight ambiguity will cause no serious harm.

Canonical forms of matrices over $\mathbf{C}(z]$

Let $A(z)$ be an m by n-matrix over the ring $\mathbf{C}(z]$ of quotients. $A(z)$ is always $\mathbf{C}(z]$-row equivalent to a matrix $B(z) \in \mathbf{C}[z]$, which can be obtained, for instance, by multiplying $A(z)$ by a common denominator of the entries of $A(z)$. Then $B(z)$ can be brought to CUT-form by elementary $\mathbf{C}[z]$-row operations. The resulting matrix can further be reduced to CUT-form where the entries in each row and above an entry which is the last nonzero entry in its column and the first nonzero entry in its row are of lower degree than the first nonzero entry in the row in question. For if this does not hold we can multiply the row in question by a sufficiently large power of z, bring the matrix to CUT-form again and thus obtain a matrix which has the desired property.

Obviously, the form of matrices constructed above is not a canonical form for the $\mathbf{C}(z]$-row equivalence. There is, however, another form, which is a canonical form. This can be constructed as follows. First write the matrix $A(z) \in \mathbf{C}(z]^{m\times n}$ to the form $\tilde{A}(1/z) \in \mathbf{C}(1/z]^{m\times n}$. Next, bring the matrix $\tilde{A}(1/z)$ to the CUT-form presented in the previous paragraph. Then multiply the rows by the least possible powers of z, bringing the first nonzero entries to elements $a(z) \in \mathbf{C}[z]$ such that $a(0) = 1$. Finally the degrees of the entries above an entry which is the last nonzero entry in its column and the first nonzero entry in its row can be reduced by the division algorithm of $\mathbf{C}[z]$.

The matrix constructed above has the following structural properties. The entries of those columns whose last nonzero entry is the first nonzero entry of its row are elements of $\mathbf{C}[z]$ and the entries above the last nonzero entry are of lower degree than the last one. Furthermore, the last nonzero entries $a(z)$ are normalized in the sense that $a(0) = 1$. The entries in other columns are not necessarily elements of $\mathbf{C}[z]$ but elements of $\mathbf{C}(1/z] = \mathbf{C}(z]$. This form of matrix over $\mathbf{C}(z]$ is also called the *canonical upper triangular form* (CUT-*form*, see Hirvonen, Blomberg, Ylinen, 1975). That it is really a canonical form for the $\mathbf{C}(z]$-row equivalence is shown by the following theorem.

69 **Theorem.** *If $A(z)$, $B(z) \in \mathbf{C}(z]^{m\times n}$ are of CUT-form and $\mathbf{C}(z]$-row equivalent, then $A(z) = B(z)$.* □

70 **Proof.** The proof can be carried out by mathematical induction in almost the same way as proof (35), and we refer to it. In particular, if $A(z)$ and $B(z)$ are of full rank m and $A(z) = P(z)B(z)$ for some $\mathbf{C}(z]$-unimodular $P(z)$, using the induction assumptions we can first show that $P(z)$ is upper triangular with diagonal entries equal to 1 and nonzero off-diagonal entries only in the last column. If i is the index of the first nonzero entry in the last row of $A(z)$, the equation corresponding to (39) obtained from the ith column is

71
$$\begin{bmatrix} a_{1i}(z) \\ \vdots \\ a_{m-1,i}(z) \\ a_{mi}(z) \end{bmatrix} = \begin{bmatrix} b_{1i}(z) + p_{1m}(z)b_{mi}(z) \\ \vdots \\ b_{m-1,i}(z) + p_{m-1,m}(z)b_{mi}(z) \\ b_{mi}(z) \end{bmatrix}.$$

Because $a_{ji}(z)$, $b_{ji}(z) \in \mathbf{C}[z]$ for each $j \in \{1, 2, \ldots, m-1\}$ and $b_{ki}(0) = 1$, $p_{jm}(z)$ must be an element of $\mathbf{C}[z]$ for each $j \in \{1, 2, \ldots, m-1\}$. Furthermore, because the entries $a_{1i}(z), a_{2i}(z), \ldots, a_{m-1,i}(z)$ are of lower degree than $a_{mi}(z)$, and $b_{1i}(z), b_{2i}(z), \ldots, b_{m-1,i}(z)$ are of lower degree than $b_{mi}(z)$, it must be $p_{1m}(z) = p_{2m}(z) = \ldots = p_{m-1,m}(z) = 0$. □

Coprime polynomial matrices

Let $A(z) \in \mathbf{C}[z]^{m\times m}$, $B(z) \in \mathbf{C}[z]^{m\times m}$ be arbitrary. According to note (A1.94) $A(z)$ and $B(z)$ are left coprime if and only if the matrix $[A(z) \vdots B(z)]$ is $\mathbf{C}[z]$- column equivalent to $[I \vdots 0]$, in which case $[I \vdots 0]$ is the Smith form of $[A(z) \vdots B(z)]$. However, there are many other possibilities for testing the coprimeness (*cf.* Rosenbrock, 1970; Barnett, 1971). In particular, there are methods based on the identification of polynomials with polynomial functions.

72 **Note.** **(i)** $A(z) \in \mathbf{C}[z]^{m\times m}$ and $B(z) \in \mathbf{C}[z]^{m\times n}$ are left coprime if and only if for each zero $c \in \mathbf{C}$ of det $A(z)$ the matrix $[A(c) \vdots B(c)]$ is of full rank m. Suppose first that $[A(z) \vdots B(z)]$ is column equivalent to $[I \vdots 0]$. Because every $\mathbf{C}[z]$-unimodular matrix $P(z)$ is such that $P(c)$ is $\mathbf{C}$-unimodular for each $c \in \mathbf{C}$, the matrices $[A(c) \vdots B(c)]$ and $[I \vdots 0]$ are $\mathbf{C}$-column equivalent and have the same rank m for each $c \in \mathbf{C}$.

Conversely, let $[A(z) \vdots B(z)]$ be column equivalent to $[L(z) \vdots 0]$ with $L(z) \in \mathbf{C}[z]^{m\times m}$. Because det $L(z)$ is an mth elementary divisor of $[A(z) \vdots B(z)]$, every zero $c \in \mathbf{C}$ of det $L(z)$ is a zero of det $A(z)$. Hence for each zero c of det $L(z)$ the matrix $[A(c) \vdots B(c)]$ and further $[L(c) \vdots 0]$ are of full rank m, which leads to a contradiction. Thus det $L(z)$ cannot have any zero, i.e. det $L(z) \in \mathbf{C}^*$ and $L(z)$ is $\mathbf{C}[z]$-unimodular.

(ii) The matrices $zI - F \in \mathbf{C}[z]^{m\times m}$ with $F \in \mathbf{C}^{m\times m}$, and $G \in \mathbf{C}^{m\times n}$ are left coprime (in $\mathbf{C}[z]^{m\times m}$) if and only if the m by kn-matrix

73 $$Q = [G \vdots FG \vdots \ldots \vdots F^{k-1}G],$$

where k is any integer not less than the degree of the minimal polynomial of F, has rank m.

Suppose first that $zI - F$ and G are not left coprime. Then there exists a zero $c \in \mathbf{C}$ of det $(zI - F)$, i.e. an eigenvalue of F such that the matrix $[cI - F \vdots G]$ is not of full rank m. Thus there exists a nonzero $d = (d_1, d_2, \ldots, d_m) \in \mathbf{C}^m$ such that

74 $$d^{\mathrm{T}}[cI - F \vdots G] = [0 \vdots 0].$$

Using this equation we obtain

75 $$d^{\mathrm{T}}[G \vdots FG \vdots \ldots \vdots F^{k-1}G] = d^{\mathrm{T}}[G \vdots cG \vdots \ldots \vdots c^{k-1}G] = [0 \vdots 0 \vdots \ldots \vdots 0],$$

which implies that the matrix (73) is not of full rank m.

Suppose then that $zI - F$ and G are left coprime and take an arbitrary nonzero $d \in \mathbf{C}^m$. If $d^{\mathrm{T}}G \neq 0$ then $d^{\mathrm{T}}Q \neq 0$. If $d^{\mathrm{T}}G = 0$, take $q \leqslant k$ and a nonzero $e = (e_1, e_2, \ldots, e_q) \in \mathbf{C}^q$ such that $d^{\mathrm{T}}, d^{\mathrm{T}}F, \ldots, d^{\mathrm{T}}F^{q-1}$ are $\mathbf{C}$-linearly independent and

76 $$d^{\mathrm{T}}F^q + \sum_{i=1}^{q} e_i d^{\mathrm{T}}F^{i-1} = 0.$$

Take $c \in \mathbf{C}$ satisfying

77 $$e_1 + e_2 c + \ldots + e_q c^{q-1} + c^q = 0$$

and let $f = (f_1, f_2, \ldots, f_q) \in \mathbf{C}^q$ be the unique solution of the equations

78 $$\begin{cases} cf_1 & = e_1, \\ cf_2 - f_1 & = e_2, \\ \vdots & \\ cf_q - f_{q-1} & = e_q, \\ \quad -f_q & = 1. \end{cases}$$

Substituting these in (76) and rearranging the terms gives

79 $$\underbrace{(f_1 d^{\mathrm{T}} + f_2 d^{\mathrm{T}} F + \ldots + f_q d^{\mathrm{T}} F^{q-1})}_{\triangleq a^{\mathrm{T}}} (cI - F) = 0.$$

Because $f_q = -1$ and d^{T}, $d^{\mathrm{T}}F$, . . . , $d^{\mathrm{T}}F^{q-1}$ are **C**-linearly independent, a is nonzero. Hence det $(cI - F) = 0$. If $d^{\mathrm{T}}Q = 0$, then $a^{\mathrm{T}}G = 0$, which together with (79) implies that $zI - F$ and G cannot be coprime, a contradiction. Thus $d^{\mathrm{T}}Q \neq 0$, and because d was arbitrary, Q must be of full rank m. □

Canonical forms for the quotient equivalence. Transfer equivalence

According to appendix A1 the pairs $(B_1(z), A_1(z))$ and $(B_2(z), A_2(z))$ as well as the partitioned matrices $[A_1(z) \vdots B_1(z)]$ and $[A_2(z) \vdots B_2(z)]$ with $A_1(z)$, $A_2(z) \in \mathbf{C}[z]^{m \times m}$ $\mathbf{C}(z)$-unimodular and $B_1(z)$, $B_2(z) \in \mathbf{C}[z]^{m \times n}$ are said to be quotient equivalent if

$$A_1(z)^{-1}B_1(z) = A_2(z)^{-1}B_2(z).$$

Let $L_1(z) \in \mathbf{C}[z]^{m \times m}$ be a GCLD of $A_1(z)$ and $B_1(z)$ and $L_2(z) \in \mathbf{C}[z]^{m \times m}$ be a GCLD of $A_2(z)$ and $B_2(z)$. Write

80 $$[A_1(z) \vdots B_1(z)] = L_1(z)[A_{11}(z) \vdots B_{11}(z)].$$

81 $$[A_2(z) \vdots B_2(z)] = L_2(z)[A_{21}(z) \vdots B_{21}(z)]$$

with $A_{11}(z)$, $B_{11}(z)$ left coprime and $A_{21}(z)$, $B_{21}(z)$ left coprime (*cf.* note (A1.94)). Then $[A_1(z) \vdots B_1(z)]$ and $[A_2(z) \vdots B_2(z)]$ are quotient equivalent if and only if $[A_{11}(z) \vdots B_{11}(z)]$ and $[A_{21}(z) \vdots B_{21}(z)]$ are $\mathbf{C}[z]$-row equivalent (*cf.* note (A1.94), (iii)).

Then bring the matrices $[A_{11}(z) \vdots B_{11}(z)]$ and $[A_{21}(z) \vdots B_{21}(z)]$ to a canonical form, say to CUT-form, for the row equivalence by multiplying them by $\mathbf{C}[z]$-unimodular matrices $P_1(z)$ and $P_2(z)$, respectively. We obtain

82 $$[A_1(z) \vdots B_1(z)] = L_1(z)P_1(z)^{-1}\underbrace{P_1(z)[A_{11}(z) \vdots B_{11}(z)]}_{\triangleq [A_{11}(z) \vdots B_{11}(z)]^*},$$

83 $$[A_2(z) \vdots B_2(z)] = L_2(z)P_2(z)^{-1}\underbrace{P_2(z)[A_{21}(z) \vdots B_{21}(z)]}_{\triangleq [A_{21}(z) \vdots B_{21}(z)]^*},$$

where $L_1(z)P_1(z)^{-1}$ and $L_2(z)P_2(z)^{-1}$ are GCLD's of $[A_1(z) \vdots B_1(z)]$ and $[A_2(z) \vdots B_2(z)]$, respectively. Thus $[A_1(z) \vdots B_1(z)]$ and $[A_2(z) \vdots B_2(z)]$ are quotient equivalent if and only if $[A_{11}(z) \vdots B_{11}(z)]^* = [A_{21}(z) \vdots B_{21}(z)]^*$.

Consequently the set of matrices $[A(z) \vdots B(z)]$ of CUT-form with $A(z)$ $\mathbf{C}(z)$-unimodular and $A(z)$, $B(z)$ left coprime constitutes a set of canonical forms for the quotient equivalence.

Note that when we have a $\mathbf{C}[z]$-module X and we are considering an input-output relation, i.e. a set of pairs $(u, y) \in X^n \times X^m$ satisfying a $\mathbf{C}[z]$-module equation of the form

84 $$A(z)y = B(z)u$$

with $\mathbf{C}(z)$-unimodular $A(z) \in \mathbf{C}[z]^{m\times m}$, $B(z) \in \mathbf{C}[z]^{m\times n}$ we call the matrices $[A(z) \vdots -B(z)]$ and $A(z)^{-1}B(z)$ a *generator* and a *transfer matrix* respectively. Furthermore the generators $[A_1(z) \vdots -B_1(z)]$ and $[A_2(z) \vdots -B_2(z)]$ are said to be *transfer equivalent* if they, or equivalently $[A_1(z) \vdots B_1(z)]$ and $[A_2(z) \vdots B_2(z)]$, are quotient equivalent.

Proper and strictly proper rational matrices

The matrices over a ring of quotients of $\mathbf{C}[z]$ are usually called rational matrices. A rational matrix $\mathscr{G}(z) \in \mathbf{C}(z)^{m\times n}$ is said to be *proper* if all of its entries are proper and *strictly proper* if all of its entries are strictly proper. By the divison algorithm every $\mathscr{G}(z) \in \mathbf{C}(z)^{m\times n}$ can be uniquely presented as a sum

85 $$\mathscr{G}(z) = \mathscr{G}^0(z) + K(z)$$

with $\mathscr{G}^0(z) \in \mathbf{C}(z)^{m\times n}$ strictly proper, $K(z) \in \mathbf{C}[z]^{m\times n}$.

As was shown in appendix A1, a rational matrix $\mathscr{G}(z) \in \mathbf{C}(z)^{m\times n}$ can always be factored as follows

86 $$\mathscr{G}(z) = A(z)^{-1}B(z) = V(z)T(z)^{-1},$$

where $A(z) \in \mathbf{C}[z]^{m\times m}$, $B(z) \in \mathbf{C}[z]^{m\times n}$, $T(z) \in \mathbf{C}[z]^{n\times n}$, $V(z) \in \mathbf{C}[z]^{m\times n}$.

87 **Note.** **(i)** Let $A(z) \in \mathbf{C}[z]^{m\times m}$ be $\mathbf{C}(z)$-unimodular and $B(z) \in \mathbf{C}[z]^{m\times n}$ and denote the row degrees of $A(z)$ and $B(z)$ by $r_1, r_2, \ldots, r_m$ and $s_1, s_2, \ldots, s_m$, respectively. If $\mathscr{G}(z) = A(z)^{-1}B(z)$ is proper (strictly proper), then $s_i \leqslant r_i$ $(s_i < r_i)$ for each $i \in \mathbf{m}$. Obviously, if $A(z)$ is row proper and $s_i \leqslant r_i$ $(s_i < r_i)$ for each $i \in \mathbf{m}$, then $\mathscr{G}(z)$ is proper (strictly proper).

Write $\mathscr{G}(z)$ in the form $\mathscr{G}(z) = 1/d(z)\, E(z)$ where $d(z) \in \mathbf{C}[z]^*$ is a common denominator of the entries of $\mathscr{G}(z)$ and $E(z) \in \mathbf{C}[z]^{m\times n}$. Then

88 $$d(z)\, B(z) = A(z)\, E(z),$$

that is

89 $$d(z)b_{ij}(z) = \sum_{k=1}^{m} a_{ik}(z)\, e_{kj}(z), \quad i \in \mathbf{m}, \quad j \in \mathbf{n},$$

which further implies that

90 $$\partial(d(z)) + \partial(b_{ij}(z)) \leqslant \max\{\partial(a_{ik}(z)) + \partial(e_{kj}(z)) \mid k \in \mathbf{m}\}, i \in \mathbf{m}, j \in \mathbf{n}.$$

Suppose first that $\mathscr{G}(z)$ is proper (strictly proper), i.e.

$$\partial(e_{kj}(z)) \overset{(<)}{\leqslant} \partial(d(z))$$

for all $k \in \mathbf{m}$, $j \in \mathbf{n}$. Thus

91 $$\partial(d(z)) + \partial(b_{ij}(z)) \overset{(<)}{\leqslant} \max\{a_{ik}(z) \mid k \in \mathbf{m}\} + \partial(d(z)) = r_i + \partial(d(z)), \; i \in \mathbf{m}, j \in \mathbf{n}$$

which implies that

$$s_i \overset{(<)}{\leqslant} r_i$$

for each $i \in \mathbf{m}$.

Conversely, suppose that $A(z)$ is row proper and

$$s_i \overset{(<)}{\leqslant} r_i$$

for each $i \in \mathbf{m}$. Suppose further that $\mathscr{G}(z)$ is not proper (strictly proper). Then there exists a column, say the jth, of $E(z)$ such that

$$\partial_{cj}(E(z)) \overset{(\geqslant)}{>} \partial(d(z)).$$

Because $A(z)$ is row proper there must exist $i \in \mathbf{m}$ such that

92 $$\partial\left(\sum_{k=1}^{m} a_{ik}(z)\, e_{kj}(z)\right) = r_i + \partial_{cj}(E(z)) \overset{(>)}{>} s_i + \partial(d(z)),$$

which contradicts the condition

93 $$\partial\left(\sum_{k=1}^{m} a_{ik}(z)\, e_{kj}(z)\right) = \partial(d(z)\, b_{ij}(z)) \leqslant \partial(d(z)) + s_i$$

obtained from (89). Hence $\mathscr{G}(z)$ is proper (strictly proper).

(ii) Let $T(z) \in \mathbf{C}[z]^{n \times n}$ be $\mathbf{C}(z)$-unimodular and $V(z) \in \mathbf{C}[z]^{m \times n}$ and denote the column degrees of $T(z)$ and $V(z)$ by $r_1, r_2, \ldots, r_n$ and $s_1, s_2, \ldots, s_n$, respectively. If $\mathscr{G}(z) = V(z)\, T(z)^{-1}$ is proper (strictly proper), then

$$r_i \overset{(>)}{\geqslant} s_i$$

for each $i \in \mathbf{n}$. Conversely, if $T(z)$ is column proper and

$$r_i \overset{(>)}{\geqslant} s_i$$

for each $i \in \mathbf{n}$, then $\mathscr{G}(z)$ is proper (strictly proper). □

Division algorithms for polynomial matrices

For matrices over $\mathbf{C}[z]$ there exist a number of different division algorithms (*cf.* Barnett, 1971). In the following note we shall consider some of them.

94 **Note.** **(i)** Let $A(z) \in \mathbf{C}[z]^{m \times m}$ have an invertible leading coefficient and $B(z) \in \mathbf{C}[z]^{m \times n}$ be arbitrary. Then there exist unique matrices $Q(z), R(z) \in \mathbf{C}[z]^{m \times n}$ such that

95 $$B(z) = A(z)\, Q(z) + R(z), \quad \partial(R(z)) < \partial(A(z)).$$

This can be proven in quite the same way as the division algorithm for polynomials.
(ii) Let $A(z) \in \mathbf{C}[z]^{m\times m}$ be $\mathbf{C}(z)$-unimodular and $B(z) \in \mathbf{C}[z]^{m\times n}$ be arbitrary. Then there exist $Q(z)$, $R(z) \in \mathbf{C}[z]^{m\times n}$ such that

96 $$B(z) = A(z)\,Q(z) + R(z),\ \partial_{ri}(R(z)) < \partial_{ri}(A(z)) \quad \text{for each } i \in \mathbf{m}.$$

If $A(z)$ is row proper, then $Q(z)$ and $R(z)$ are unique. This can be shown as follows. Write

97 $$A(z)^{-1}B(z) = \mathcal{G}^0(z) + Q(z),$$

where $\mathcal{G}^0(z) \in \mathbf{C}(z)^{m\times n}$ is strictly proper and $Q(z) \in \mathbf{C}[z]^{m\times n}$. Then $R(z) \in \mathbf{C}[z]^{m\times n}$ defined by

98 $$R(z) = B(z) - A(z)Q(z)$$

satisfies

99 $$\mathcal{G}^0(z) = A(z)^{-1}R(z).$$

Hence note (87), (i) implies that $\partial_{ri}(R(z)) < \partial_{ri}(A(z))$ for each $i \in \mathbf{m}$.

If $A(z)$ is row proper, then the condition for row degrees implies that $A(z)^{-1}R(z)$ is strictly proper. This further implies that there is exactly one pair $Q(z)$, $R(z)$ satisfying

100 $$A(z)^{-1}B(z) = R(z)^{-1}B(z) + Q(z).$$

Note that if we add to (95) the condition that $A(z)^{-1}R(z)$ should be strictly proper, which is equivalent to the condition

101 $$\partial(\operatorname{adj} A(z))\,R(z) < \partial(\det A(z)),$$

we always have unique $Q(z)$, $R(z)$ irrespective of whether $A(z)$ is row proper or not.
(iii) The division algorithms in items (i) and (ii) are left ones and $Q(z)$ and $R(z)$ in

102 $$B(z) = A(z)\,Q(z) + R(z)$$

are a left quotient and a left remainder, respectively, of $B(z)$ by $A(z)$. Of course, we have also the corresponding right division algorithms. In particular, if $A(z) \in \mathbf{C}[z]^{n\times n}$ is $\mathbf{C}(z)$-unimodular and $B(z) \in \mathbf{C}[z]^{m\times n}$, then there exist $Q(z)$, $R(z) \in \mathbf{C}[z]^{m\times n}$ such that

103 $$B(z) = Q(z)\,A(z) + R(z),\ \partial_{ci}(R(z)) < \partial_{ci}(A(z)) \quad \text{for each } i \in \mathbf{n},$$

and if $A(z)$ is column proper then $Q(z)$ and $R(z)$ are unique. □

A3

Polynomials and rational forms in an endomorphism

Let X be a nontrivial vector space over the field $\mathbf{C}$ of complex numbers. The set $L(X, X)$ of all vector space endomorphisms (denoted also by $\mathrm{end}_{\mathbf{C}}(X)$) together with the pointwise addition

1 $$(\phi + \psi)(x) = \phi(x) + \psi(x)$$

and the multiplication by composition

2 $$\phi\psi = \phi \circ \psi$$

is a ring with the identity mapping 1 as its unity. The field $\mathbf{C}$ can be considered as a subring of $L(X, X)$ by identifying each $a \in \mathbf{C}$ with the corresponding mapping $x \mapsto ax$. Hence $L(X, X)$ with scalar multiplication $(a, \phi) \mapsto a\phi$ is a vector space over $\mathbf{C}$, and because each $a \in \mathbf{C}$ commute with the elements of $L(X, X)$, it holds that

3 $$a(\phi\psi) = (a\phi)\psi = \phi(a\psi)$$

for all $\phi, \psi \in L(X, X)$. This implies that $L(X, X)$ with the operations above is an *algebra* over $\mathbf{C}$ (see Greub, 1963).

Let $\sigma \in L(X, X)$ be arbitrary. The ring $\mathbf{C}[\sigma]$ of polynomials in σ over $\mathbf{C}$ is a commutative subring of $L(X, X)$ and contains the unity 1 of $L(X, X)$. Because $\mathbf{C}$ is a subring of $\mathbf{C}[\sigma]$, $\mathbf{C}[\sigma]$ is also a subspace of $L(X, X)$, thus it is even a subalgebra of $L(X, X)$.

Suppose then that the endomorphism σ is an indeterminate over $\mathbf{C}$, which is equivalent to saying that the set of powers $\{\sigma^0, \sigma^1, \sigma^2, \ldots\}$ of σ is linearly independent over $\mathbf{C}$. Then $\mathbf{C}[\sigma]$ is an integral domain satisfying the division algorithm, hence a principal ideal domain. Furthermore, every subset which is closed under multiplication and does not contain the zero of $\mathbf{C}[\sigma]$ is a denominator set. In addition to the set of all nonzero elements $\mathbf{C}[\sigma]^*$ and the set of all powers of σ, for instance the sets of all epimorphisms, of all monomorphisms and especially of all automorphisms belonging to $\mathbf{C}[\sigma]$ are considerable denominator sets.

$\mathbf{C}[\sigma]$-module

Because X is a vector space over $\mathbf{C}$, X is an additive abelian group. Furthermore, the scalar multiplication $\mathbf{C}[\sigma] \times X \to X$ defined by

4 $$(a(\sigma), x) \mapsto a(\sigma)x$$

satisfies the axioms (A1.27), . . . , (A1.30) for a module. Hence X is a left module over $\mathbf{C}[\sigma]$. For each $a(\sigma) \in \mathbf{C}[\sigma]$ the endomorphism of the $\mathbf{C}[\sigma]$-module X defined by the scalar multiplication $x \mapsto a(\sigma)x$ is an epimorphism, a monomorphism or an isomorphism of $\mathbf{C}[\sigma]$-modules if and only if $a(\sigma)$ is an epimorphism, a monomorphism or an isomorphism, respectively, of vector spaces over $\mathbf{C}$.

Let S be a denominator set in $\mathbf{C}[\sigma]$ and $\mathbf{C}[\sigma]_S$ the corresponding ring of quotients. According to appendix A1 X is a module over $\mathbf{C}[\sigma]_S$ such that for each $s(\sigma)a(\sigma)/s(\sigma)) \in \mathbf{C}[\sigma]_S, x \in X$, the scalar multiplication satisfies

5 $$(s(\sigma)a(\sigma)/s(\sigma))x = a(\sigma)x$$

if and only if every $s(\sigma) \in S$ is an automorphism, i.e. belongs to $\mathrm{aut}_{\mathbf{C}}(X)$. In particular, X is a module over the ring of quotients of $\mathbf{C}[\sigma]$ by the set $\mathbf{C}[\sigma] \cap \mathrm{aut}_{\mathbf{C}}(X)$. Further, X is a vector space over the field $\mathbf{C}(\sigma)$ of quotients of $\mathbf{C}[\sigma]$ if and only if every $a(\sigma) \in \mathbf{C}[\sigma]^*$ is an automorphism (*cf.* Blomberg and Salovaara, 1968).

If S is a denominator set in $\mathbf{C}[\sigma]$ such that every $a(\sigma) \in S$ is an automorphism of X, then each $b(\sigma)/a(\sigma) \in \mathbf{C}[\sigma]_S$ defines an endomorphism of X by the assignment $x \mapsto b(\sigma)/a(\sigma)x$. This endomorphism is equal to the endomorphism $a(\sigma)^{-1} \circ b(\sigma)$, where $a(\sigma)^{-1}$ is the inverse automorphism of X. The assignment $b(\sigma)/a(\sigma) \mapsto a(\sigma)^{-1} \circ b(\sigma)$ defines an isomorphism of rings between $\mathbf{C}[\sigma]_S$ and the subring

6 $$\{\phi \mid \phi \in L(X, X) \text{ and there exists } a(\sigma) \in S, b(\sigma) \in \mathbf{C}[\sigma] \text{ such that } \phi = a(\sigma)^{-1} \circ b(\sigma)\}$$

of $L(X, X)$ (*cf.* Blomberg and Salovaara, 1968).

If every element of a denominator set S is a monomorphism but not an automorphism, then X can be identified with a $\mathbf{C}[\sigma]$-submodule of X_S by the assignment $x \mapsto x/1$. In this case the extension of X to X_S makes the elements of S surjective.

In general, the kernel of the morphism $x \mapsto x/1$ consists of all elements $x \in X$ such that $a(\sigma)x = 0$ for some $a(\sigma) \in S$. Thus, if some element of S is not a monomorphism, the morphism $x \mapsto x/1$ cannot be a monomorphism.

A4

The space $\mathfrak{D}$ of generalized functions

In this paragraph we shall try to give the reader a working knowledge of a specific class of generalized functions of particular importance in this context. The presentation basically follows the lines of the first few paragraphs of the introductory chapter I in Gelfand and Shilov, 1964.

The space $\mathcal{K}'$ of all complex-valued generalized functions on **R**

Let $\mathcal{K}$ denote the set of all complex-values infinitely continuously differentiable functions with bounded support defined on **R** (the support of a continuous function ϕ on **R** is the closure of the set $\{t | t \in \mathbf{R} \text{ and } \phi(t) \neq 0\}$). $\mathcal{K}$ is called a space of "test functions". It is a vector space over **C** with respect to the natural pointwise operations.

The set $\mathcal{K}'$ of all *complex-valued generalized functions* on **R** is then taken to mean the set of all linear continuous functionals $\mathcal{K} \to \mathbf{C}$. Linearity and continuity are understood in the sense explained in sections 1.2 and 1.3 of chapter I in Gelfand and Shilov, 1964.

$\mathcal{K}'$ is again a vector space over **C** with respect to the natural pointwise operations.

Let us collect a number of relevant properties of the elements of $\mathcal{K}'$ in the following note.

1 **Note.** (i) Let $f: \mathbf{R} \to \mathbf{C}$ be an ordinary function which is locally summable (i.e. f is absolutely integrable on every bounded interval of **R**). Then there is a unique corresponding element of $\mathcal{K}'$—call it also f—such that

2 $$f(\phi) = \int_{\mathbf{R}} f(\tau)\phi(\tau)\, d\tau$$

for every $\phi \in \mathcal{K}$.

Any generalized function $f \in \mathcal{K}'$ which allows a representation of the form (2) is said to be "regular". In the sequel we shall feel free to identify regular elements of $\mathcal{K}'$ with the corresponding ordinary functions $\mathbf{R}\to\mathbf{C}$. It will always be clear from the context which interpretation is meant.

The space C^∞ of all infinitely continuously differentiable complex-valued ordinary functions on $\mathbf{R}$ is an example of a well-behaved function space which can be identified with the corresponding subspace of regular elements of $\mathcal{K}'$. The same holds more generally for the space of all piecewise continuous functions $\mathbf{R}\to\mathbf{C}$. This space contains in particular the step functions also.

Let the (left continuous) *unit step* at $t\in\mathbf{R}$, $U_t:\mathbf{R}\to\mathbf{C}$ be given by

3 $$U_t(\tau) = \begin{cases} 0 \text{ for } \tau\in(-\infty, t], \\ 1 \text{ for } \tau\in(t,\infty). \end{cases}$$

There is then a corresponding regular generalized unit step $U_t\in\mathcal{K}'$ given by (*cf.* (2))

4 $$U_t(\phi) = \int_{\mathbf{R}} U_t(\tau)\phi(\tau)\,d\tau = \int_t^\infty \phi(\tau)\,d\tau$$

for every $\phi\in\mathcal{K}$.

(ii) Let $t\in\mathbf{R}$ and $k\in\{0,1,2,\ldots\}$ and consider the functional $\delta_t^{(k)}:\mathcal{K}\to\mathbf{C}$ given by

5 $$\delta_t^{(k)}(\phi) = (-1)^k p^k\phi(t)$$

for every $\phi\in\mathcal{K}$, where $p^k\phi\in\mathcal{K}$ denotes the ordinary kth-order derivative of ϕ (note that $p^k\phi(t)$ means $(p^k\phi)(t)$).

It is found that $\delta_t^{(k)}\in\mathcal{K}'$ for every t and k. $\delta_t^{(k)}$ is called the "delta function of order k at t". The delta function of order zero at t, i.e. $\delta_t^{(0)}$, is also called the Dirac delta function. For $\delta_t^{(0)}$ the simpler notation δ_t is also used.

Now $\delta_t^{(k)}$ does not allow a representation of the form (2). It is therefore called a "singular" generalized function. Even if there is consequently no ordinary function $f:\mathbf{R}\to\mathbf{C}$ such that $\delta_t^{(k)}(\phi) = (-1)^k p^k\phi(t)$ would be equal to $\int_{\mathbf{R}} f(\tau)\phi(\tau)d\tau$ for every $\phi\in\mathcal{K}$ it is customary to write *symbolically*

6 $$\delta_t^{(k)}(\phi) = (-1)^k p^k\phi(t) \triangleq \int_{\mathbf{R}} \delta_t^{(k)}(\tau)\phi(\tau)\,d\tau$$

for every $\phi\in\mathcal{K}$.

The integral sign appearing in (6) is said to mean "integration in the generalized sense".

The equalities (5) and (6) indicate the well-known "sifting property" of $\delta_t^{(k)}$ (*cf.* Zadeh and Desoer, 1963, A.2).

Note that the value $p^k\phi(t)$ appearing in (5) and (6) can be written as

7 $$p^k\phi(t) = -\int_t^\infty p^{k+1}\phi(\tau)\,d\tau = -\int_{\mathbf{R}} U_t(\tau)p^{k+1}\phi(\tau)\,d\tau.$$

As a result, (5) can be replaced by

8 $$\delta_t^{(k)}(\phi) = (-1)^{k+1}\int_{\mathbf{R}} U_t(\tau)p^{k+1}\phi(\tau)\,d\tau$$

for every $\phi\in\mathcal{K}$.

(iii) Let $x\in\mathcal{K}'$ be an arbitrary generalized function. Then the "generalized kth-order derivative" of x, denoted by p^kx, is defined as the functional $\mathcal{K}\to\mathbf{C}$ given by

9 $$p^k x(\phi) = (-1)^k x(p^k \phi)$$

for every $\phi \in \mathcal{K}$.

It is found that $p^k x \in \mathcal{K}'$, i.e. every generalized function has a generalized derivative of any order, and any such derivative is again a generalized function.

Suppose that the ordinary function $f: \mathbf{R} \to \mathbf{C}$ is continuous and that the ordinary kth-order derivative of f, $p^k f: \mathbf{R} \to \mathbf{C}$ is piecewise continuous. f and $p^k f$ are then locally summable, and they thus determine the corresponding regular generalized functions f and $p^k f \in \mathcal{K}'$ according to (2). We have

10 $$p^k f(\phi) = \int_{\mathbf{R}} (p^k f(\tau))\, \phi(\tau)\, d\tau$$

for every $\phi \in \mathcal{K}$.

Repeated application of integration by parts to the right-hand member of (10) yields

11 $$\int_{\mathbf{R}} (p^k f(\tau))\, \phi(\tau)\, d\tau = (-1)^k \int_{\mathbf{R}} f(\tau) p^k \phi(\tau)\, d\tau.$$

But the right-hand member of (11) is, by definition (*cf.* (2)), equal to $(-1)^k f(p^k \phi)$, and so (10) can be replaced by

12 $$p^k f(\phi) = (-1)^k f(p^k \phi)$$

for every $\phi \in \mathcal{K}$. Comparison of (12) with (9) then shows that the generalized function $p^k f \in \mathcal{K}'$ as given by (12) and determined by the ordinary kth-order derivative of the ordinary function $f: \mathbf{R} \to \mathbf{C}$ in this case coincides with the generalized kth-order derivative as given by (9) of the generalized function $f \in \mathcal{K}'$ determined by the ordinary function $f: \mathbf{R} \to \mathbf{C}$. Ordinary and generalized derivatives are thus compatible in a certain sense (for further details, see Gelfand and Shilov, 1964, chapter I).

Consider then the ordinary unit step at $t \in \mathbf{R}$, $U_t: \mathbf{R} \to \mathbf{C}$ as given by (3) and the corresponding generalized unit step $U_t \in \mathcal{K}'$ as given by (4). The ordinary kth-order derivative of the ordinary unit step does not exist, but the generalized kth-order derivative $p^k U_t \in \mathcal{K}'$ of the generalized unit step $U_t \in \mathcal{K}'$ does exist and it is given by (*cf.* (9))

13 $$p^k U_t(\phi) = (-1)^k U_t(p^k \phi)$$

for every $\phi \in \mathcal{K}$. Using (4), the right-hand member of (13) can be written as

14 $$(-1)^k U_t(p^k \phi) = (-1)^k \int_{\mathbf{R}} U_t(\tau) p^k \phi(\tau)\, d\tau,$$

and (13) can thus be replaced by

15 $$p^k U_t(\phi) = (-1)^k \int_{\mathbf{R}} U_t(\tau) p^k \phi(\tau)\, d\tau$$

for every $\phi \in \mathcal{K}$. From (8) and (15) it then follows that

16 $$\delta_t^{(k)} = p^{k+1} U_t,$$

i.e. the delta function of order k at t is identical with the generalized $(k+1)$th-order derivative of the generalized unit step at t. It also follows that in the generalized sense

17 $$\delta_t^{(k+1)} = p\delta_t^{(k)}$$

for every $k \in \{0, 1, 2, \ldots\}$, and

18 $$\delta_t = pU_t.$$

(iv) It is also possible, under certain conditions, to define a generalized convolution of generalized functions as well as a product of a generalized function and an infinitely differentiable ordinary function. We shall, however, not use these operations in this context.
(v) The space $\mathcal{K}'$ of generalized functions introduced above is invariant with respect to the generalized differentiation operation defined in (9). If the elements of $\mathcal{K}'$ are regarded, in a generalized sense, as complex-valued functions defined on the time interval $T = \mathbf{R}$, then it follows that $\mathcal{K}'$ qualifies as a signal space as specified in section 4.1. □

The space $\mathcal{K}'$ is too rich

The space $\mathcal{K}'$ of generalized functions discussed above is, as signal space, much too rich for our purposes. The main difficulty here is that if x is an arbitrary element of $\mathcal{K}'$, and $t \in \mathbf{R}$, then it is generally not possible to give the term "the value of x at t" a reasonable and useful meaning, neither is it possible to divide x into two nonoverlapping segments x^- and x^+—with x^- representing x on the time interval up to the point t, and x^+ representing x on the time interval from t onwards—such that x is uniquely given as x^- followed by x^+. This difficulty can be avoided by choosing a suitable subspace of $\mathcal{K}'$ as the basic signal space. This will now be done.

The space $\mathcal{D}$ of piecewise infinitely regularly differentiable complex-valued generalized functions on **R**

We shall be interested in the subset $\mathcal{D}$ of $\mathcal{K}'$ characterized in the following way.

19 $\mathcal{D} \triangleq \{x | x \in \mathcal{K}'$, x and every generalized derivative of x of any order are of the general form $f + \Sigma_{i \in I} c_i \delta_{t_i}^{(k_i)}$, where the various quantities have the following meanings:
I is a suitable index set,
f is a regular element of $\mathcal{K}'$ corresponding to an ordinary piecewise continuous function $\mathbf{R} \to \mathbf{C}$,
$\delta_{t_i}^{(k_i)}$, $i \in I$, $t_i \in \mathbf{R}$, $k_i \in \{0, 1, 2, \ldots\}$ denotes a delta function of order k_i at t_i,
c_i, $i \in I$, is a complex number,
the set $\{i | i \in I, t_i \in \theta, c_i \neq 0\}$ contains, for every bounded interval $\theta \subset \mathbf{R}$, at most a finite number of elements$\}$.

Any element of $\mathcal{D}$ has thus at most a finite number of delta functions of various order on any bounded time interval.

As an example, let $x \in \mathcal{D}$ be of the form stated above, i.e.

20 $$x = f + \sum_{i \in I} c_i \delta_{t_i}^{(k_i)},$$

and let f have its discontinuities at the points $t_j \in \mathbf{R}$ for $j \in J$. Then the generalized derivative px of x is of the form

21 $$px = f_1 + \sum_{j \in J} (f(t_j+) - f(t_j-))\,\delta_{t_j} + \sum_{i \in I} c_i \delta_{t_i}^{(k_i+1)},$$

where f_1 again corresponds to an ordinary piecewise continuous function $\mathbf{R} \to \mathbf{C}$. It represents the ordinary derivative of f on $\mathbf{R} - \{t_j | j \in J\}$. $f(t_j+)$ and $f(t_j-)$ denote in the usual way the right-hand and left-hand limits respectively of the ordinary function f at t_j.

It is readily concluded that $\mathcal{D}$ is a subspace of $\mathcal{K}'$. The elements of $\mathcal{D}$ have generalized derivatives of any order, and the generalized differentiation operator, denoted by p, can be regarded as a linear mapping $\mathcal{D} \to \mathcal{D}$. Furthermore, the space C^∞ of infinitely continuously differentiable ordinary functions $\mathbf{R} \to \mathbf{C}$ can, and will, be identified with the corresponding subspace of $\mathcal{D}$. The zero of $\mathcal{D}$ is identified with the zero of C^∞.

The space $\mathcal{D}$ is here, for obvious reasons, called the space of "piecewise infinitely regularly differentiable complex-valued generalized functions on **R**" (not a very convenient term!).

$\mathcal{D}$ is a signal space

The above properties mean that $\mathcal{D}$ qualifies as a signal space in a generalized sense with regard to the conditions specified in section 4.1.

$\mathcal{D}$ is a signal space possessing the richness property (5.1.5)

Consider then a polynomial operator $a(p) : \mathcal{K}' \to \mathcal{K}'$ of the form

22 $$a(p) \triangleq a_0 + a_1 p + \ldots + a_n p^n, \quad a_n \neq 0$$

as given by (4.1.1).

Now it is known that the set of all solutions $y \in \mathcal{K}'$ to $a(p)y = 0$ forms an n-dimensional subspace of C^∞ (considered as a subspace of $\mathcal{K}'$; *cf.* Gelfand and Shilov, 1964, chapter I, section 2.6, and note (5.1.6), (i)), i.e. the extension of the signal space from C^∞ to $\mathcal{D}$ and further to $\mathcal{K}'$ does not bring about any new solutions y to $a(p)y = 0$. Hence

23 $$\ker a(p)_{\text{with respect to } \mathcal{D}} = \ker a(p)_{\text{with respect to } \mathcal{K}'}.$$

Regarding $\mathcal{K}'$ as the "largest signal space of complex-valued functions on **R**" we see that $\mathcal{D}$ possesses the richness property (5.1.5) in a generalized sense.

$\mathcal{D}$ is a regular signal space

$\mathcal{D}$ is clearly a nontrivial signal space. It is, moreover, also regular, because every polynomial operator $a(p) : \mathcal{D} \to \mathcal{D}$ of the form (22) is a surjection with respect to $\mathcal{D}$ (*cf.* note (5.1.4), (i)). This can be seen in the following way.

Let $x \in \mathcal{D}$ be arbitrary and given in the form

24 $$x = f + \sum_{i \in I} c_i \delta_{t_i}^{(k_i)}$$

corresponding to the specification given in (19), and consider the differential equation

25
$$a(p)y = x.$$

It is asserted that there is a solution $y \in \mathfrak{D}$ to (25) of the form

26
$$y = y_0 + \sum_{i \in I} y_i,$$

where y_0 satisfies $a(p)y_0 = f$, and where the y_i's satisfy $a(p)y_i = c_i\delta_{t_i}^{(k_i)}$.

Note first that if $n = 0$, then $a(p) = a_0$, a constant, and a solution y of the form (26) to (25) can be trivially found. So let us suppose that $n \geqslant 1$.

Now f was regular and it can be identified with an ordinary piecewise continuous function on $\mathbf{R}$. It is therefore clear that a regular $y_0 \in \mathfrak{D}$ can always be found such that $a(p)y_0 = f$. Such a y_0 can be determined using any standard method for solving ordinary linear differential equations.

Then let a $t_0 \in \mathbf{R}$ be fixed, pick one of the i's, $i \in I$, and consider the differential equation $a(p)y_i = c_i\delta_{t_i}^{(k_i)}$. It is clear that if $\eta_i \in \mathfrak{D}$ is such that $a(p)\eta_i = \delta_{t_i}$, then $y_i = c_i p^{k_i}\eta_i \in \mathfrak{D}$ satisfies $a(p)y_i = c_i\delta_{t_i}^{(k_i)}$. Thus it remains to find a solution $\eta_i \in \mathfrak{D}$ to $a(p)\eta_i = \delta_{t_i}$. This case is considered in Gelfand and Shilov, 1964, chapter I, section 5.3, and there it is shown that a regular and piecewise continuous solution η_i exists as given by

27
$$\eta_i = \begin{cases} \alpha_1 e_1 + \alpha_2 e_2 + \ldots + \alpha_n e_n \text{ on } (-\infty, t_i), \\ \beta_1 e_1 + \beta_2 e_2 + \ldots + \beta_n e_n \text{ on } (t_i, \infty), \end{cases}$$

where $\{e_1, e_2, \ldots, e_n\}$ is a regular basis for $\ker a(p)$, and where $\alpha_1, \alpha_2, \ldots, \alpha_n$ and $\beta_1, \beta_2, \ldots, \beta_n$ are suitable complex numbers. Denoting $\beta_i - \alpha_i \triangleq \gamma_i$ for $i = 1, 2, \ldots, n$, the γ_i's are required to satisfy

28
$$\begin{aligned} e_1(t_i)\gamma_1 + e_2(t_i)\gamma_2 + \ldots + e_n(t_i)\gamma_n &= 0, \\ pe_1(t_i)\gamma_1 + pe_2(t_i)\gamma_2 + \ldots + pe_n(t_i)\gamma_n &= 0, \\ &\cdots \\ p^{n-2}e_1(t_i)\gamma_1 + p^{n-2}e_2(t_i)\gamma_2 + \ldots + p^{n-2}e_n(t_i)\gamma_n &= 0, \\ p^{n-1}e_1(t_i)\gamma_1 + p^{n-1}e_2(t_i)\gamma_2 + \ldots + p^{n-1}e_n(t_i)\gamma_n &= \frac{1}{a_n}. \end{aligned}$$

The γ_i's are uniquely determined by (28), because the coefficient matrix is always nonsingular.

The solution η_i according to (27) can thus be constructed by choosing either $\alpha_1, \alpha_2, \ldots, \alpha_n$ or $\beta_1, \beta_2, \ldots, \beta_n$ freely, and then determining the other set so that the γ_i's with $\gamma_1 \triangleq \beta_1 - \alpha_1$ satisfy (28). We shall make the following choice:

29
$$\begin{aligned} &\text{If } t_i \geqslant t_0, \text{ then } \alpha_1 = \alpha_2 = \ldots = \alpha_n = 0, \text{ and} \\ &\text{if } t_i < t_0, \text{ then } \beta_1 = \beta_2 = \ldots = \beta_n = 0. \end{aligned}$$

We have thus determined a unique solution $y_i \in \mathfrak{D}$ to $a(p)y_i = c_i\delta_{t_i}^{(k_i)}$ for an arbitrary $i \in I$. The same construction can be applied for all $i \in I$.

Next consider the sum (26). It is seen that with the α_i's and β_i's chosen as explained above, the sum $y = y_0 + \Sigma_{i \in I} y_i$ contains only a finite number of nonzero terms on every bounded interval of $\mathbf{R}$. The sum therefore exists and it clearly satisfies $a(p)y = u$. Consequently, $\mathfrak{D}$ is a regular signal space.

Restrictions of generalized functions

The restriction of an ordinary function $\mathbf{R} \to \mathbf{C}$ to an arbitrary interval $T \subset \mathbf{R}$ is a well-known concept in the theory of functions. A corresponding concept can be defined for generalized functions, but generally only for open intervals $T \subset \mathbf{R}$. Ordinary and generalized restrictions are mutually compatible as far as regular generalized functions and their ordinary counterparts are concerned.

So let $T \subset \mathbf{R}$ be an open interval, and let $\mathcal{K}_T \subset \mathcal{K}$ denote the subspace of the space $\mathcal{K}$ of test functions consisting of all the elements of $\mathcal{K}$ having their support in T. The restriction of a generalized function $x \in \mathcal{K}'$ to $\mathcal{K}_T$ is said to be the "restriction of x to T", it is denoted by $x|T$ and it is defined as the functional $\mathcal{K}_T \to \mathbf{C}$ given by

30 $$x|T(\phi) = x(\phi)$$

for every $\phi \in \mathcal{K}_T$. Accordingly, we obtain the restricted spaces $C^\infty|T$, $\mathcal{D}|T$, and $\mathcal{K}'|T$ of generalized functions as the sets of all restrictions of the elements of C^∞, $\mathcal{D}$, and $\mathcal{K}'$ respectively to T. These sets are all vector spaces over $\mathbf{C}$ with respect to the operations of $\mathcal{K}'$ appropriately restricted. They possess essentially the same properties as the corresponding unrestricted spaces. $C^\infty|T$ can be identified with the corresponding space of ordinary functions restricted to T. The spaces $\mathcal{D}$ and $\mathcal{D}|T$, with $T \subset \mathbf{R}$ some open interval, are further discussed in section 8.2. Note that the interval T is considered as arbitrary but fixed. The notations C^∞ and $\mathcal{D}$ are therefore also used for $C^\infty|T$ and $\mathcal{D}|T$ respectively.

References

The reference list given below contains only those papers, books, etc. which have actually been used as references in writing the book.

Aracil, J. and Montes, C. G. (1976). "External description of multivariable systems," *Int. J. Control*, 1976, **23**, no. 3, 409–420.

Arbib, M. A. (1966). "Automata theory and control theory—a rapprochement," *Automatica*, **3**, 161–189.

Arbib, M. A. and Zeiger, H. P. (1969). "On the relevance of abstract algebra to control theory," *Automatica*, **5**, 589–606.

Barnett, S. (1971). *Matrices in control theory.* Van Nostrand Reinhold.

Beghelli, S. and Guidorzi, R. (1976). "A new input-output canonical form for multivariable systems," *IEEE Trans. Automat. Contr.* **AC-21**, 692–696.

Blomberg, H. (1972a). "On set theoretical and algebraic systems theory." In Pichler, F. and Trappl, R. (ed.), *Advances in cybernetics and systems research, Vol. I.* Transcripta Books.

Blomberg, H. (1972b). "Large-scale systems as interconnections of subsystems." Helsinki University of Technology, Systems Theory Laboratory, **B15**.

Blomberg, H. (1974). "Note on the feedback control of multivariable linear systems." Helsinki University of Technology, Systems Theory Laboratory, **B21**.

Blomberg, H. (1975). "Systems and interconnections." Helsinki University of Technology, Systems Theory Laboratory, **B24**.

Blomberg, H., Sinervo, J., Halme, A., and Ylinen, R. (1969). "On algebraic methods in systems theory." Acta Polytechnica Scandinavica, **Ma 19**.

Blomberg, H. and Salovaara, S. (1968). "On the algebraic theory of ordinary linear time-invariant differential systems." Finland's Institute of Technology, Scientific Researches No. 27.

Blomberg, H. and Ylinen, R. (1976). "An operator algebra for analysis and synthesis of feedback and other systems." Helsinki University of Technology, Systems Theory Laboratory, **B31**.

Blomberg, H. and Ylinen, R. (1978). "Foundations of the polynomial theory for linear systems." *Int. J. General Systems*, **4**, 231–242.

Brocket, R. W. (1970). *Finite dimensional linear systems.* Wiley.

Fisher, G. D. (1961). "Laplacesche Transformation, Anfangswerte und unstetige Funktionen, physikalische Dimensionen." *Arch. f. Elektrotechn*, **XLVI**, 295–311.

Forney, G. D. (1975). "Minimal bases of rational vector spaces with applications to multivariable linear systems." *Siam J. Control*, **13**, no. 3, 493–520.

Fuhrmann, P. A. (1977). "On strict system equivalence and similarity." *Int. J. Control*, **25**, no. 1, 5–10.

Gantmacher, F. R. (1959). *The theory of matrices, Vols. I and II.* Chelsea.
Gelfand, I. M. and Shilov, G. E. (1964). *Generalized functions, Vol. 1.* Academic Press.
Greub, W. H. (1963). *Linear algebra.* Springer.
Guidorzi, R. (1975). "Canonical structures in the identification of multivariable systems." *Automatica*, **11**, 361–374.
Hirvonen, J., Blomberg, H., and Ylinen, R. (1975). "An algebraic approach to canonical forms and invariants for linear time-invariant differential and difference systems." *Int. J. Systems Sci.*, **6**, no. 12, 1119–1134.
Jacobson, N. (1951). *Lectures in abstract algebra, vol I—basic concepts.* Van Nostrand.
Jacobson, N. (1953). *Lectures in abstract algebra, vol. II—linear algebra.* Van Nostrand.
Jacobson, N. (1964). *Lectures in abstract algebra, vol. III—the fields and Galois theory.* Van Nostrand.
Kalman, R. E. (1960). "On the general theory of control systems." *Proc. 1st IFAC Congress, Moscow.* Butterworths.
Kalman, R. E., Falb, P. L., and Arbib, M. A. (1969). *Topics in mathematical system theory.* McGraw-Hill.
Kamen, E. W. (1975). "On an algebraic theory of systems defined by convolution operations." *Math. Systems Theory*, **9**, no. 1, 57–74.
Klir, G. (1969). *An approach to general systems theory.* Van Nostrand Reinhold.
Kučera, V. (1979). *Discrete linear control: The polynomial equation approach.* Wiley.
Lang, S. (1965). *Algebra.* Addison–Wesley.
Lipschutz, S. (1964). *Set theory and related topics.* Schaum.
MacDuffee, C. C. (1933). *The theory of matrices.* Springer.
MacLane, S. and Birkhoff, G. (1967). *Algebra.* Macmillan.
Maeda, H. (1974). "On the duality for large-scale systems." *Int. J. Control*, **19**, no. 2, 315–322.
Mesarovic, M. D. (1960). *The control of multivariable systems.* Wiley.
Mesarovic, M. D. (1969). Mathematical theory of general systems. In: Hammer, P. (ed.). *Advances in mathematical systems theory.* The Pennsylvania State University Press.
Mesarovic, M. D., Macko, D., and Takahara, Y. (1970). *Theory of hierarchical multilevel systems.* Academic Press.
Mesarovic, M. D. and Takahara, Y. (1975). *General systems theory: mathematical foundations.* Academic Press.
Mikusinski, J. (1959). *Operational calculus.* Pergamon Press.
Morf, M. (1975). *Extended system matrices,—transfer functions and system equivalence.* Decision and Control Conference, Houston.
Ogata, K. (1967). *State space analysis of control systems.* Prentice-Hall.
Orava, P. J. (1973). "Causality and state concepts in dynamical systems theory." *Int. J. Systems Sci.*, **4**, no. 4, 679–691.
Orava, P. J. (1974). "Notion of dynamical input-output systems: causality and state concepts." *Int. J. Systems Sci.*, **5**, no. 8, 793–806.
Padulo, L. and Arbib, A. (1974). *Systems theory.* Saunders.
Pernebo, L. (1977). "Notes on strict system equivalence." *Int. J. Control*, **25**, no. 1, 21–38.
Pernebo, L. (1978). "Algebraic control theory for linear multivariable systems."

Lund Institute of Technology. Dept. of Automatic Control, Report: LUTFD2/(TFRT-1016)1-307/(1978).
Polak, E. (1969). "Linear time-invariant systems." In Zadeh, L. A. and Polak, E. (ed.). *System theory*. McGraw-Hill.
Pontryagin, L. S., Boltyanskii, V. G., Gamkrelidze, R. V., and Mishchenko, E. F. (1962). *The mathematical theory of optimal processes*. Wiley.
Pugh, A. C. and Shelton, A. K. (1978). "On a new definition of strict system equivalence." *Int. J. Control*, **27**, no. 5, 657–672.
Ramar, K., Ramaswami, B., and Murti, V. G. K. (1972). Note on "On the definition of transfer function". *Int. J. Control*, **16,** no. 3, 599–602.
Rosenbrock, H. H. (1970). *State-space and multivariable theory*. Nelson.
Rosenbrock, H. H. (1974a). "Order, degree, and complexity." *Int. J. Control*, **19**, no. 2, 323–331.
Rosenbrock, H. H. (1974b). "Structural properties of linear dynamical systems." *Int. J. Control*. **20**, no. 2, 191–202.
Rosenbrock, H. H. (1974c). "Redundancy in linear time-invariant, finite-dimensional systems." Third IFAC Symposium on Multivariable technological systems, Manchester.
Rosenbrock, H. H. (1977a). "The transformation of strict system equivalence." *Int. J. Control*. **25**, no. 1, 11–19.
Rosenbrock, H. H. (1977b). "A comment on three papers." *Int. J. Control*. **25**, no. 1, 1–3.
Rosenbrock, H. H. and Hayton, G. E. (1974). "Dynamical indices of a transfer function matrix." *Int. J. Control*. **20**, no. 3, 177–189.
Salovaara, S. (1967). "On set theoretical foundations of system theory." Acta Polytechnica Scandinavica **Ma 15**.
Schwarz, H. (1967). *Mehrfachregelungen, Grundlagen einer Systemtheorie. Erster Band*. Springer.
Sinervo, J. (1972). "A note concerning minimality of ordinary linear time-invariant differential systems." In F. Pichler, R. Trappl (ed), *Advances in Cybernetics and Systems Research, vol. I*. Transcripta Books.
Sinervo, J. and Blomberg, H. (1971). "Algebraic theory for ordinary linear time-invariant difference systems." Acta Polytechnica Scandinavica **Ma 21**.
Sontag, E. D. (1976). "On linear systems and noncommutative rings." *Math. Systems Theory*. **9**, no. 4, 327–344.
Timonen, J. (1974). "On the stability of large-scale systems." (Dipl. Eng. thesis; in Finnish). Helsinki University of Technology.
Truxal, J. G. (1955). *Automatic feedback control system synthesis*. McGraw-Hill.
Willems, J. C. (1971). *The analysis of feedback systems*. The MIT Press.
Windeknecht, T. G. (1967). "Mathematical systems theory: causality." *J. Math. Syst. Theory* **1**, 279–288.
Windeknecht, T. G. (1971). *General dynamical processes, a mathematical introduction*. Academic Press.
Wolovich, W. A. (1973). "The determination of state-space representations for linear multivariable systems." *Automatica*, **9**, 97–106.
Wolovich, W. A. (1974). *Linear multivariable systems*. Springer.
Wolovich, W. A. (1977). "Skew prime polynomial matrices." Brown University, Providence, R.T. Division of Engineering, Eng. OC-1.
Wolovich, W. A. and Guidorzi, R. (1977). "A general algorithm for determining state-space representations." *Automatica*, **13**, 295–299.

Wymore, A. W. (1967). *A mathematical theory of systems engineering: the elements.* Wiley.

Ylinen, R. (1975). "On the algebraic theory of linear differential and difference systems with time-varying or operator coefficients." Helsinki University of Technology, Systems Theory Laboratory, **B23**.

Ylinen, R. (1980). "An algebraic theory for analysis and synthesis of time-varying linear differential systems." Acta Polytechnica Scandinavica **Ma 32.**

Zadeh, L. A. and Desoer, C. A. (1963). *Linear system theory, the state space approach.* McGraw-Hill.

Index

Page references in **bold face** indicate definitions and explanations

Abstract algebra, 13, 301, 303–320, 321–339, 340–341
Abstract input–output system, 6, **50–55**
  block diagram representation of, 51, 57
Adder, 30, 123
Adjacency matrix, 59
Algebra, 340
Algebraic, **321**
Algebraic equation, 25, 110, 183, 239
Anticausality, **285**
Associate, **305**
Augmented matrix, 15, 172, 174
Automorphism of groups, **304**
Automorphism of vector spaces, 341

Binary operation, **303**
  associative, **303**
  commutative, **303**
Block diagram algebra, 120
Boolean matrix, 58

Cancellation problem, 1
Candidate for feedback compensator, 34–44, 191
  satisfactory, 35
Candidate for observer, 179
Causality, 49, **52**, 66, 73, 284
  generalized, 205, **209–216**, 230
  strict, **285**
Causal input–output relation, **275**
Causal system, **53**, 73, 231, 275
Cause–effect relationship, 49, 50, 52, 56, 59

Characteristic polynomial, **23**, 29, 70, 185, **292**
Coefficient list, **221**, 298
Coefficient matrix, **221**, 222
  augmented, 221, 222
Column equivalence, **14**, **311**
Comaximal, **305**, *see also* Matrix; Polynomial matrix
Commutative diagram, 93, 97
Composition of family of difference input–output relations, 288
  causal, 289
  feedback, 291
  input–output relation determined by, 288
  observable, 289
  order of, 289
  parallel, 291
  reconstructible, 289
  regular, 289
Composition of family of differential input–output relations, **7**, 82, 90, 106, **110–122**, 202, 243
  block diagram representation of, 10, 113
  controllability aspects of, 121–122
  description of, 115
  equivalent, 114, 115, 118
  feedback, 12, 27, 184–190
  input–output equivalent, 107, **117**
  input–output relation determined by, **112–117**, 250–251
  observable, **24**, **117**, 119
    partially, **118–120**
  order of, **117**

Composition of family of differential input–output relations—*cont.*
  parallel, 25, 123
    controllability aspects of, 125–126
    observable, 125
    transfer matrix of, 126
  regular, **9**, **114**, 116, 250
  series, 9, 11, 26, 137–143
    controllability aspects of, 142–143
    observable, 140–143
    transfer matrix of, 139
  series–parallel, 141, 143–145
  system determined by, **250–259**
  transfer equivalent, 120
  transfer matrix determined by, 120, 252
Composition of projection and delay operator, 296
Composition of projection and differential operator, 216–225
Concatenation property, 208
Control input, 30
Control problem, 236, 240
Controllability, **25**, 55, **100**, 159, 202, 236, 286
Controllability index, 194
Converse relation, **89**
Coprime, **305**, *see also* Matrix; Polynomial matrix

Decentralized control, 45
Decomposition of difference input–output relation, 288
  parallel, 290
Decomposition of differential input–output relation, 106, **118**, 160–176
  observable, **118**
  parallel, 126–133
  Rosenbrock representation, 160–176
  series, 140–145, 197
  series–parallel, 141, 143–145
  state-space, 160
Decoupling theory, 45
Delta function, 206, 207, **343–348**
  Dirac, **343**
  Kronecker, **309**
Denominator, **305**
Denominator set, **306**, 325
Determinateness, 47, 56, **63–66**, 71–72
Diagnostics, 45

Difference equation, linear time-invariant, 271, 274, 275, 282, 284
Difference input–output system, *see* Difference system
Difference system, 3, 271, 273, **275**, 293
  causal, 275
  linear, 294
Differential equation, linear time-invariant, 5, 73, 201
Differential input–output system, *see* Differential system
Differential operator, **4**, 75, **76**
  causal, 210
Differential system, 3, 75, **79**, 90, 231
  causal, 230
  controllable, **231**
  dimension of, 231
  linear, 230
  proper, **232**
  strictly proper, **232**
Differentiation operator, **4**
  generalized, **346**
Distinguishable by a single input–output pair, **55**
Distribution, *see* Generalized function
Distributive, **303**
Divide, **305**
Division algorithm, 17, **323–325**
Divisor, **305**, *see also* Matrix; Polynomial matrix
  greatest common, 19, **305**
Dynamic system, 52, **53**, 64, 69–72, 90, 107, 201, 230

Electrical network, 259
Elementary column operation, 18, **312**
Elementary matrix, 15, **311**
Elementary row operation, 14, **311**
Elimination procedure, 9, **23**, **107–110**, 116, 176
Endomorphism of groups, **304**
Endomorphism of vector spaces, 340
Epimorphism of groups, **304**
Equivalence of matrices, **311**
  canonical forms for, **315**
Estimator, 3, 107
Euclidean algorithm, **324–325**

Factor set, **306**

Feedback compensation, 27–44, 184–199
Feedback compensator, 27–44, 184–199, 292
  candidate for, **34–44**, 191–199
    satisfactory, **35**
Feedback composition, 12, 27–44, 184, 187, 291
Feedback control, 2, 27, 106, 107, 184
Feedback interconnection, 69
Field, 13, **304**
Field of quotients, 14, 84, 210, 278, **307**
Final segment, 204, 226
First element, 99
Fixed point, 63
Fourier transform, 1, 67
Fundamental theorem of algebra, **324**

Generalized convolution, 345
Generalized differentiation operator, 206, **346**
Generalized derivative, 206, **343**
Generalized function, complex-valued, 203, 205, **342–348**
  piecewise infinitely regularly differentiable, 5, 76, 203, 205–210, **345–348**
    concatenation property of, 208
    subspaces of, 208
    value at *t*, 206
  regular, **342**
  restriction of, **348**
  singular, **343**
Generator, **337**
  quotient equivalent, 337
  transfer equivalent, 337
Generator for difference input–output relation, **282**
  anticausal, **285**
  causal, **285**
  input–output equivalent, **283**
  order of, **282**
  proper, 285
  regular, **282**
  row equivalent, 283
  strictly causal, **285**
  transfer equivalent, **285**
Generator for differential input–output relation, **6**, 88, **90–94**
Generator for differential input–output relation—*cont.*
  characteristic polynomial of, **23**
  input–output equivalent, **23**, 79, 88, **92–94**, 96, 106
  order of, **6**, **90**
  proper, **21**, **95**
  quotient equivalent, **337**
  regular, **6**, 88, **90**, 91–92
  row equivalent, **22**, **92–94**, 95
  strictly proper, **22**, **95**
  transfer equivalent, 88, **95–100**
Green's function, 215
Group, **303**
  abelian, **303**, 304
  additive, **303**
  commutative, **303**
  multiplicative, **303**

Hierarchical control, 45

Ideal, two-sided, **305**
  principal, **305**
Identification, 23, 44
Impulse response function, 215
Indeterminate over ring, 13, 81, **84**, 277, **321**
Infinitely continuously differentiable complex-valued function, 5, 76, 206, 343
Initial condition list, **221**
Initial condition mapping, **217–225**, 297
  for polynomial matrix operator, 219, 220, 297
  for polynomial operator, 219, 297
Initial segment, 78, 204, 226, 275
Input, 50
  overall, 58, 62
Input–output equivalence, **23**, 79, 88, **92–94**, 107, **117**, 145, 161–170, 283
  canonical form for, 88, 93–94, 104, 162, 284
  complete invariant for, 93–94, 162
Input–output mapping, **50**, 201
  causal, **52**, 209, 214–216, 231
  parametric, **50–52**, 64, 227, 230, 241, 246, 247, 249, 252
    equivalent, 51

Input–output relation, **51**, 73, 201
difference, **275**
causal, **275**
characteristic polynomial of, **292**
controllable, **286**
decomposition of, 288
generation of, 274, *see also* Generator for difference input–output relation
minimal, 286
order of, **282**
reachable, **286**
regular, **282**
transfer equivalent, **286**
differential, **5**, **78**, 88, 89–105
asymptotically stable, **23**
block diagram representation of, 111
characteristic polynomial of, **23**
controllable, **25**, **100**, 121
decomposition of, **118**, 197
generation of, 78, 89, *see also* Generator for differential input–output relation
minimal, 97, **99**, 129
order of, **6**, **90**
proper, **21**, **95**
regular, **6**, 88, **90–91**
strictly proper, **22**, **95**
transfer equivalent, 88, **98–100**
internal, **9**, **113**
reduced, **147**
internal-overall, **8**, **115**
overall, **8**, 9, 112, **115**
Input set, 50
Input signal, 6, 78
Input-state relation, **158**
Insensitivity with respect to parameter variations, 34–35, 44, 178, 191, 258
Integral domain, 13, 84, **304**
Integration in generalized sense, 225, 343
Integration by parts, 225
Interconnection constraint, **8**, **57–59**, 111, 113, 245
Interconnection of family of differential systems, 243–269
determinate, 246, 250
equivalent, 255, 258
Interconnection of family of differential systems—*cont.*
input–output mapping determined by, 246–250
restricted state space, 245–246
stability properties of, 258
system determined by, 247–259
dimension of, 250
transfer matrix determined by, 247
unrestricted state space, 244–245
Interconnection of family of systems, **56–61**, 61–72, 110, 201
block diagram representation of, 62
determinate, 59, **63–67**, 71–72, 114
indeterminate, 66, 72
input–output mapping determined by, 64, 65–67, 70–72
input–output relation determined by, **63**, 65–67, 70–72
system determined by, **62–64**, 66–67, 71–72
Interconnection matrix, **7**, **57**, 61, 111
Internal system, 62, **63–64**, 65–67, 247–259
Inverse, **303**
Irreducible, **305**
Isomorphism of groups, **304**
Isomorphism of rings, 84

Kernel, **304**

Laplace transform, 67–68, 80, 94, 149, 203, 216, 224–226
generalized, **225**, 229
Leading coefficient, 35, **322**
Leading term, **322**
Linear mapping, **307**
Linear mappings, algebra of, 84, 210, 340
Linear dynamic system, 79, 204, 226–242
finite dimensional, 227–242
Linear system, **53–55**, 201
dimension of, **54**
finite dimensional, **54**, 73
linear description of, **54**
List, **308**

Matrix difference equation, 279, 275
Matrix differential equation, 13, 78, 89

Matrix over principal ideal domain, 312–320
determinantal divisor of, **314**
diagonal reduction of, **314**
greatest common divisor of, 316–320
invariant factor of, **314**
least common multiple of, 316
rank of, **315**
secondary, **313**
secondary row operation on, **313**
Smith (canonical) form of, **315**
triangular reduction of, **313**
Matrix over ring, **309**
adjoint of, **310**
column equivalent, **311**
comaximal, left, **311**
right, **310**
coprime, left, **310**
right, **311**
determinant of, **309**
determinantal rank of, **310**
diagonal, **309**
elementary, **311**
elementary column operation on, **312**
elementary row operation on, **311–312**
equivalent, **311**
identity, **309**
invertible, **309**
left divisor of, **310**
greatest common, **310**
left multiple of, **310**
least common, **310**
lower (left) triangular, **309**
minor of, **310**
partitioned form of, **309**
polynomial, *see* Polynomial matrix
rational, *see* Rational matrix
right divisor of, **310**
greatest common, **310**
right multiple of, **310**
least common, **310**
row equivalent, **311**
unimodular, **309**
unit, **309**
upper (right) triangular, **309**
Matrix over ring of quotients, **318–320**
canonical upper triangular form of, **334**
Matrix over ring of quotients—*cont.*
quotient equivalent, **319**
Smith–McMillan form of, **320**
Mikusinski operational calculus, 204
Mode, 43, 44, 151, 155
Model, abstract, 49
mathematical, 4, 49
Module, 13, 73, 80, 84, 210–211, **307–308**
cyclic, **307**
finitely generated, **307**
free, **308**
basis for, **308**
dimension of, **308**
rank of, **308**
strengthening of, 85, 211–216
trivial, **307**
zero, **307**
Module of quotients, 278, 295, **308–309**
Monoid, **303**
Monomorphism of groups, **304**
Morphism of groups, **304**
Morphism of modules, **307**
Morphism of rings, **304**
Morphism of vector spaces, 53
Multiple, **303**, **305**, *see also* Matrix; Polynomial matrix
least common, **305**

Negative, **303**
Numerator, **305**

Observability, **24**, 55, 106, **117**, 159, 177, 289
Observability index, 239
Observable component, **120**
Observer, 106, 107, **177**
Output, 50
overall, 58, 61
Output controllability, 160
Output set, 50
Output signal, 6, 78
Overall system, 7, 58, 62, **63–64**, 65–67, 247–259

Pole assignment problem, 29–30
Polynomial, 3, 13, 76, 321
leading coefficient of, 35, **322**
monic, 18

Polynomial over field of complex numbers, **324**
  multiple zero of, **324**
Polynomial function, **323–324**
Polynomial matrix, 3, 14–21, 76, 85–86, 279, **326–327**
  canonical form of, 14, 104
    lower triangular, **18**, **328**
    row proper, 105, 233, **333**
    upper triangular, **18**, 93, **328–329**
  column degree of, **326**
  column equivalent, 14
  column permuted canonical form of, **329**
  column proper form of, **327**
  common left divisor of, **19**
    greatest, **19**, 97
  common right divisor of, **19**
    greatest, **19**
  coprime, left, **19**, 100, 335–336
    right, **19**, 108–109, 117
    skew, 132
  degree of, **326**
  determinant of, 77
  division algorithm for, 338–339
  elementary, **15**
  elementary column operation on, 20
  elementary row operation on, **14**
  invertible, **14**
  leading coefficient of, **326**
  minor of, **327**
  rank of, 35, **85**
  row degree of, 37, 110, 326
  row equivalent, **14**, **92**, 145
  row proper form of, 35, 104, 109, 223, **327**, 329–333
  triangular, 14–17
  triangularization of, 14–17
  unimodular, **14**, **87**, 90, 326
Polynomial matrix operator, **5**, **77**, 78, 85, 86–87, 274, 279, *see also* Polynomial matrix
  causal, **210**
Polynomial operator, **5**, 13, 77, 81–82, 84–85, 274, *see also* Polynomial over field of complex numbers; Polynomial over ring
  causal, **210**
  degree of, **81**
Polynomial over ring, **321–322**
Polynomial over ring—*cont.*
  degree of, **322**
  leading coefficient of, **322**
  leading term of, **322**
  monic, **322**
  root of, **323**
  zero of, **323**
Polynomial systems theory, 2, 73, 80, 106, 145, 176, 243, 249, 252
Power, **303**
Prime, **305**
Principal ideal domain, **305**
Projection mapping, **207–210**, **294**
Projection method, 203, **232–243**, 248, 252, 298

Quotient, **305**, **323**
Quotient equivalence, **319–320**, 336
  canonical form for, 320, 325, 336
  invariant for, 320

Range, **304**
Rational, **325**
Rational form, **325**
  proper, **325**
  strictly proper, **325**
Rational function, **326**
Rational matrix, **21**, 94, **212**, 279, 337
  factorization of, 98, 337
  proper, **21**, **95**, **337**
  strictly proper, **22**, **95**, **337**
  strictly proper part of, **95**
Rational matrix operator, **212–216**, 295, *see also* Rational matrix
Rational operator, **295**
  causal, **295**
Reachability, 55, **286**
Realizability condition, **52**, 63–64, **64–67**, 69, 73, 202, 210
Realization problem, 176, 237–242
Real system, 4
  large-scale, 47
Reconstructibility, 55, **289**
Remainder, **323**
Response space, zero input, 25, 79, 83, 231, 253
Response, zero input, 25, **54**
  zero state, **55**

Richness condition, 52, 66, 72
Ring, **304**
 commutative, 13, **304**
 trivial, **304**
Ring of quotients, 278, **305–307**
Ring of quotients of polynomials, 278–279, **325–326**
Ring with unity, **304**
Rosenbrock representation, **145–160**, 160–176
 controllability aspects of, 151–155
 input–output equivalent, 162
 observable, 150, 162, 174
 order of, **149**, 175
 strictly system equivalent, **173**
 transfer matrix of, 150, 151, 153, 175
Rosenbrock system matrix, 6, 90, 107, **148–149**
 modified, **148–149**
  augmented, **171**, 172, 174
  input–output equivalent, **162–176**
  observable part of, **165**, 166
 state-space modified, 158, 175
 strictly system equivalent, 163, 172–176
Row equivalence, 14, **92–94**, 145, 283, 311–315, *see also* Input–output equivalence
 canonical form for, 93–94, 284, 328, 333
 invariant for, **314**

Scalar, **307**
Scalar multiplication, **307**
Semigroup, **303**, 304
Sensitivity analysis, 45
Shift operator, 273
Sifting property, **343**
Signal, 52, 273
Signal space, 3, 4, 5, 73, 75, **76–77**, **81–83**, 89, 107, 204–225, **273**
 generalized, 206, 208, **346**
 regular, **81**, **276**, 346
 sufficiently rich, 80, 82, 86, 206, 277, 346
Similarity transformation, 175
Skew coprime, 132
Smith (canonical) form of matrix, 104, 134, 135, **315**
Smith–McMillan form of matrix, **320**
Stability, **23**, 28–29, 36, 178, 190, 292
Stabilizability, 106
State, **50**, **158**, 201
 distinguishable by a single input–output pair, **55**
 equivalent, **50**
 initial, **53**
State constraint, **59**, 244–246
State-output mapping, **158**
State set, 50, 201
 minimal, 51, 235, 241
State space, 159
 dimension of, 159
 restricted, 245–246
 unrestricted, 244–245
State-space approach, 2, 106
State-space representation, 2, 90, 144, **158–160**, 160–176
 controllability aspects of, 100, 159
 observable, 100, 159
 order of, **159**
Stochastic dynamic system, 53
Strict system equivalence, 163, **172–176**
transformation of, **172**
Subalgebra, 340
Subdomain, **305**
Subfield, **305**
Subgroup, **304**
Submodule, **307**
Subring, **304**
Subspace, 76
Subsystem, 7
System, **51–52**, 201
 block diagram representation of, 51, 57
System description, **51**, 57
 equivalent, **51**
 linear, **54**
System matrix, *see* Rosenbrock system matrix
Systems theory, basic concepts of, **47–48**, 201
 polynomial, 2, 80, 106

Test function, **342–348**
 improper, 224
Time, initial, **53**
Time evolution, 78, 204, 226

Time function, **52**
Time set, **52**, 68, 271, 273
Time system, **52**, 64, 68
  uniform, **52**, 53
Transcendental, **321**
Transfer equivalence, 88, **95–100**, 120, 285, 286, **337**
  canonical form for, 88, 99, 105, 286
  complete invariant for, 98
Transfer function, 94
Transfer function technique, 1, 2
Transfer matrix, **21**, 88, **94–95**, 96–100, 120, 126, 175, 212, **285**, **337**, *see also* Rational matrix
  factorization of, 98, 144, 194, 198, 337
  minimal input–output relation determined by, **99**, 286
  pole of, **154**
  proper, **21**, **95**, 285
  strictly proper, **22**, **95**, 285
  strictly proper part of, **95**
Transfer matrix operator, **212**
Transfer ratio, **21**, **95**

Unique factorization domain, **305**
Unit, **304**
Unit delay operator, **273**
Unit element, **303**
Unit prediction operator, **273**
Unit step, **343–345**
  regular generalized, **343**
Unity, **303**
Unobservable component, **120**

Vector space, 13–14, 210–216, **307**
Vector space of quotients, 279, 295

Weighting function, 215

Z-transform, 1, 2
Zero, decoupling, 152, 175
  input–decoupling, **154–158**
  input–output-decoupling, **154–158**
  output-decoupling, 150–158
Zero divisor, **304**
Zero element, **303**